Expanding Nanobiotechnology

This book will cover recent and advanced concepts in nanobiotechnology, with deep focus on their applications and commercialization.

Expanding Nanobiotechnology: Applications and Commercialization covers topics like green synthesis, microbial synthesis and their applications in medicine, imaging and nanotheranostics. It also includes success stories in food, energy and environment. The book explains the mechanistic insights in the biological synthesis of nanoparticles and includes discussion on the toxicity and interaction with the natural systems in detail. The chapters also examine synthesis and fabrication aspects in a concise manner. The book also highlights the possible therapeutic and imaging applications of biogenic nanomaterials in cancer therapies and explains the diverse mechanisms. The authors also present the hurdles, issues and concerns that hinder the large-scale commercial applications of the biogenic nanoparticles. Finally, apart from detailed discussion on the advanced futuristic applications of nanobiotechnology, specific contributions covering important topics like sustainability, ethics and dual use research concerns are included.

This book is intended for researchers, biotechnologists, microbiologists and entrepreneurs spanning across diverse areas, including medicine, energy and the environment.

Expanding Nanobiotechnology

Applications and Commercialization

Edited by

Ali Talha Khalil and Arshad Islam

CRC Press
Taylor & Francis Group
Boca Raton London New York

CRC Press is an imprint of the
Taylor & Francis Group, an **informa** business

Designed cover image: Shutterstock ©

First edition published 2025
by CRC Press
2385 NW Executive Center Drive, Suite 320, Boca Raton FL 33431

and by CRC Press
4 Park Square, Milton Park, Abingdon, Oxon, OX14 4RN

CRC Press is an imprint of Taylor & Francis Group, LLC

ISBN: 978-1-032-45516-7 (hbk)
ISBN: 978-1-032-45758-1 (pbk)
ISBN: 978-1-003-37856-3 (ebk)

DOI: 10.1201/9781003378563

Typeset in Times
by KnowledgeWorks Global Ltd.

Contents

About the Editors

Ali Talha Khalil, a native of Peshawar (Pakistan), is working as Assistant Professor and consultant molecular biologist at the Lady Reading Hospital Medical Teaching Institution (LRH-MTI), Peshawar, which is one of the largest tertiary care hospitals in the region. Besides his major responsibilities, Dr. Khalil is in-charge of the COVID-19 diagnostic facility at LRH. He serves as a secretary and member of the Institutional Review Board (IRB) and Institutional Research Committee (IRC) of LRH that are mandated for reviewing the ethical, technical and financial aspects of research work conducted at the Institute. Dr. Khalil's major expertise and research interests are in molecular biology and materials sciences. He is using eco-friendly and sustainable avenues for the synthesis of novel nanomaterials and nanocomposites and uses them for different nanomedicine and environmental applications.

Dr. Arshad Islam is currently working as Assistant Professor and Consultant of Biochemistry, at the Department of Pathology, Government Lady Reading Hospital Medical Teaching Institution, Peshawar, Pakistan. Dr. Islam has earned his PhD degree in Biological Chemistry/Inorganic Biochemistry from Universidade Federal de Minas Gerais (UFMG), Brazil (2014), and Master of Philosophy in Biochemistry/Molecular from Quaid-i-Azam University (QAU), Islamabad, Pakistan (2009). During his PhD and postdoctoral research, he has been working, in general, with "the development of metal-based organometallic and nanoscale drugs and drug delivery systems, and their mechanism of action" for the purpose of improved drug's potency and efficacy, aiming at the treatment of parasitic diseases such as leishmaniasis and Chagas disease as well as bacterial infections and cancer, exploring both the in vitro and in vivo model of the diseases.

Contributors

Banzeer Ahsan Abbasi
Department of Botany
Rawalpindi Women University
Rawalpindi, Pakistan

Latifat Abdulsalam
Department of Bioengineering
King Fahd University of Petroleum and Minerals (KFUPM)
Dhahran, Saudi Arabia

Irshad Ahmad
Department of Bioengineering
King Fahd University of Petroleum and Minerals (KFUPM)
Dhahran, Saudi Arabia

and

Interdisciplinary Research Center for Membranes and Water Security
King Fahd University of Petroleum and Minerals (KFUPM)
Dhahran, Saudi Arabia

Washma Aimen
Sustainable Bioenergy and Biorefinery Lab
Department of Microbiology
Islamabad, Pakistan

Ikrimah A. Alhanbali
Department of Bioengineering
King Fahd University of Petroleum and Minerals (KFUPM)
Dhahran, Saudi Arabia

Iftikhar Ali
Center for Plant Sciences and Biodiversity
University of Swat Charbagh
Charbagh, Pakistan

and

Department of Genetics and Development
Columbia University Irving Medical Center
New York, New York

Sidra Ali
Sustainable Bioenergy and Biorefinery Lab
Department of Microbiology
Islamabad, Pakistan

Nargis Aman
Nazar College of Pharmacy
DAKSON Institute of Health Science
Islamabad, Pakistan

Deepak Anand
Department of Bioengineering
King Fahd University of Petroleum and Minerals (KFUPM)
Dhahran, Saudi Arabia

Zaman Ashraf
Department of Chemistry
Rawalpindi Women University
Rawalpindi, Pakistan

Bushra Begum
Department of Chemistry
Shaheed Benazir Bhutto Women University
Peshawar, Pakistan

Hussan Ara Begum
Department of Botany
Abdul Wali Khan University
Mardan, Pakistan

Naheed Bibi
Department of Chemistry
Shaheed Benazir Bhutto Women University
Peshawar, Pakistan

Nurcan Dogan
Istanbul Technical University
Istanbul, Türkiye

Hajera Gul
Department of Chemistry
Shaheed Benazir Bhutto Women University
Peshawar, Pakistan

Abdul Haq
Pakistan Council of Scientific and Industrial Research (PCSIR)
Peshawar, Pakistan

Muhammad Bahlool Hassan
Sustainable Bioenergy and Biorefinery Lab
Department of Microbiology
Islamabad, Pakistan

Shumaila Ijaz
School of Biomedical Engineering
Shenzhen University Medical School
Shenzhen, China

Naila Ijaz
Nano composite lab
Department of Physics
University of agriculture Faisalabad, Faisalabad 38000, Pakistan

Muhammad Imran
Department of Pulmonology
Lady Reading Hospital Medical Teaching Institution
Peshawar, Pakistan

Javed Iqbal
Department of Botany
Bacha Khan University
Chārsadda, Pakistan

Rashid Iqbal
Department of Agronomy
Faculty of Agriculture and Environment
The Islamia University of Bahawalpur
Bahawalpur, Pakistan

Muhammad Irfan
Gebze Technical University
Kocaeli, Türkiye

Arshad Islam
Department of Pathology
Lady Reading Hospital Medical Teaching Institution
Peshawar, Pakistan

Muhammad Jahangir
Department of Food Science and Technology
The University of Haripur
Haripur, Pakistan

Nargis Jamila
Shaheed Benazir Bhutto Women University
Peshawar, Pakistan

Aisha Kanwal
Key Laboratory of Applied Surface and Colloid Chemistry
School of Chemistry & Chemical Engineering
Shaanxi Normal University
Xi'an, China

Sobia Kanwal
Department of Biology and Environmental Sciences
Allama Iqbal Open University
Islamabad, Pakistan

Mohsin Kazi
Department of Botany
Bacha Khan University
Chārsadda, Pakistan

and

Department of Pharmaceutics
College of Pharmacy
King Saud University
Riyadh, Saudi Arabia

Asmat Ullah Khan
Department of Zoology
Shaheed Benazir Bhutto University
Sheringal, Pakistan

Imran Khan
Department of Food Science and Technology
The University of Haripur
Haripur, Pakistan

Tariq Mahmood
Department of Plant Sciences
Faculty of Biological Sciences
Quaid-i-Azam University
Islamabad, Pakistan

El-Hassan M. Mahmoud
Department of Bioengineering
King Fahd University of Petroleum and Minerals (KFUPM)
Dhahran, Saudi Arabia

Shaila Mehwish
Department of Health Biotechnology
Women University Swabi
Swabi, Pakistan

Sumaira Miskeen
Department of Food Science and Technology
The University of Haripur
Haripur, Pakistan

James Mordecai
Department of Bioengineering
King Fahd University of Petroleum and Minerals (KFUPM)
Dhahran, Saudi Arabia

El-Hussein Moustafa
Faculty of Pharmacy
Minia University
Minya, Egypt

Ghulam Murtaza
Faculty of Environmental Science and Engineering
Kunming University of Science and Technology
Kunming, China

Sara Najeeb
Department of Food Science and Technology
The University of Haripur
Haripur, Pakistan

Raina Aman Qazi
Department of Chemistry
Shaheed Benazir Bhutto Women University
Peshawar, Pakistan

Fakhra Riaz
SA-Centre for Interdisciplinary Research in Basic Sciences
International Islamic University
Islamabad, Pakistan

Fatima Riaz
SA-Centre for Interdisciplinary Research in Basic Sciences
International Islamic University
Islamabad, Pakistan

Ikram Ullah
Department of Biotechnology and Genetic Engineering
Hazara University
Mansehra, Pakistan

Nabi Ullah
Department of Inorganic and Analytical Chemistry
Faculty of Chemistry
University of Lodz
Lodz, Poland

Saleem Ullah
Biomaterial processing and products
VTT Technical Research Centre
Espoo, Finland

and

Department of Transdisciplinary Science and Engineering
School of Environment and Society
Tokyo Institute of Technology
Tokyo, Japan

Zakir Ullah
Department of Plant Sciences
Faculty of Biological Sciences
Quaid-i-Azam University
Islamabad, Pakistan

Zahida Wasil
Shaheed Benazir Bhutto Women University
Peshawar, Pakistan

Tabassum Yaseen
School of Biomedical Engineering
Shenzhen University Medical School
Shenzhen, China

1 Nano-Vaccines and Therapeutics

Emerging Horizons to Better Fight Infectious Diseases

Arshad Islam, Abdul Haq, Sidra Ali, Muhammad Imran, Fakhra Riaz, Fatima Riaz, Washma Aimen, and Muhammad Bahlool Hassan

1.1 INTRODUCTION

For centuries, infectious diseases have posed a threatening challenge to human health, necessitating unending efforts to mitigate their impact. Despite the outstanding efficacy of traditional vaccines and pharmacotherapies in the prophylaxis and management of infectious conditions, these conventional modalities are not without limitations. The emergence of antimicrobial resistance (AMR), the transient nature of immunity conferred by booster vaccinations, and the challenges associated with targeted drug delivery emphasize the imperative for the exploration of novel therapeutic strategies (1).

Nanotechnology has recently emerged as a novel paradigm in the arsenal against infectious diseases, thereby warranting the consideration of nanoparticles within this field (2). This chapter aims to provide a comprehensive overview of the emerging field of nano-vaccines and nano-therapeutics and explore the unique properties of nanoparticles and their potential to revolutionize disease prevention and treatment strategies. Also, the potential of nanoparticles as vaccine antigen carriers will be explored, highlighting their capacity to encapsulate antigens within or alongside their structures, thereby eliciting potent immune responses (3).

In addition to vaccines, nanoparticles also offer great potential for other therapeutic applications. Nanoparticles can, for example, be designed to carry drugs into infected cells or tissues, helping to minimize adverse side effects on healthy tissues and organs (4). Targeted drug delivery increases bioavailability of drug at its action site and minimizes dosage. Furthermore, nanoparticles, in a controlled and sustained manner, can deliver drugs to immune cells, ensuring continued activity and improved patient compliance (5).

This chapter will explore the various types of nanoparticles employed in vaccine and therapeutic development, and the mechanisms by which these nanoparticles

DOI: 10.1201/9781003378563-1

interact with the immune system and enhance drug delivery. Moreover, the current research efforts directed towards utilizing nanotechnology for combating a wide range of infectious diseases, from established threats like human immunodeficiency virus (HIV) and tuberculosis to emerging pathogens like coronaviruses, will be explored.

The journey of nano-vaccines and therapeutics is still in its early stages, yet the potential for transformative change is undeniable. By utilizing the potential of nanotechnology, a new era of effective and targeted interventions is underway to be introduced, ultimately leading to a healthier future for all (6).

1.2 INFECTIOUS DISEASES AND THEIR GLOBAL IMPACT

Microbial pathogens, defined as bacteria, viruses, parasites, and fungi, continue to pose a significant threat to global health, responsible for considerable morbidity and mortality. In 2020, the World Health Organization estimated that these threats led to 13.9 million deaths worldwide, accounting for about 24% of all global mortality, as depicted in Figure 1.1 (7). Since low- and lower-middle-income countries face the impact of this burden, microbial pathogens magnify existing inequalities.

Over 5 million children under the age of five die as a result of infectious diseases due to their juvenile immune systems. This shocking statistic reflects the disproportionate burden these microscopic pathogens place on young populations. The spectrum of severity varies considerably, ranging from mild sicknesses like the common cold to life-threatening conditions such as those caused by HIV and tuberculosis (8). Transmission dynamics also differ, with some pathogens readily communicable, while others exhibit geographical restrictions, as exemplified by neglected tropical diseases such as malaria, leishmania, dengue, Chagas disease, and others (9).

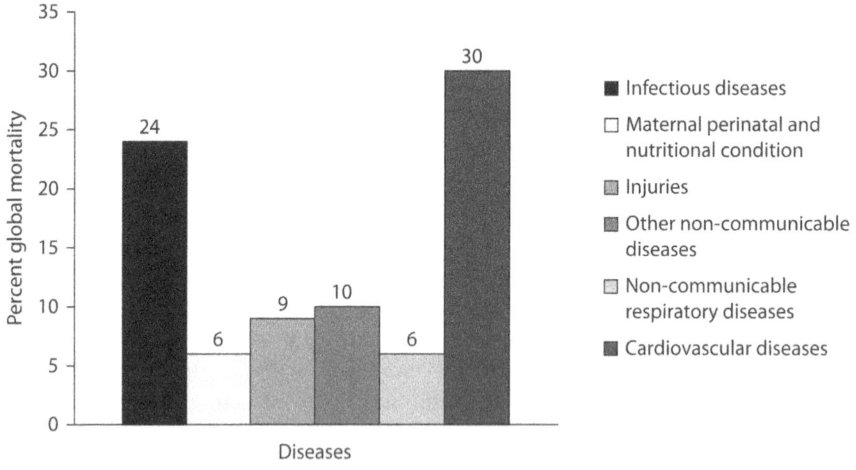

FIGURE 1.1 Global burden of infectious diseases.

Despite remarkable progress in prevention and treatment, infectious diseases remain a petrifying global challenge, demanding persistent innovation. Traditional antibiotics are ineffective due to the emergence of multi-drug-resistant organisms and existing vaccines to keep pace with rapidly evolving infectious pathogens such as malaria. The disastrous coronavirus disease 2019 (COVID-19) outbreak served as a wake-up call for the modern world for faster, more innovative, and sustainable solutions (10).

Nanotechnology offers a transformative approach in the fight against infectious diseases. By manipulating materials at the atomic level, scientists are developing novel nano-vaccines and pharmaceuticals with the potential to revolutionize healthcare. These advancements hold the promise of targeted delivery of potent medications directly to the infected cells, long-lasting immunization strategies, and ultra-precise diagnostic techniques (11). The arsenal of nanotechnological developments provide the platform for regaining control over various fatal infectious diseases.

1.3 LIMITATIONS OF TRADITIONAL VACCINES AND THERAPEUTICS

Despite their numerous advantages, traditional vaccines and therapeutics, unfortunately, are associated with severe side effects and limitations. These limitations and shortcomings may hinder the effectiveness of these vaccines in certain situations. For example, their limited duration of immunity, ineffectiveness against rapidly evolving pathogens (e.g., influenza virus mutates frequently and rendering existing vaccines less effective over time and necessitating frequent reformation), mild side and more serious reactions, and inefficacy in immunocompromised patients. Understanding and considering these shortcomings is critical for developing more effective strategies to fight infectious diseases (12). Figure 1.2 shows drawbacks and risks associated with conventional vaccinations and therapeutics.

One of the most significant disadvantages of most traditional vaccines is their narrow spectrum of protection. Many vaccines require several injections to achieve broad-spectrum immunity (13). Furthermore, even after initial immunization, an additional vaccination may be required because the immunity level declines over time. Attempts to create universal vaccines that would protect against different pathogenic strains as long as possible still pose serious scientific challenges (14).

Safety is another vital factor that is not always possible to preserve. Even though live attenuated vaccines have high efficiency, they also have a small chance of causing the very disease they are designed to combat. Some adjuvant groups can cause some mild repercussions as well, despite their ability to boost immune response. The delicate balance between preserving maximum efficacy and minimum safety risks continues to be sought. Also, the ingredients of some traditional vaccines are harmful to the planet (15).

The traditional dependence on vaccinations and other therapeutics is often inhibited due to resource limitations. A perceived primary downside of vaccines is the cold chain requirements: very few vaccines have no less than interruptions in

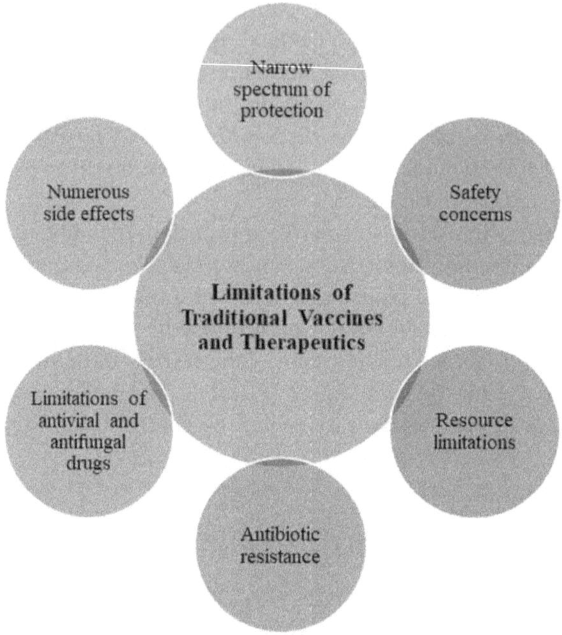

FIGURE 1.2 Limitations of traditional vaccines and therapeutics.

temperature control. Predominantly, the developing world often lacks proper infrastructure to conduct such a process, which makes the administration unattainable. In addition, the need for multiple doses is antithetical to the ability to pay, meaning that costs seldom can be allocated in such a way as to attain promising affordability (16).

As for the bacteria, the once-presumed invincible antibiotics are rapidly losing to antibiotic-resistant strains. Naturally, the established multiple drug-resistant (MDR) bacteria are unable to be treated with antibiotics, which presents an acute issue. This poses a significant threat, particularly in hospital settings where MDR bacterial infections can be life-threatening (17).

Antiviral and antifungal drugs also face similar challenges. Not all viruses or fungi are pharmacologically viable targets, and the current treatments are often limited by numerous side effects, and the growing threat of drug resistance, a well-recognized challenge with extensive research (18). Therefore, the development of new drugs with a broader spectrum of action and the ability to evade resistance is the only possible approach.

Although classical methods of fighting infectious diseases have saved millions of lives, their limitations made them inefficient in some situations (19). The existing potential for further research and innovative practices, such as personalized drugs, nanotechnology, and expanded-spectrum drugs, allows for a much more efficient envisioning of the future. Therefore, understanding the limitations of current treatment and prevention approaches can cause future improvements in preventing microscopic enemies from harming humans (20).

1.4 NANOTECHNOLOGY-BASED VACCINE: A POTENTIAL SOLUTION

Traditional vaccines and small-molecule therapeutics face significant limitations in combating infectious diseases. These limitations include narrow targeting specificity, potential safety concerns, challenges with accessibility, and the emergence of drug resistance. These drawbacks necessitate the development of more sophisticated strategies. The past few decades have witnessed the accelerated development of novel tools in the fight against pathogens at the nanoscale, with nanotechnology emerging as a key player (21). While the concept of nanobots directly destroying infected cells remains hypothetical, nanotechnology offers a more realistic approach. It allows for the development of targeted nanocarriers that deliver potent therapeutic agents specifically to infected cells, minimizing exposure to healthy tissues. These "smart" delivery systems potentially enable effective treatment at lower doses, reducing human toxicity. Additionally, they may induce more durable immune responses, potentially leading to fewer vaccinations and simplifying immunization schedules, particularly beneficial in resource-limited settings with inadequate healthcare infrastructure (22).

Nanomedicine offers a promising strategy to combat AMR. Nanoparticles can be engineered for targeted bacterial eradication, potentially enhancing the efficacy of traditional antibiotics and overcoming existing resistance mechanisms. Furthermore, nanomedicines with potent antifungal and antiviral properties can be developed, leveraging similar design principles to combat a broad spectrum of pathogens (23).

Beyond individual-level interventions, nanotechnology presents transformative possibilities for diagnostic and therapeutic applications. The development of highly sensitive diagnostic nanosystems capable of rapid pathogen detection within minutes or hours holds immense promise. Furthermore, personalized nanomedicines tailored to an individual's genetic makeup could revolutionize healthcare by enabling precision medicine at the molecular level (24).

Consequently, widespread clinical application of nanotechnology necessitates substantial optimization for safety and efficacy. Nonetheless, the potential for nanomedicine to combat infectious diseases is undeniable. The key question lies in the timeframe for translating these discoveries into effective therapeutic strategies. In essence, nanotechnology offers a compelling avenue for mitigating the burden of infectious diseases and enhancing human health (25).

1.5 KEY FEATURES OF NANOMATERIALS FOR VACCINE DEVELOPMENT

Traditional vaccines, despite their effectiveness, often suffer from limitations such as short-lived immunity and reduced efficacy against specific pathogens. Nanomaterials offer a promising avenue to address these shortcomings. Their unique properties, including high surface area for efficient antigen presentation, targeted delivery to immune cells, and controlled release of immunological components, hold significant promise for the development of next-generation vaccines capable of inducing

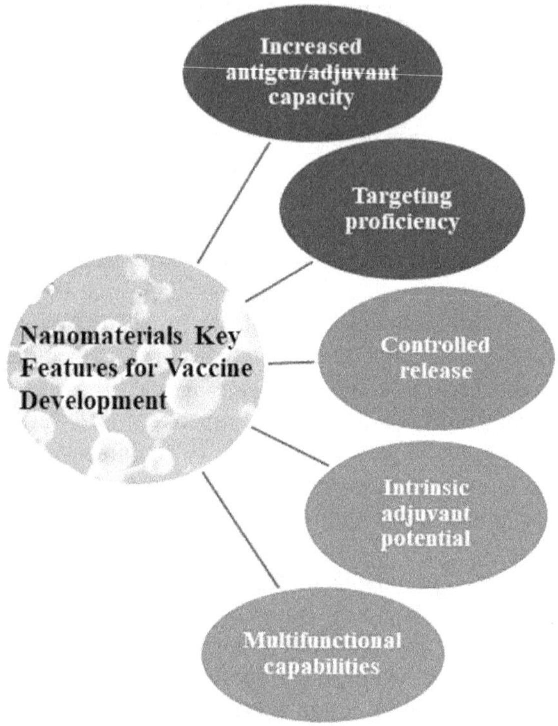

FIGURE 1.3 Key features of nanomaterials for vaccine development.

robust and durable immune responses while minimizing off-target effects (11). Key features of nanomaterials beneficial for the development of nano-vaccine are shown in Figure 1.3.

1.5.1 High Surface Area for Enhanced Antigen Presentation

Nanoparticles possess a remarkably high surface area to volume ratio. This translates to the ability to carry significantly more antigens or adjuvants per particle compared to traditional vaccines. This increased capacity for antigen presentation is crucial for inducing a robust immune response. The optimized geometry of nanoparticles facilitates efficient interaction with immune cells, particularly antigen-presenting cells (APCs). This enhanced interaction allows APCs to more effectively process and present antigens to T-cells, ultimately leading to their full activation (26).

1.5.2 Targeting Delivery via Ligand Conjugation

Nanoparticle surface can be functionalized with specific ligands that can bind to receptors on desired cells. In the context of vaccines, these ligands can be designed to target APCs, the immune system's workhorses responsible for presenting antigens to T-cells. This targeted delivery significantly reduces off-target effects and minimizes the risk of side effects by ensuring the nanoparticles reach and interact with the intended cells (27). Furthermore, nanoparticles can be engineered for even more precise targeting. By incorporating responsive elements, they can be designed to release their cargo only in response to specific environmental cues. For instance, nanoparticles could be programmed to release their antigen payload upon encountering a specific pH level (indicative of an infected environment) or in response to inflammatory signals. This approach further enhances targeting by ensuring the vaccine components are released only at the site of infection, minimizing systemic release and maximizing efficacy (28).

1.5.3 Controlled Release of Immunogens

Unlike traditional vaccines with burst antigen delivery, causing an initial widespread immune response, nanomaterials enable sustained and programmable release of immunogens, mimicking the kinetics of natural infection and promoting the development of robust immunological memory. This controlled release strategy can improve vaccination adherence by reducing the number of doses required, thereby enhancing cost-effectiveness (29).

1.5.4 Intrinsic Adjuvant Potential

Another distinct characteristic that a few nanomaterials have is intrinsic adjuvant potential. This translates to their ability to stimulate a more potent immune response compared to their bulk counterparts, even without the addition of explicit immunostimulatory agents. This intrinsic property could potentially reduce the need for traditional adjuvants, which may be associated with inflammatory side effects. Additionally, these nanoparticles can be further tailored to selectively induce Th1 or Th2 immune responses, depending on the desired immunological profile (30).

1.5.5 Multifunctional Platform Design

Nanoparticles offer a unique platform for integrating multiple functionalities. They can be engineered to co-deliver antigens, adjuvants, and even imaging agents within a single carrier. This versatility allows for targeted delivery to immune cells while simultaneously enabling real-time monitoring of vaccine activity (31).

1.5.6 Enhanced Thermostability

Another significant advantage of nanomaterials lies in their improved stability and storage properties, particularly relevant for resource-limited settings. Traditional vaccines often require stringent cold chain logistics, which can be challenging to maintain and limit accessibility (32). In contrast, some nanomaterials exhibit exceptional thermostability, allowing for storage and transportation at higher temperatures. This translates to reduced costs associated with refrigeration and broader vaccine reach due to less stringent storage requirements. Examples include liposomes and other promising candidates like polymer-based nanoparticles or virus-like particles (VLPs) being actively developed for various infectious diseases (33).

In conclusion, the unique properties of nanomaterials, including high surface area, targeted delivery, controlled release, and intrinsic adjuvant activity, offer a promising platform for the development of next-generation vaccines. These multifunctional platforms can be engineered to deliver antigens, adjuvants, and even imaging agents, enabling targeted immune cell stimulation and real-time monitoring of vaccine efficacy. Furthermore, enhanced thermostability of certain nanomaterials facilitates storage and transportation at higher temperatures, particularly beneficial for resource-limited settings. While challenges in production and regulatory approval remain, the potential of nanotechnology to revolutionize vaccine design and ultimately combat infectious diseases is undeniable. Continued advancements in this field hold significant promise for improving global health outcomes (34).

1.6 TYPES OF NANO-VACCINES

The field of vaccinology is undergoing a significant transformation with the emergence of novel nano-engineered vaccines. These vaccines leverage the unique properties of nanoparticles to enhance antigen delivery, immune stimulation, and overall vaccine efficacy. While a diverse range of nano-vaccines are under development, this section focuses on the leading platforms: polymer-based nanoparticles, liposomes, dendrimers, VLPs, DNA vaccines, and RNA vaccines (35). Figure 1.4 enlists the various types of nano-vaccines and also highlights their key features.

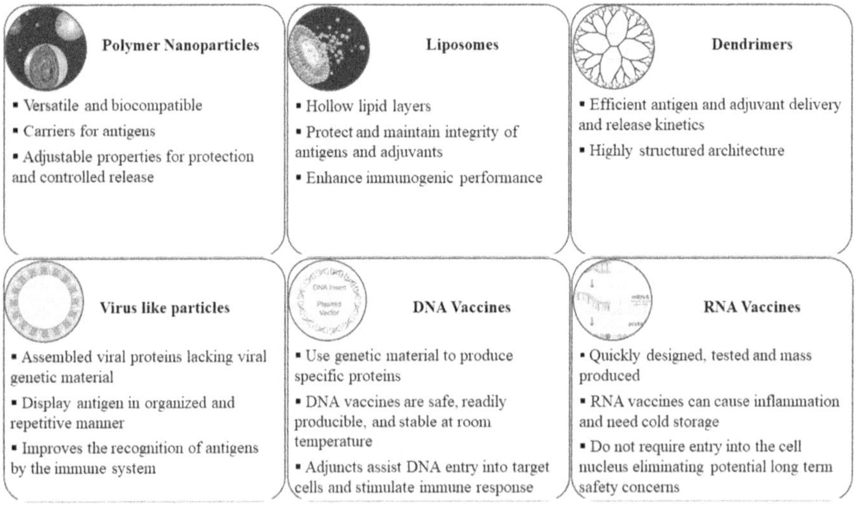

FIGURE 1.4 Types of nanoparticle-based vaccines.

1.6.1 Nanoparticle-Based Vaccines

Nanoparticle-based vaccines represent a versatile approach to vaccine formulation. Their diverse nanostructures offer distinct advantages, including the following:

a. ***Targeted Delivery*:** Nanoparticles can be engineered to specifically target antigen delivery to immune cells, enhancing the immune response (36).
b. ***Controlled Release*:** The adjustable properties of nanoparticles allow for precise control over antigen release kinetics, optimizing the immune response (37).
c. ***Biocompatibility and Antigen Protection*:** Polymer nanoparticles, a prominent example, offer biocompatibility and the ability to shield antigens from degradation, ensuring their stability and efficacy (38).

1.6.2 Virus-Like Particles (VLPs)

VLPs are non-infectious, self-assembling structures composed of viral proteins that mimic the natural viral structure but lack genetic material. This design offers several advantages:

a. ***Safety*:** VLPs are devoid of viral genetic material, eliminating the risk of replication and infection (39).
b. ***Antigen Presentation*:** VLPs present antigens in a highly organized and repetitive manner, mimicking the natural viral structure. This facilitates efficient activation of the immune system by B-cells, leading to the production of high-affinity antibodies against the targeted pathogen (39).
c. ***Scaffolding for Vaccine Development*:** VLPs provide a reliable platform for displaying antigens, making them a promising strategy for vaccines against various infectious diseases (40).

1.6.3 DNA Vaccines

DNA vaccines utilize plasmid DNA encoding specific antigens. Upon delivery, these plasmids enter host cells and instruct them to produce the encoded antigen, triggering an immune response. Key advantages of DNA vaccines include the following:

a. ***Safety*:** DNA vaccines lack infectious potential as they do not contain replicating viral or bacterial genetic material (41).
b. ***Rapid Production and Stability*:** DNA vaccines can be readily manufactured and exhibit stability at room temperature, facilitating rapid deployment and storage, particularly beneficial in resource-limited settings (42).

1.6.4 RNA Vaccines

RNA vaccines operate in a manner like DNA vaccines by introducing genetic information into human cells to produce viral or bacterial proteins that generate an immune response. They do not require entry into the cell nucleus to begin protein synthesis, simplifying their operation (43). mRNA vaccines, a form of RNA vaccination, have attracted much consideration in the struggle against illness like COVID-19 due to their high efficacy and rate of development (44). Additionally, while RNA vaccines often cause inflammation and need cold storage, DNA vaccines are stable at room temperature and readily transportable. For this purpose, DNA vaccines are a distinctively appealing alternative for infectious disease prevention and immunization. Ongoing advancements in RNA vaccine technology continue to improve their efficacy and applicability in preventing infectious diseases (45).

1.7 ADVANTAGES OF NANO-VACCINES OVER TRADITIONAL VACCINES

Nano-vaccines offer several advantages over traditional vaccines, potentially revolutionizing the field of immunization such as high precision and efficiency, and versatility. These benefits include enhanced immunogenicity and efficacy, improved stability and delivery, multivalency (targeting multiple antigens), and adjuvanticity (boosting immune response) (46).

1.7.1 Enhanced Immunogenicity and Efficacy

Nano-vaccines exhibit superior immunogenicity compared to conventional vaccines. Their ability to present antigens in an organized and repetitive manner facilitates efficient uptake by APCs. Soluble antigens in traditional vaccines often struggle with endocytosis by APCs, leading to reduced immune response (27). The high surface area and organized antigen display on nano-vaccines enhance recognition by the immune system, resulting in a more robust and targeted response. Additionally, repetitive antigen presentation promotes the development of strong and long-lasting immunological memory, leading to improved protection against a wider range of pathogens. This enhanced immunogenicity paves the way for more effective and durable immunization strategies (47).

1.7.2 Improved Stability and Delivery

Nano-vaccine platforms, such as polymer nanoparticles and liposomes, offer significant advantages in stabilizing antigen formulations and controlling their delivery. These structures shield antigens from degradation during transport, ensuring their structural integrity and efficacy upon reaching target cells (48). The controlled release kinetics of nano-vaccines further enhances their impact on the immune system. By gradually releasing antigens over time, they allow for sustained immune system stimulation and ensure efficient interaction between different immune components. This controlled release strategy also facilitates the migration of macrophages to the site of action, further amplifying the immune response (49).

1.7.3 Multivalency: Targeting Multiple Antigens

One of the most remarkable advantages of nano-vaccines is their multivalency, allowing them to target multiple antigens simultaneously. Protein nanoparticles, for example, can present a diverse array of antigens in a manner that closely mimics their natural conformation. This enables them to effectively stimulate a broader immune response against a wider range of potential pathogens (50). This multivalent targeting is particularly valuable for complex diseases or rapidly evolving pathogens where a single antigen-based vaccine may be insufficient. The consistent performance of nano-vaccines across various antigen combinations highlights their versatility in achieving optimal immune responses. These features make them promising candidates for developing next-generation, broadly protective vaccines (51).

1.7.4 Adjuvanticity: Boosting Immune Response

Another key advantage of nano-vaccines is their inherent adjuvanticity. Certain nanomaterials, such as dendrimers and lipid nanoparticles, possess intrinsic adjuvant properties that stimulate and regulate the immune response to antigens. These adjuvants enhance the recognition of antigens by immune system cells, particularly dendritic cells, leading to a more potent and targeted response against pathogens (52). This amplified response translates to the development of long-lasting immunity, effectively protecting the body from infectious agents. By incorporating adjuvants, nano-vaccines offer a promising strategy for developing more effective and personalized immunization approaches. The adjuvanticity of nano-vaccines underscores their potential as a transformative tool in the fight against infectious diseases and the advancement of public health goals (53).

1.8 EXAMPLES OF NANO-VACCINES IN DEVELOPMENT

Nano-vaccines offer several advantages over traditional vaccines, including improved immune response, better antigen delivery, and the ability to target multiple antigens. These next-generation vaccines are being developed for various diseases like HIV/AIDS, malaria, influenza, and tuberculosis, with some showing promise in early trials. Their unique properties have the potential to revolutionize vaccinology and improve global health outcomes.

1.8.1 Acquired Immunodeficiency Syndrome (HIV/AIDS)

A promising nano-vaccine for HIV/AIDS utilizes liposomes (nanoparticles around 150 nm) to target immune cells and stimulate a cell-mediated immune response against the virus. This approach shows promise in preclinical studies and may soon enter Phase I trials. The nano-vaccine design offers broader coverage and potentially achieves a 60% efficacy rate, surpassing traditional HIV vaccines (54, 55).

1.8.2 Malaria

Nano-vaccines for malaria, parasitic disease caused by *Plasmodium falciparum*, target the sexual stages of the plasmodium parasite, aiming to block transmission by mosquitoes. These vaccines use nanoparticles to deliver antigens and potentially boost the production of antibodies that disrupt parasite development (56). Additionally, gold nanoparticles act as adjuvants, enhancing the immune response. This promising approach is in the advanced preclinical stages and may soon progress to Phase II trials. The design offers the potential for a long-lasting effect and may overcome limitations of strain-specific immunity observed in some malaria vaccines (57, 58).

1.8.3 Influenza

Influenza, commonly known as the flu, is a contagious respiratory illness caused by influenza viruses. In order to treat influenza, a bivalent nano-vaccine using self-assembling VLPs is in early clinical development. This design has the potential for a universal influenza vaccine by stimulating both cellular and humoral immune responses against a broad range of influenza strains (59, 60).

1.8.4 Tuberculosis

Tuberculosis is an infectious disease caused by the bacteria *Mycobacterium tuberculosis*. Promising preclinical research explores chitosan nanoparticles (300–400 nm) as a delivery system for DNA vaccines against *Mycobacterium tuberculosis*. This approach aims to induce both cell-mediated and humoral immune responses for improved protection. The vaccine is expected to enter Phase I human trials soon (61, 62).

1.9 NANO-THERAPEUTICS FOR INFECTIOUS DISEASES

1.9.1 Applications in Infectious Diseases

Nano-therapeutics represent a revolutionary approach to combating infectious diseases by utilizing the power of nanotechnology, especially in the fight against infectious diseases. These innovative interventions exploit the unique physicochemical properties of nanoparticles to achieve several key advancements:

a. ***Targeted Delivery*:** Nanoparticles can be meticulously engineered based on size and surface characteristics to precisely target pathogens within the body. This enhanced targeting allows for drug delivery directly to infected cells or tissues, minimizing off-target effects and maximizing therapeutic efficacy (63).
b. ***Optimized Pharmacokinetics*:** Nano-therapeutics enable improved bioavailability and controlled drug release. This translates to increased drug

concentrations at the site of infection, ultimately leading to more potent treatment outcomes.

c. ***Overcoming Resistance Mechanisms*:** Traditional therapies often face challenges like diminishing drug activity due to bacterial or viral resistance and poor drug deposition at the intended site. Nano-therapeutics offer a potential solution by bypassing these resistance mechanisms and delivering drugs more effectively (64).

By utilizing nanocarriers for drug delivery and vaccine development, nano-therapeutics hold immense promise for revolutionizing infectious disease management. This cutting-edge field aims to eliminate the limitations of existing therapies, ultimately leading to more targeted and effective treatments with improved patient recovery rates (64).

Applications of nano-therapeutics in combating infectious diseases are listed in Figure 1.5.

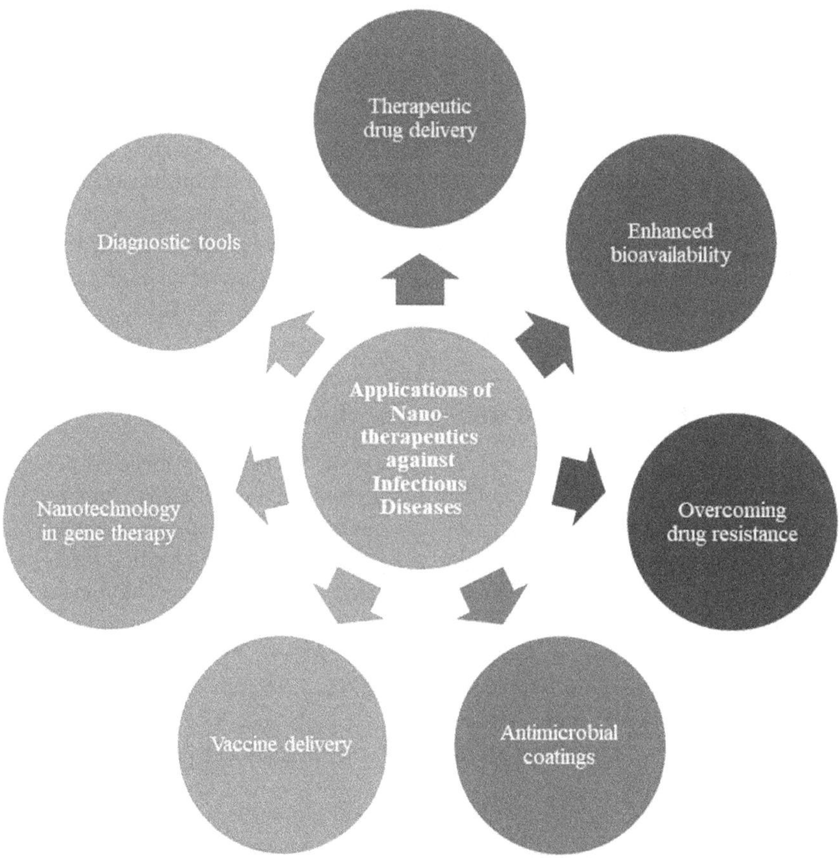

FIGURE 1.5 Applications of nano-therapeutics in fighting infectious diseases.

1.9.2 Types of Nano-Therapeutics for Infectious Diseases

The field of nano-therapeutics compiles a wide range of novel technologies aiming to transform the delivery of healthcare through the application of nanotechnology to fight diseases. Studies and diverse research have explored several types of this novel way of treatment, each with unique capabilities and possibilities. They include, for example, antimicrobial nanoparticles developed to fight resistant pathogens, as well as not less complicated drug delivery systems that can target specific cells with medication. Moreover, antibiofilm and other applications developed to fight and disrupt biofilms seek to solve a known issue in sepsis prevention. Additionally, nano-therapeutic applications for diagnostics and treatment enrichment enhance sensitivity and enable early detection. The diverse types of existing and potential nano-therapeutics portray a vast field of opportunity in existing medicine development. They dramatically improve the coverage and quality of treatment, enabling a new strategy for care allocation and patient involvement (65).

1.9.3 Antimicrobial Nanoparticles

Antimicrobial nanoparticles such as silver and gold nanoparticles exert their antimicrobial effects on a broad spectrum of pathogens and are thus critical in nanotherapeutics. The physicochemical properties of these nanoparticles specifically enable them to inhibit the growth of a wide range of pathogens such as bacteria, viruses, and fungi. As a result, these unique characteristics of antimicrobial nanotherapeutics present a remarkable approach to infectious disease treatment and the challenge of evolving AMR. The ability of nanoparticles to target and kill pathogens is a significant increase in the effectiveness of antimicrobial treatment (66). Therefore, this reveals that nanoparticles are a potential saviour to infectious disease and can significantly improve treatment.

1.9.4 Drug Delivery Systems for Targeted Therapy

Drug delivery systems employ nano-therapeutic mechanisms for targeted delivery of therapeutic agents to particular cells or tissues using nanoparticles to enhance the therapeutic ratio of treatment. These systems encase the drugs in nanoparticles and attach to targeting ligands that ensure the drug is released in a controlled (66).

Generally, the properties of nanoparticles, such as penetration of biological barriers and specific areas, can be varied and applied to various diseases and therapies. Moreover, nano-therapeutic drug delivery systems referring to specifically tailored delivery of drug medications within the desired site of therapy are getting recognition as they decrease the probable toxicity because of inaccurate administration of drugs (67). Summary of the drug therapies for the targeted delivery of nano-therapeutics is stated in Table 1.1.

1.9.5 Antibiofilm Strategies

A major challenge to combating microbial biofilms is that they are non-responsive to available treatments, necessitating the need for nano-therapeutics-based antibiofilm

TABLE 1.1
Advantages and Disadvantages of Drug Delivery Systems for Targeted Therapy

Targeting Strategy	Description	Advantages	Disadvantages
Passive targeting	Exploits the enhanced permeability and retention (EPR) effect to passively accumulate nanoparticles within diseased tissues. Leaky blood vessels and impaired lymphatic drainage in these areas promote nanoparticle extravasation and retention, making this approach particularly effective for solid tumours.	• No need for specific targeting ligands • Applicable to a broad range of diseases with leaky vasculature	• Limited targeting specificity • Potential accumulation in healthy tissues with similar vascular characteristics
Active targeting	Employs functionalized nanoparticles decorated with ligands or antibodies that bind to specific receptors or biomarkers expressed on target cells. This strategy enhances therapeutic efficacy by facilitating active interaction with target areas, overcoming biological barriers, and enabling precise drug delivery, thereby reducing systemic toxicity.	• Improved targeting specificity • Reduced off-target effects • Enhanced therapeutic efficacy	• Requires identification and conjugation of appropriate targeting ligands • May be hindered by heterogeneity in receptor expression
Stimuli-responsive targeting	Utilizes drug delivery systems designed to release therapeutic cargo in response to specific environmental cues associated with the diseased state. These stimuli, such as variations in pH, temperature, or enzymatic activity, can trigger drug release at the target site, promoting localized action and minimizing off-target effects.	• Controlled drug release at the target site • Reduced systemic exposure and off-target effects	• Complex design and engineering required • Potential for premature drug release during circulation

strategies. Biofilms cause persistent infections in healthcare facilities, and their eradication is problematic. The use of new antibiofilm strategies involving nano-therapeutic interventions with nanoparticles edited with antibiofilm agents is an effective solution to problems related to biofilm. These strategies include targeting genes of biofilm formation, disrupting the formed biofilm, and treating biofilm-related infections (68). Goals intended to be realized by nano-therapeutic antibiofilm strategies include the utilization of the advantages of nanoparticles such as the penetration between cells of microbes and the specific targeting of the microbial cells. As a result, the need for non-responsiveness to biofilms and enhanced care of infection will be an added advantage. This advancement marks a significant stride in infection treatment related to biofilms and the utilization of clinical care protocols across healthcare fields (69).

1.9.6 Diagnostics and Imaging Tools

In nano-therapeutics, nanoparticles play a vital role in promoting diagnostics and imaging tools. These nanoparticles have unique optical and magnetic properties, which help to ensure that the tools are highly sensitive to all diseases and monitoring. For instance, quantum dots and iron oxide nanoparticles have properties that help in imaging and tracking biomarkers and other molecules relating to diseases with high precision in biological systems. These tools have immense capabilities in promoting early detection of diseases, ensuring that medicine is personalized, and facilitating monitoring of responses to therapeutics in real time (70). Therefore, the use of nanoparticles and their capabilities to sensitive detection and imaging significantly improve healthcare outcomes; therefore, patients receive accurate information regarding their disease; providers offer personalized care and ensure that they monitor the effect of their interventions. The use of nanoparticles and the creation of images through diagnostics in nano-therapeutics is thus an innovative approach that increases precision and sensitivity (71).

1.10 CHALLENGES AND OPPORTUNITIES

Nanotechnology-based vaccines and therapeutics could transform the sector and result in more powerful solutions for a range of diseases. Nevertheless, the area in which they are most likely to succeed has a variety of difficulties that will need to be resolved to exploit their core competencies.

1.10.1 Regulatory Hurdles

First and foremost, significant lab testing and powerful clinical trials would be able to outlast regulatory systems. As with any other diet part, effective functioning of the Food and Drug Administration (FDA) and European Medicine Agency (EMA) require these nano-vaccines and therapeutics to be completely safe for public health use (72). Another major issue in the effort to commercialize nano-vaccines is the manufacturing process, which is a much more complicated process that requires process-specific and high-tech hardware. Even though this could increase manufacturing costs and scaling may become problematic, the ability to sustain manufacturing consistency is also vital (73).

1.10.2 Characterization and Stability

The nano-vaccines and therapeutics could be novel and complex, making it hard to characterize and instability. The small size and complexity of the structures in nano-vaccines and therapeutics can make them impossible to characterize correctly and impossible to stop from becoming denatured and denaturant over time. It is critical for patient safety and success (74).

1.10.3 Cost Effectiveness

The high costs of developing and commercializing nano-vaccines and therapeutics are a disadvantage. They entail high research and development expenditures to produce novel medical products and maybe difficult to obtain the necessary financing to develop them. In addition, manufacturing costs and difficulties scaling production can also offset the high costs, making nano-vaccines and other therapeutics more expensive and less attainable (53).

1.10.4 Robust Delivery System

A critical challenge that arises with nano-vaccines and therapeutics in the development and commercialization of the product is the dearth of robust delivery systems. Therefore, since these nano-sized particles require a different delivery mechanism to ensure they travel to the desired body location. Therefore, a delivery system for nano-vaccines and therapeutics should be developed to ensure effective and efficient results (75).

1.10.5 Safety Concerns

A significant risk is the safety and potential toxicity risks associated with nano-vaccines and therapeutics. For both further regulatory approval and broad acceptance by the populace, understanding of the bio-distribution and pharmacokinetics of nano-sized materials must be further carried out over time. In the end, bio-nano R&D will take humanity to more advanced medical products that are safer to use (76).

1.10.6 Public Awareness and Acceptance

Besides the scientific and technical challenges, there are substantial market acceptance and reimbursement challenges to making nano-vaccines and therapeutics profitable. In this regard, it is vital to focus on the education of healthcare professionals and the public concerning the benefits and perspectives for nano-vaccines and therapeutics. Moreover, it is important to consider the complexity and ambiguity of reimbursement and pricing strategies for innovative medical products given the innovative potential of the novel technology (77).

Several challenges need to be addressed to fully realize the potential of developing and commercializing nano-vaccines and therapeutics in reshaping medical and healthcare practices. Regulatory hurdles, manufacturing and scalability challenges, product characterization and stability, development costs, safety and toxicity, and market acceptance and reimbursement are necessary to be considered. Nevertheless, continued investment in research and collaboration with regulatory authorities and policymakers will help to overcome these challenges and leverage the utilization of nano-vaccines and therapeutics for the benefits of medical treatment and healthcare.

1.11 FUTURE OPPORTUNITIES IN THE FIELD OF NANO-VACCINES

The development of nano-vaccines holds several promising opportunities for future healthcare breakthroughs. Some of these include the following.

1.11.1 Personalized Medicine

Nano-vaccines developed based on their genetic profiles, health histories, and specific pathogen exposure could dramatically reduce these vaccine failures. Consequently, nano-vaccines that target an individual's immune response's specific characteristics could lead to a new era in preventative medicine and lower the cost of vaccine development (78).

1.11.2 Combination Therapies

Nano-vaccines can be combined with other pharmaceutical interventions to enhance and better some treatments. For example, if one vaccine is engineered to work with therapy and enhance the assault on a disease, the outcome is better since different pathways have been attacked or multiple disease antigens have been targeted at a go (79).

1.11.3 Artificial Intelligence in Drug Design

AI can take a considerable part in uncovering new antigens for nano-vaccines and optimizing their design. This would include creating machine learning models that could simulate how different formulations of nanoparticles could interact with the immune system. Artificial intelligence could help fasten the time it takes to develop a vaccine drug and bring it to the market from the research stage to the clinical trials and the market stage. This, in turn, would drive the public health domain in a trivial way. Despite the potential available, implementing these options would also require a significant amount of research and integration between the separate scientific domains to prepare medicine and the relevant regulatory bodies (80).

1.12 CONCLUSIONS

The prospects of nanotechnology have overhauled the landscape for the global endeavour to prevent, treat, and ultimately eliminate infectious diseases in the foreseeable future. Nano-vaccines are a valuable component of this emerging technology. Nano-vaccines include nanoparticles composed of or manufactured with substances that can stimulate the host's immune response. Apart from the vehicle, such nanocarriers were shown to have intrinsic adjuvant properties, in the role of immune cell stimulants. Thus, nano-vaccines can provoke rapid and sustained humoral and cell-based immunity. Firstly, they have the potential to deliver antigens to the desired site. Secondly, they can raise the bioavailability of antigens. Thirdly, the risk of adverse reactions could decrease. To realize the potential, numerous nanoparticle-based vaccines are under development, including VLPs, liposomes, polymeric NPs, nanogels, lipid NPs, and dendrimers. Researchers are actively investigating nano-vaccines for a variety of applications, including cancer immunotherapy and infectious diseases. The technology, however, poses several challenges, including longer time stability, safety concerns, cost challenges, and scalability for marketing. The projections are promising, with a range of possibilities such as personalized vaccines, needle-free formulations, or combinatorial nano-vaccines, with a wide range of potential candidate vaccines. Nonetheless, this is an avant-garde technology that requires thorough investigation in order to utilize our civilization in real pandemics.

REFERENCES

1. Celis-Giraldo CT, López-Abán J, Muro A, Patarroyo MA, Manzano-Román R. Nanovaccines against animal pathogens: The latest findings. Vaccines. 2021;9(9):988. https://doi.org/10.3390/vaccines9090988.
2. Pati R, Shevtsov M, Sonawane A. Nanoparticle vaccines against infectious diseases. Frontiers in Immunology. 2018;9:385476. https://doi.org/10.3389/fimmu.2018.02224.
3. Cifuentes-Rius A, Desai A, Yuen D, Johnston AP, Voelcker NH. Inducing immune tolerance with dendritic cell-targeting nanomedicines. Nature Nanotechnology. 2021; 16(1):37–46. https://doi.org/10.1038/s41565-020-00810-2.
4. Zazo H, Colino CI, Lanao JM. Current applications of nanoparticles in infectious diseases. Journal of Controlled Release. 2016;224:86–102. https://doi.org/10.1016/j.jconrel.2016.01.008.
5. Chuang ST, Conklin B, Stein JB, Pan G, Lee K-B. Nanotechnology-enabled immunoengineering approaches to advance therapeutic applications. Nano Convergence. 2022;9(1):19. https://doi.org/10.1186/s40580-022-00310-0.
6. Harun-Ur-Rashid M, Jahan I, Imran AB. Recent advancement of nanostructured materials for clinical challenges in vaccinology. Nanovaccinology: clinical application of nanostructured materials research to translational medicine: Springer; 2023. p. 135–60. https://doi.org/10.1007/978-3-031-35395-6_8.
7. Omar RF, Boissinot M, Huletsky A, Bergeron MG. Tackling infectious diseases with rapid molecular diagnosis and innovative prevention. Infectious Disease Reports. 2024;16(2):216–27. https://doi.org/10.3390/idr16020017.
8. Sanders D. The struggle for health: medicine and the politics of underdevelopment: Oxford University Press; 2023.
9. Ramírez-Rendon D, Passari AK, Ruiz-Villafán B, Rodríguez-Sanoja R, Sánchez S, Demain AL. Impact of novel microbial secondary metabolites on the pharma industry. Applied Microbiology and Biotechnology. 2022;106(5):1855–78. https://doi.org/10.1007/s00253-022-11821-5.
10. Parkins MD, Lee BE, Acosta N, Bautista M, Hubert CR, Hrudey SE, et al. Wastewater-based surveillance as a tool for public health action: SARS-CoV-2 and beyond. Clinical Microbiology Reviews. 2024;37(1):e00103–22. https://doi.org/10.1128/cmr.00103-22.
11. Pal K. Nanovaccinology: clinical application of nanostructured materials research to translational medicine: Springer Nature; 2023.
12. Montero DA, Vidal RM, Velasco J, Carreño LJ, Torres JP, Benachi O MA, et al. Two centuries of vaccination: historical and conceptual approach and future perspectives. Frontiers in Public Health. 2024;11:1326154. https://doi.org/10.3389/fpubh.2023.1326154.
13. Wei C-J, Crank MC, Shiver J, Graham BS, Mascola JR, Nabel GJ. Next-generation influenza vaccines: Opportunities and challenges. Nature Reviews Drug Discovery. 2020;19(4):239–52. https://doi.org/10.1038/s41573-019-0056-x.
14. Weinberger B. Vaccines and vaccination strategies for older adults. Healthy Longevity and Immune System. 2022:119–64. https://doi.org/10.1007/978-3-030-87532-9_7.
15. Yadav DK, Yadav N, Khurana SMP. Vaccines: present status and applications. Animal biotechnology: Elsevier; 2020. p. 523–42. https://doi.org/10.1016/B978-0-12-811710-1.00024-0.
16. Sarkar A, Maitra T, Neogy S. Blockchain in healthcare system: security issues, attacks and challenges. Blockchain technology: applications and challenges: Springer; 2021. p. 113–33. https://doi.org/10.1007/978-3-030-69395-4_7.

17. Marimani M. Combination therapy against multidrug resistance. Combination therapy against multidrug resistance: Elsevier; 2020. p. 39–64. https://doi.org/10.1016/B978-0-12-820576-1.00002-3.
18. Catalano A, Iacopetta D, Ceramella J, Scumaci D, Giuzio F, Saturnino C, et al. Multidrug resistance (MDR): A widespread phenomenon in pharmacological therapies. Molecules. 2022;27(3):616. https://doi.org/10.3390/molecules27030616.
19. Thomas S, Dilbarova R, Rappuoli R. Future challenges for vaccinologists. Vaccine design: methods and protocols: Volume 1: Vaccines for human diseases: 2016. p. 41–45. https://doi.org/10.1007/978-1-4939-3387-7_2.
20. Gopikrishnan M, Haryini S. Emerging strategies and therapeutic innovations for combating drug resistance in Staphylococcus aureus strains: A comprehensive review. Journal of Basic Microbiology. 2024. https://doi.org/10.1002/jobm.202300579.
21. Contera S, Bernardino de la Serna J, Tetley TD. Biotechnology, nanotechnology and medicine: Portland Press Ltd.; 2020. p. 551–54. https://doi.org/10.1042/ETLS20200350.
22. Chen S, Huang X, Xue Y, Álvarez-Benedicto E, Shi Y, Chen W, et al. Nanotechnology-based mRNA vaccines. Nature Reviews Methods Primers. 2023;3(1):63. https://doi.org/10.1038/s43586-023-00246-7.
23. Mubeen B, Ansar AN, Rasool R, Ullah I, Imam SS, Alshehri S, et al. Nanotechnology as a novel approach in combating microbes providing an alternative to antibiotics. Antibiotics. 2021;10(12):1473. https://doi.org/10.3390/antibiotics10121473.
24. Trivedi S, Kamdar A, Bhoyar V, Belgamwar V. Nanodiagnostics and targeted drug delivery: integrated technologies. Nanotechnology principles in drug targeting and diagnosis: Elsevier; 2023. p. 353–93. https://doi.org/10.1016/B978-0-323-91763-6.00011-4.
25. Lateef A, Elegbede JA, Akinola PO, Ajayi VA. Biomedical applications of green synthesized-metallic nanoparticles: a review. Pan African Journal of Life Sciences. 2019;3: 157–82. https://doi.org/10.36108/pajols/9102/30(0170).
26. Singha S, Shao K, Ellestad KK, Yang Y, Santamaria P. Nanoparticles for immune stimulation against infection, cancer, and autoimmunity. ACS Nano. 2018;12(11):10621–35. https://doi.org/10.1021/acsnano.8b05950.
27. Wang F, Ullah A, Fan X, Xu Z, Zong R, Wang X, Chen G. Delivery of nanoparticle antigens to antigen-presenting cells: from extracellular specific targeting to intracellular responsive presentation. Journal of Controlled Release. 2021;333:107–28. https://doi.org/10.1016/j.jconrel.2021.03.027.
28. Mitchell MJ, Billingsley MM, Haley RM, Wechsler ME, Peppas NA, Langer R. Engineering precision nanoparticles for drug delivery. Nature Reviews Drug Discovery. 2021;20(2):101–24. https://doi.org/10.1038/s41573-020-0090-8.
29. Liu Y, Yang G, Jin S, Xu L, Zhao CX. Development of high-drug-loading nanoparticles. ChemPlusChem. 2020;85(9):2143–57. https://doi.org/10.1002/cplu.202000496.
30. Garg A, Dewangan HK. Nanoparticles as adjuvants in vaccine delivery. Critical Reviews™ in Therapeutic Drug Carrier Systems. 2020;37(2). https://doi.org/10.1615/CritRevTherDrugCarrierSyst.2020033273.
31. Kim D, Shin K, Kwon SG, Hyeon T. Synthesis and biomedical applications of multifunctional nanoparticles. Advanced Materials. 2018;30(49):1802309. https://doi.org/10.1002/adma.201802309.
32. Singh AK, Yadav TP, Singh G, Pritam M, Pandey B, Ansari MI, et al. Nanosize carriers for drug and vaccine delivery: Advances and challenges. Nanoscience & Nanotechnology-Asia. 2021;11(6):64–86. https://doi.org/10.2174/2210681211666210505115217.
33. Rosales-Mendoza S, González-Ortega O, Rosales-Mendoza S, González-Ortega O. The mucosal immune system: an outlook for nanovaccines development. Nanovaccines: an innovative technology to fight human and animal diseases: 2019: 15–35. https://doi.org/10.1007/978-3-030-31668-6_2.

34. Durán-Lobato M, López-Estévez AM, Cordeiro AS, Dacoba TG, Crecente-Campo J, Torres D, Alonso MJ. Nanotechnologies for the delivery of biologicals: Historical perspective and current landscape. Advanced Drug Delivery Reviews. 2021;176:113899. https://doi.org/10.1016/j.addr.2021.113899.
35. Facciolà A, Visalli G, Laganà P, La Fauci V, Squeri R, Pellicanò G, et al. The new era of vaccines: The "nanovaccinology". European Review for Medical & Pharmacological Sciences. 2019;23(16):7163–82.
36. Petkar KC, Patil SM, Chavhan SS, Kaneko K, Sawant KK, Kunda NK, Saleem IY. An overview of nanocarrier-based adjuvants for vaccine delivery. Pharmaceutics. 2021; 13(4):455. https://doi.org/10.3390/pharmaceutics13040455.
37. Al-Halifa S, Gauthier L, Arpin D, Bourgault S, Archambault D. Nanoparticle-based vaccines against respiratory viruses. Frontiers in Immunology. 2019;10:435389. https://doi.org/10.3389/fimmu.2019.00022.
38. Diaz-Arévalo D, Zeng M. Nanoparticle-based vaccines: opportunities and limitations. Nanopharmaceuticals: Elsevier; 2020. p. 135–50. https://doi.org/10.1016/B978-0-12-817778-5.00007-5.
39. Nooraei S, Bahrulolum H, Hoseini ZS, Katalani C, Hajizade A, Easton AJ, Ahmadian G. Virus-like particles: Preparation, immunogenicity and their roles as nanovaccines and drug nanocarriers. Journal of Nanobiotechnology. 2021;19:1–27. https://doi.org/10.1186/s12951-021-00806-7.
40. Gupta R, Arora K, Roy SS, Joseph A, Rastogi R, Arora NM, Kundu PK. Platforms, advances, and technical challenges in virus-like particles-based vaccines. Frontiers in Immunology. 2023;14:1123805. https://doi.org/10.3389/fimmu.2023.1123805.
41. Kisakova LA, Apartsin EK, Nizolenko LF, Karpenko LI. Dendrimer-mediated delivery of DNA and RNA vaccines. Pharmaceutics. 2023;15(4):1106. https://doi.org/10.3390/pharmaceutics15041106.
42. Sefidi-Heris Y, Jahangiri A, Mokhtarzadeh A, Shahbazi M-A, Khalili S, Baradaran B, et al. Recent progress in the design of DNA vaccines against tuberculosis. Drug Discovery Today. 2020;25(11):1971–87. https://doi.org/10.1016/j.drudis.2020.09.005.
43. Gregorio NE, Levine MZ, Oza JP. A user's guide to cell-free protein synthesis. Methods and Protocols. 2019;2(1):24. https://doi.org/10.3390/mps2010024.
44. To KK, Cho WC. An overview of rational design of mRNA-based therapeutics and vaccines. Expert Opinion on Drug Discovery. 2021;16(11):1307–17. https://www.tandfonline.com/doi/abs/10.1080/17460441.2021.1935859.
45. Uddin MN, Roni MA. Challenges of storage and stability of mRNA-based COVID-19 vaccines. Vaccines. 2021;9(9):1033. https://doi.org/10.3390/vaccines9091033.
46. Ang MJY, Chan SY, Goh Y-Y, Luo Z, Lau JW, Liu X. Emerging strategies in developing multifunctional nanomaterials for cancer nanotheranostics. Advanced Drug Delivery Reviews. 2021;178:113907. https://doi.org/10.1016/j.addr.2021.113907.
47. Kishimoto TK, Maldonado RA. Nanoparticles for the induction of antigen-specific immunological tolerance. Frontiers in Immunology. 2018;9:319730. https://doi.org/10.3389/fimmu.2018.00230.
48. Yadav HK, Dibi M, Mohammad A, Srouji AE. Nanovaccines formulation and applications – A review. Journal of Drug Delivery Science and Technology. 2018;44:380–87. https://doi.org/10.1016/j.jddst.2018.01.015.
49. Ahmad MZ, Ahmad J, Haque A, Alasmary MY, Abdel-Wahab BA, Akhter S. Emerging advances in synthetic cancer nano-vaccines: Opportunities and challenges. Expert Review of Vaccines. 2020;19(11):1053–71. https://pubmed.ncbi.nlm.nih.gov/33315512/.
50. Li J, Ren H, Zhang Y. Metal-based nano-vaccines for cancer immunotherapy. Coordination Chemistry Reviews. 2022;455:214345. https://doi.org/10.1016/j.ccr.2021.214345.

51. Khojini JY, Babaei B, Shakarami M, Mofidi M, Tahershamsi Z, Arjmand TF, et al. Biomimetic nanovaccines: A novel approach in immunization. Current Pharmaceutical Design. 2023;29(18):1391–1408. https://doi.org/10.2174/1381612829666230529094128.
52. Moni SS, Abdelwahab SI, Jabeen A, Elmobark ME, Aqaili D, Ghoal G, et al. Advancements in vaccine adjuvants: The journey from alum to nano formulations. Vaccines. 2023;11(11):1704. https://doi.org/10.3390/vaccines11111704.
53. Nath R, Mandal A, Bhowmik R, Roy R, Biswas R, Bhattacharyya S. A pervasive review on new advancements of nano vaccines on Covid-19 pandemic. International Journal of Pharmaceutical Sciences Review and Research. 2021;70(2). http://dx.doi.org/10.47583/ijpsrr.2021.v70i02.015.
54. Yenkoidiok-Douti L, Jewell CM. Integrating biomaterials and immunology to improve vaccines against infectious diseases. ACS Biomaterials Science & Engineering. 2020; 6(2):759–78. https://doi.org/10.1021/acsbiomaterials.9b01255.
55. Liu S, Hu M, Liu X, Liu X, Chen T, Zhu Y, et al. Nanoparticles and antiviral vaccines. Vaccines. 2023;12(1):30. https://doi.org/10.3390/vaccines12010030.
56. Duffy PE. Transmission-blocking vaccines: Harnessing herd immunity for malaria elimination. Expert Review of Vaccines. 2021;20(2):185–98. https://pubmed.ncbi.nlm.nih.gov/33478283/.
57. Huang WC, Sia ZR, Lovell JF. Adjuvant and antigen systems for malaria transmission-blocking vaccines. Advanced Biosystems. 2018;2(10):1800011. https://doi.org/10.1002/adbi.201800011.
58. Deshmukh R, Dewangan B, Harwansh RK, Agrawal R, Garg A, Chopra H. Current trends in nanotechnology-based drug delivery systems for the diagnosis and treatment of malaria: A review. Current Drug Delivery. 2024. https://doi.org/10.2174/0115672018291253240115012327.
59. Stepanova E, Isakova-Sivak I, Rudenko L. Options for the development of a bivalent vaccine against SARS-CoV-2 and influenza. Expert Review of Vaccines. 2022;21(11): 1533–35. https://pubmed.ncbi.nlm.nih.gov/36004567/.
60. Teng Z, Meng L-Y, Yang J-K, He Z, Chen X-G, Liu Y. Bridging nanoplatform and vaccine delivery, a landscape of strategy to enhance nasal immunity. Journal of Controlled Release. 2022;351:456–75. https://doi.org/10.1016/j.jconrel.2022.09.044.
61. Giri TK, Ghosh B, Patra F. Chitosan-based nanoparticulate system for pulmonary drug delivery. Polysaccharide based nano-biocarrier in drug delivery: CRC Press; 2018. p. 273–89.
62. Das A, Ali N. Nanovaccine: An emerging strategy. Expert Review of Vaccines. 2021;20(10):1273–90. https://pubmed.ncbi.nlm.nih.gov/34550859/.
63. Zaidi S, Misba L, Khan AU. Nano-therapeutics: A revolution in infection control in post antibiotic era. Nanomedicine: Nanotechnology, Biology and Medicine. 2017;13(7): 2281–301. https://doi.org/10.1016/j.nano.2017.06.015.
64. Ghosal A, Vashist A, Tiwari S, Sharmin E, Ahmad S, Bhattacharya J. Nanotechnology for therapeutics. Advances in personalized nanotherapeutics: 2017. p. 25–40. https://doi.org/10.1007/978-3-319-63633-7_3.
65. Chavda VP. Nanotherapeutics and nanobiotechnology. Applications of targeted nano drugs and delivery systems: Elsevier; 2019. p. 1–13. https://doi.org/10.1016/B978-0-12-814029-1.00001-6.
66. Harish V, Tewari D, Gaur M, Yadav AB, Swaroop S, Bechelany M, Barhoum A. Review on nanoparticles and nanostructured materials: Bioimaging, biosensing, drug delivery, tissue engineering, antimicrobial, and agro-food applications. Nanomaterials. 2022;12(3):457. https://doi.org/10.3390/nano12030457.
67. Manzari MT, Shamay Y, Kiguchi H, Rosen N, Scaltriti M, Heller DA. Targeted drug delivery strategies for precision medicines. Nature Reviews Materials. 2021;6(4):351–70. https://doi.org/10.1038/s41578-020-00269-6.

68. Mohanta YK, Chakrabartty I, Mishra AK, Chopra H, Mahanta S, Avula SK, et al. Nanotechnology in combating biofilm: A smart and promising therapeutic strategy. Frontiers in Microbiology. 2023;13:1028086. https://doi.org/10.3389/fmicb.2022.1028086.
69. Subhaswaraj P, Syed A, Siddhardha B. Novel nanotherapeutics as next-generation anti-infective agents: Current trends and future prospectives. Current Drug Discovery Technologies. 2020;17(4):457–68. https://doi.org/10.2174/1570163816666190715120708.
70. Woźniak M, Płoska A, Siekierzycka A, Dobrucki LW, Kalinowski L, Dobrucki IT. Molecular imaging and nanotechnology—Emerging tools in diagnostics and therapy. International Journal of Molecular Sciences. 2022;23(5):2658. https://doi.org/10.3390/ijms23052658.
71. Salih S, Elliyanti A, Alkatheeri A, AlYafei F, Almarri B, Khan H. The role of molecular imaging in personalized medicine. Journal of Personalized Medicine. 2023;13(2):369. https://doi.org/10.3390/jpm13020369.
72. Cordaillat-Simmons M, Rouanet A, Pot B. Live biotherapeutic products: The importance of a defined regulatory framework. Experimental & Molecular Medicine. 2020; 52(9):1397–406. https://doi.org/10.1038/s12276-020-0437-6.
73. Aqib AI, Zaheer T, Usman M, Arslan M, Atta K. Commercialization of nanovaccines: Utopia or a reality? Nanovaccinology as targeted therapeutics: 2022. p. 267–82. https://doi.org/10.1002/9781119858041.ch17.
74. Zaheer T, Pal K, Zaheer I. Topical review on nano-vaccinology: Biochemical promises and key challenges. Process Biochemistry. 2021;100:237–44. https://doi.org/10.1016/j.procbio.2020.09.028.
75. Halwani AA. Development of pharmaceutical nanomedicines: From the bench to the market. Pharmaceutics. 2022;14(1):106. https://doi.org/10.3390/pharmaceutics14010106.
76. Zielińska A, Costa B, Ferreira MV, Miguéis D, Louros J, Durazzo A, et al. Nanotoxicology and nanosafety: Safety-by-design and testing at a glance. International Journal of Environmental Research and Public Health. 2020;17(4657). https://doi.org/10.3390/ijerph17134657.
77. Shan X, Gong X, Li J, Wen J, Li Y, Zhang Z. Current approaches of nanomedicines in the market and various stage of clinical translation. Acta Pharmaceutica Sinica B. 2022;12(7):3028–48. https://doi.org/10.1016/j.apsb.2022.02.025.
78. Clack K, Soda N, Kasetsirikul S, Mahmudunnabi RG, Nguyen NT, Shiddiky MJ. Toward personalized nanomedicine: The critical evaluation of micro and nanodevices for cancer biomarker analysis in liquid biopsy. Small. 2023;19(15):2205856. https://doi.org/10.1002/smll.202205856.
79. Abusalah MAH, Chopra H, Sharma A, Mustafa SA, Choudhary OP, Sharma M, et al. Nanovaccines: A game changing approach in the fight against infectious diseases. Biomedicine & Pharmacotherapy. 2023;167:115597. https://doi.org/10.1016/j.biopha.2023.115597.
80. Alshawwa SZ, Kassem AA, Farid RM, Mostafa SK, Labib GS. Nanocarrier drug delivery systems: Characterization, limitations, future perspectives and implementation of artificial intelligence. Pharmaceutics. 2022;14(4):883. https://doi.org/10.3390/pharmaceutics14040883.

2 Mechanistic Insights of Natural Plants and Microbial Extract Synthesis of Metal Nanoparticles and Nanocomposites

Naheed Bibi, Aisha Kanwal, Bushra Begum, Zahida Wasil, Nargis Jamila, Raina Aman Qazi, and Nargis Aman

2.1 INTRODUCTION TO NANOBIOTECHNOLOGY

Nanobiotechnology plays a vital role in constructing atomic or molecular-level nanomachines by incorporating principles of biological systems for producing a variety of materials with improved physicochemical properties. It is also used for evaluating the efficacy of nanomaterials and establishing new or modified existing processes to find solutions to different systems that are expected to revolutionize their applications in numerous arenas (Pal 2021, Show et al. 2023).

Multifunctional nanomachines with enhanced performance, selectivity, and sensitivity have been made possible by this unique fusion of nanotechnology and life sciences. Besides, the advances in this field have rendered it convenient for researchers to work on cutting-edge research to further expand our understanding of how to apply nanobiotechnology to diverse functional devices and nanobiosystems (Show et al. 2023). Henceforth, nanobiotechnology offers simple, unconventional, rapid, cost-efficient, and innovative solutions for tackling the existing different challenges worldwide (Amin, Hwang, and Park 2011).

2.2 METAL-BASED NANOMATERIALS

The rapid progressions in nanobiotechnology pave the way for generating novel inorganic nanosystems, particularly, metal-based nanomaterials (Adeyemi et al. 2022, Li et al. 2023). Metal-based nanomaterials take different forms; (1) monometallic

 DOI: 10.1201/9781003378563-2

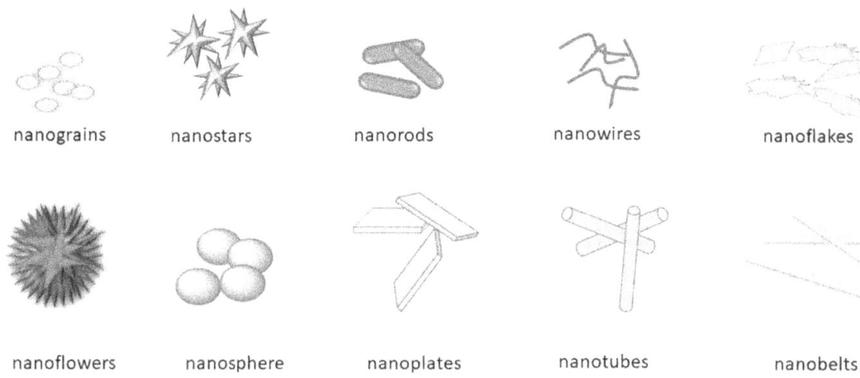

FIGURE 2.1 Various shapes of metal-based nanostructures.

nanoparticles (NPs), (2) bimetallic NPs/nanoalloys (composed of two different metals), (3) metal oxide NPs, (4) metal sulfides NPs, and (5) metal nanocomposites (combination of metal with metal oxide/sulfide/etc.) (Adeyemi et al. 2022, Berta et al. 2021, Vijayaram et al. 2023). In addition to NPs, these nanostructures are not limited to only nanofibers, nanowires, nanoflakes, nanoflowers, nanograins, and nanoplates, but they have also been designed in the form of nanospheres, nanorods, nanostars, nanotubes, nanobelts, etc. as shown in Figure 2.1. These shapes of metal-based nanomaterials are strongly reliant on reaction parameters (Sharwani et al. 2022, Vijayaram et al. 2023).

Monometallic NPs which have garnered significant attention from researchers are gold (Au), silver (Ag), zinc (Zn), platinum (Pt), palladium (Pd), selenium (Se), iron (Fe), copper (Cu), etc. (Adeyemi et al. 2022, Berta et al. 2021). Most notable metal oxide NPs include zinc oxide (ZnO), iron oxide (Fe_2O_3), nickel oxide (NiO), cerium oxide (CeO_2), titanium dioxide (TiO_2), and so on (Adeyemi et al. 2022). Likewise, metal sulfide NPs such as zinc sulfide (ZnS), silver sulfide (Ag_2S), copper sulfide (CuS), and cadmium sulfide (CdS) have been reported (Sharma et al. 2015). Bimetallic NPs for example Ag-Au, Ag-Co, Ag-Pd, Au-Pt, Cu-Zn, Au-Pd, Ag-Mn, Pt-Pd, Fe-Ag, Ag-Ni, Ag-Cu, etc. have also been reported. The bimetallic NPs are prepared in a wide range of structural geometries (spherical, hexagonal, triangular, cubical, and so on) with dimensions of 3–80 nm (Berta et al. 2021).

Furthermore, metal-based nanocomposites have been designed by a combination of metal with other materials like metal oxides and sulfide (cobalt-nickel sulfide ($CoNi_2S_4$)). Metal-based nanocomposites with the other metal oxides include binary (Ag-ZnO, Zn-SiO_2, Au/Fe_3O_4, Cu-Fe_3O_4, MgO-CuO, CdO-ZnO, Pd/Fe_3O_4, NiO-ZnO, CuO-Bi_2O_3, CuO-ZnO, Ni-TiO_2, $CuWO_4$, ZnO-SnO_2, etc.), ternary ($CuFe_2O_4$@Ag, CdO-NiO-ZnO, CdO-MgO-Fe_2O_3, CeO_2/ZnO/$ZnAl_2O_4$, CeO_2.CuO. ZnO, $CuWO_4$/NiO, Pd/Au/Fe_3O_4, $CoTiO_3$/$CoFe_2O_3$), and quaternary (ZnO-ZnS-CdO-CdS) nanocomposites (Berta et al. 2021, Khan et al. 2023, Sharwani et al. 2022). Compared to other metal-based nanomaterials, nanocomposites are preferred due to their well-understood physical, chemical, electrochemical, optical, and structural features (Azhar et al. 2021). Metal-based nanocomposites, in accurate design, are expected to provide innovative prospects for the upgradation of state-of-the-art technologies. However, further exploration is necessary in designing metal nanocomposites with control size, porosity, and morphology along with improved surface area. Also, the assessment of the probable toxicity, biosafety, and biocompatibility of these metal nanocomposites is highly required (Yadav, Rani, and Saini 2022).

The size and shape of the nanomaterial are important parameters that must be carefully monitored in designing and tailoring the functions of a metal-based nanomaterial. Numerous studies in literature reveal that the reduction of size to nanometer scale and new structural changes induces special features into the metal-based nanomaterials (Adeyemi et al. 2022, Dikshit et al. 2021). Metal-based nanomaterials exhibit several valuable physicochemical properties including electrical conductance, eco-friendliness, and uniformity. Moreover, they show exceptional catalytic activity, relatively better chemical and thermal stability, and enhanced optical properties, which makes them more appropriate. The above-mentioned striking features of metal-based nanomaterials is catalyzing the world of nanobiotechnology for their application in biosensors, drug-delivery devices, neural circuits, disease diagnostics, tissue engineering, catalysis, cosmetic agents, etc. (Dutt et al. 2023, Singh et al. 2023). Nevertheless, other most important and extensive uses of these nanomaterials are in the field of nanomagnets, food packaging, bioremediation, and antimicrobial studies (Amin, Hwang, and Park 2011, Dikshit et al. 2021).

2.3 SYNTHESIS PROCEDURES OF METAL-BASED NANOMATERIALS

Though some metal-based nanomaterials are naturally occurring (Ag, Au, Fe, Mn, etc.), yet many have been fabricated with desirable functionality which enable one to fine-tune the intrinsic characteristics of the materials (Dikshit et al. 2021, Sharma et al. 2015). The unique features of metal-based nanomaterials strongly depend upon the synthesis procedure. There are different methods to prepare metal-based nanomaterials namely physical methods, chemical methods, and biological methods. Figure 2.2 shows various methodologies of the preparation of variety of metal-based nanomaterials (Dikshit et al. 2021, Guleria et al. 2022).

These methods are usually carried out via two approaches named as top-down approach (physical methods) and bottom-up approach (chemical and biological methods) (Dutt et al. 2023). The top-down method includes size reduction or destruction involving fragmentation of bulk of a material into smaller particles leading to the production of nanomaterials. Conversely, bottom-up is a constructive approach and takes place by self-combining of smaller particles into nanomaterials (Figure 2.3) (Chandrakala, Aruna, and Angajala 2022).

Physical methods involve various techniques including mechanical ball milling, thermal decomposition, spray pyrolysis, sputtering, photoirradiation, ultrasonication, laser ablation, and lithography (Guleria et al. 2022).

The commonly adopted chemical methods for the fabrication of metal-based nanomaterials include chemical reduction, solvothermal, pyrolysis, photochemical, co-precipitation, sonochemical, chemical vapor deposition, electrolysis, and sol-gel methodologies (Dikshit et al. 2021).

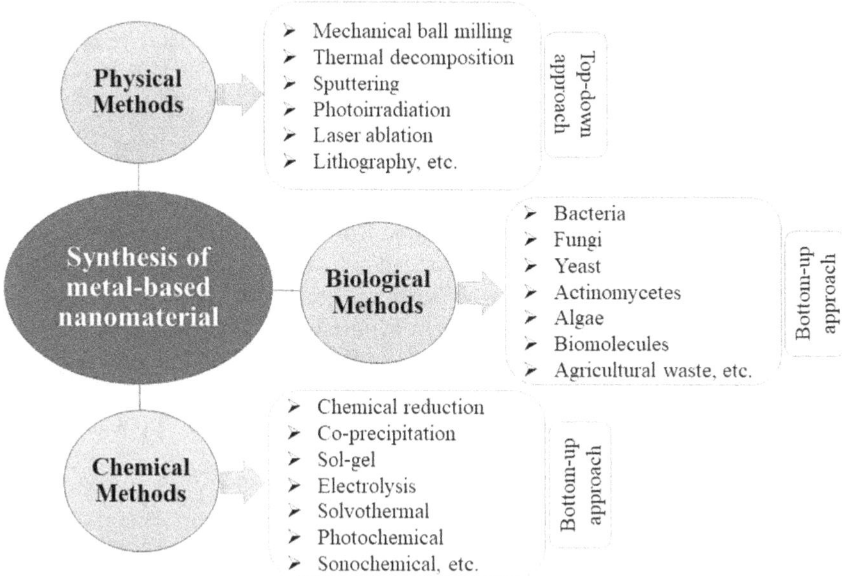

FIGURE 2.2 Synthesis methods for metal-based nanomaterials.

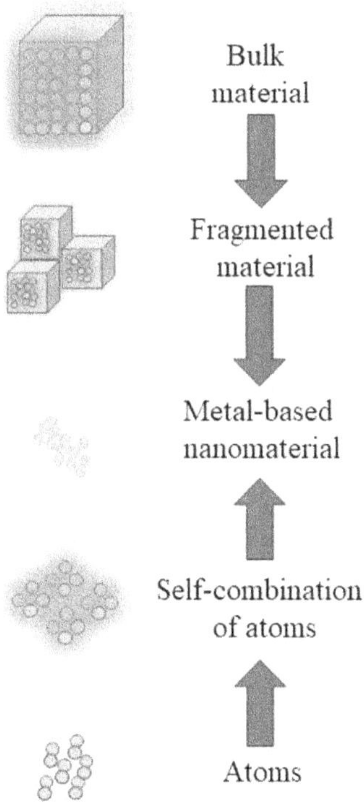

FIGURE 2.3 Schematic illustration of various approaches to metal-based nanomaterials synthesis.

Both these physical and chemical techniques are less efficient, expensive, time-consuming, and need intensive labor. They also pose severe ecological threats owing to the use of harmful chemicals along with the discharge of toxic side products. Besides, these methods are also constrained by the lack of predictability of the shape and size of the nanomaterial (Chandrakala, Aruna, and Angajala 2022). Henceforth, it calls for alternative cleaner, greener, simple, efficient, and scalable methods, which alleviate the above-mentioned drawbacks in the conventional synthesis techniques.

An alternative method is biological method which comprises biogenic materials, for instance, biomolecules, plants, microorganisms (bacteria, fungi, Actinomycetes, yeast, algae), and industrial wastes (Vijayaram et al. 2023). Biological synthesis methods, colloquially named as "green synthesis" or "biogenic synthesis" are advantageous in terms of their non-toxicity, simplicity, high yield, and well-defined morphology. Concurrently, it allows easy tuning of the electronic and structural properties of the nanomaterials to enable them for a wide range of applications.

2.4 GREEN SYNTHESIS

Nanobiotechnology involves the production of NPs utilizing biosynthetic and non-polluting methods and is one of the most significant and rapidly developing areas of study. NPs typically measure less than 100 nm in size and have a huge surface-volume ratio. Due to their captivating mechanical, sensing, physical, magnetic, electronic, and optical capabilities, NPs have drawn the attention of the researchers worldwide and differ greatly from their bulk counterparts in many ways (Karthik et al. 2018). Numerous industries, including energy, food, engineering, cosmetics, agriculture, and medicine, have extensively utilized NPs due to their unique natural, biological, and physical features (Wang, Fang, and Megharaj 2014). The biosynthesis of NPs, also known as the green synthesis, has got attention due to the biocompatibility, low toxicity, and sustainability of the process (Khan and Bano 2016). Figure 2.4 depicts the biological synthesis of metal NPs (MNPs) and nanocomposites from various sources like plants extracts, bio-polymers, algae, bacteria, and fungi as precursors.

2.5 PRECURSOR MATERIAL

2.5.1 Green Source

2.5.1.1 From Plants Extracts

An important area of biosynthesis of MNPs is utilizing plant resources. Using plants rather than other environmentally friendly biological systems like bacteria and fungi has a number of benefits, such as low cost and quicker isolation as well as facile handling procedures. Besides this, plant-based biosynthesis can easily be scaled up for the mass production of NPs. Plants contain a variety of bioactive compounds, such as enzymes, proteins, amino acids, terpenoids, polysaccharides, phenolics, vitamins, alkaloids, flavones, and organic acids, which possess the ability to reduce metal ions (Mustapha et al. 2022). An extensive literature is available where the synthesis of MNPs from leaves, flowers, seeds, bark (Giri et al. 2022), fruits, and roots has been done (Chopra et al. 2022). The plant extracts function as stabilizing and

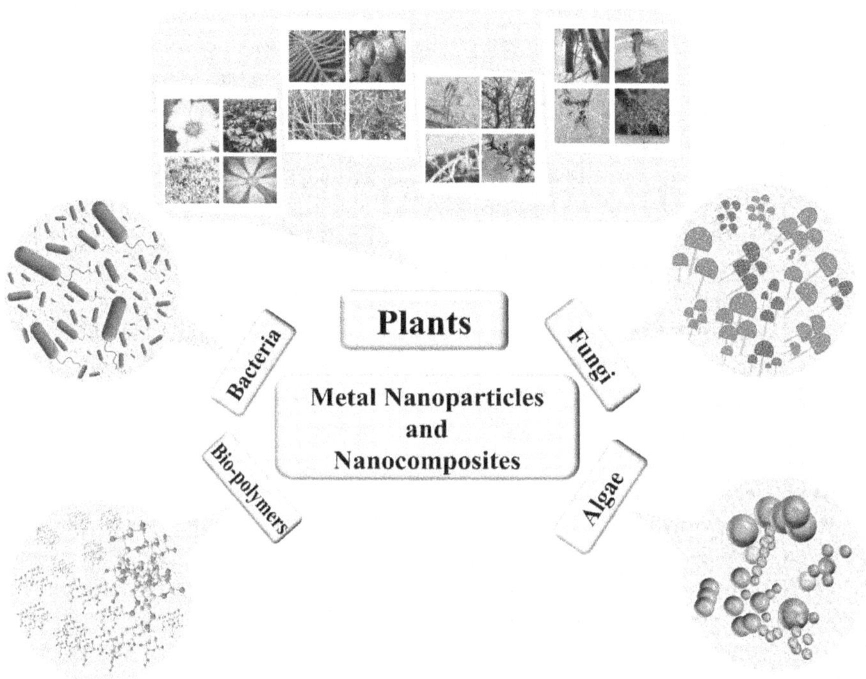

FIGURE 2.4 Biological synthesis of metal nanoparticles and nanocomposites from various sources.

reducing agents. Alfalfa sprouts have been used for the fabrication of metallic NPs (Gardea-Torresdey et al. 2003). Alfalfa roots have the capacity to take Ag^+ from agar media and transport it into plant shoots in the same oxidation state. These ions in the shoots organize themselves to form AgNPs. In order to synthesize gold and silver NPs of various shapes, Sasidharan et al. used embelin (3-undecyl-2,5-dihydroxy-1,4-benzoquinone), an ancient medicinal substance extracted from the *Embelia tsjeriam-cottam* fruit. Embelin forms complexes with metals and has antioxidant potential, making it a potential agent for reducing metal ions (Sasidharan et al. 2018). *Eugenia roxburghii*-leaf-extract-based AgNPs were discovered to be highly effective against bacterial biofilm development (Giri et al. 2022). The *Orobanche aegyptiaca* stem extract was used in the fabrication of CuNPs, for antibacterial activity against *Escherichia coli* and *Staphylococcus aureus* as well as nematicidal characteristics against *Meloidogyne incognita* (Akhter et al. 2020). Anthocyanins, flavonoids, oxalic acids, sugars, and other phenolic substances that can act as antioxidants and reducing agents are found in fruit extracts. The Cornelian cherry fruit extracts based on Fe_2O_3 NPs have been found to enhance the growth of barley (Rostamizadeh et al. 2020). *Lycium chinense* fruit/flower extracts have been reported to facilitate the production of Au and Ag NPs, that demonstrated an inhibitory effect toward *S. aureus* and *E. coli* infections (Chokkalingam et al. 2019). Despite the benefits of MNPs, agglomeration is sometimes unavoidable.

2.5.1.2 From Bacteria

Microorganisms serve as mini-factories to facilitate the economical and environmentally friendly synthesis of NPs since they are low-cost, sustainable reducing agents. The microbial source, intracellular or extracellular fabrication, metal ion concentration, incubating temperature, production time/incubation time, pH, and simplicity of extraction are important factors in microorganism-assisted NP production. This approach is limited by factors such as culture contamination, long procedures, and poor control over NP size (Rafique et al. 2017). Metals and metalloids-based NPs are typically synthesized by extracellular enzymes generated by various bacterial species. When some donors, such as–OH groups, donate electrons to metal ions, metallic NPs are formed. Extracellular enzymes such as nitrate reductase can aid in this process. It has been discovered that the functional groups of bacterial proteins, such as –OH, $-NH_2$, –COOH, and –SH, help to stabilize NPs by serving as interaction sites for metal ions, followed by reduction, and hence forming NPs in the cell-wall or periplasm of the cell. Intracellular enzymes such as cytochrome oxidases have also been identified to assist the reduction process of metallic ions via electron transfer, including cytoplasmic substances (e.g., nicotinamide adenine dinucleotide + hydrogen [NADH] and nicotinamide adenine dinucleotide phosphate + hydrogen [NADPH]), nutrients, and organic acids (Lahiri et al. 2021). These bacteria utilize the enzymatic reductive approach to convert metallic ions into elemental states by encasing the metal ions within them and adsorbing them on the surface (Khan et al. 2018). Bacteria are recognized to have an exceptional ability for reducing heavy metal ions, making them one of the finest alternatives for NPs formation.

The formation of NPs is caused by the presence of numerous interacting pathways within bacterial cells. Another advantage of exploiting bacterial cells is their ability to produce high yields of stable NPs (Fariq, Khan, and Yasmin 2017). Because of their substantial surface-to-volume ratio, CuNPs and their oxides are favored as noble and desirable metals in numerous fields. CuONPs with different sizes have been produced using *Streptomyces* species. Results showed that CuONPs have excellent antibacterial action against prokaryotic and eukaryotic microbial cells (Hassan et al. 2019). Agglomerated PdNPs with hexagonal shape have been synthesized from secondary metabolites found in *Saccharomyces cerevisiae*. The yeast extract contains bioactive substances that function as secondary metabolites to aid in the synthesis of PdNPs (Sriramulu and Sumathi 2018). Exploring marine resources is crucial for the formation of various NPs since marine microorganisms are easily able to adapt to harsh environmental circumstances. Patil et al. studied the ability of the naturally occurring marine bacteria *Paracoccus haeundaensis* BC74171^T to synthesize AuNPs extracellularly (Patil et al. 2019).

2.5.1.3 From Fungi

Different cellular parts such as cell wall, cellular membrane, enzymes, proteins, and various additional intracellular elements of the fungal body are essential to the synthesis of NPs. Fungi employ both extracellular and intracellular enzymes in the formation of NPs (Fariq, Khan, and Yasmin 2017). Metallic NPs were synthesized by enzymes such as NADH-dependent reductase (Guilger-Casagrande and de Lima 2019). Nitrate reductase and anthraquinones from *Fusarium oxysporum* have been found to mediate the reduction of Ag^+. A NADH-dependent oxidoreductase that originates from fungi also produces AuNPs (Kitching, Ramani, and Marsili 2015). The production of NPs is often larger than that of bacterial cells due to the much higher amount of biomass found in fungi. Fungi may be beneficial owing to the presence of mycelia, which offer greater contact through its wide surface area, despite the fact that bacteria are used more commonly in the fabrication of metallic NPs. Fungi produce more enzymes than bacteria, thereby accelerating the transformation. The cell wall in particular plays an important role in the absorption and reduction of metal ions (Khandel and Shahi 2018). Additionally, certain functional groups, including C-O-C, C-O-R, =C=O, and =C=C=, generated by heterocyclic groups found in fungal extracts function as capping ligands (Sanghi and Verma 2009). Higher surface-to-volume ratio AgNPs have demonstrated superior antibacterial activity than conventional metallic silver. The endophytic fungi offer a wide range of bioactive secondary metabolites with distinctive structures that can be studied for their capacity to produce AgNPs through a reliable, environmentally friendly technique. The ability of an endophytic fungus connected to the ethnomedicinal plant *Potentilla fulgens* L. to produce AgNPs have also been investigated (Devi and Joshi 2015). According to Khandel et al. the main candidates for the formation of NPs are nitrate reductases and NADPH-dependent reductases. In addition, a variety of high-molecular weight proteins and enzymes produced by the fungus itself stabilize NPs upon extracellular production (Khandel and Shahi 2018).

2.5.1.4 From Algae

Due to their low toxicity and strong capacities to bio-accumulate and reduce metals, algae are important eukaryotic organisms from an economic and ecological standpoint that are being employed more and more in the field of nanotechnology. Algae are well-known for being capable of capturing heavy metal ions and changing into more malleable forms (Fawcett et al. 2017). Physical parameters, including pH, reactant concentration, exposure time, reaction duration, and temperature, influence the nucleation, growth, and stabilization of nanomaterials. The shape, size, and aggregation of the cells can all be altered by changing these variables. Algal extracts (phycocyanin and phycoerythrin) contain a variety of nutrients, including carbohydrates, vitamins, minerals, lipids, fatty acids, antioxidants (polyphenols and tocopherols), carotenoids (carotene and xanthophyll), chlorophylls, and phycobilin, in varying concentrations. Theoretically, these active compounds have been recognized as stabilizers and reducers of NPs. Thakkar et al. used *Sargassum wightii* to produce very stable extracellular AuNPs for the first time (Thakkar, Mhatre, and Parikh 2010). According to the findings of the Fourier transformed infrared (FTIR) study, the presence of biomolecules (polysaccharides, pigments, and peptides) in algae, with related functional groups i.e., carboxyl, hydroxyl, cysteine, and amine group promote reduction and coating of the freshly generated NPs. According to research by Dahoumane et al. bioreduction takes place inside the thylakoidal membranes of *Cosmarium impressulum*, followed by release into the cytoplasm (Dahoumane et al. 2014). The reduced metals are eventually transported outside the cell membrane and become encased in the polysaccharide web of the cell wall before being discharged as a stable gold colloid (Jena et al. 2014). For instance, polyphenols and proteins containing carbonyl groups found in an extract obtained from the red seaweed *Corallina officinalis* may help produce and stabilize AuNPs (El-Kassas and El-Sheekh 2014). In another study, Khanehzaei et al. also provided a description of how CuONPs were created utilizing red seaweed extracts (*Kappaphycus alvarezii*) (Khanehzaei et al. 2014). Several hydroxide and sulfur groups and pairs of electrons from the water-soluble sulfated polysaccharides found in seaweed cell walls were also found to cap the surfaces of NPs. Rajathi et al. used a brown alga called *Stoechospermum marginatum* for the formation of gold NPs. The study also discovered that the hydroxyl groups present in the diterpenoids contained in brown seaweed are specifically responsible for reducing gold ions. Additionally, the NPs exhibited strong antibacterial activity toward a variety of chosen bacterial pathogens (Rajathi et al. 2012).

2.5.1.5 From Biopolymers

In general, biopolymers are polymeric biomolecules composed of monomeric units that are covalently attached to form larger molecules (Zinoviadou et al. 2022). The term "bio" implies that these are in fact biodegradable substances derived from microorganisms. The term "biopolymer" refers to a category of materials typically derived from organic natural resources, such as microorganisms, vegetation, or even animals. Due to their natural compositions, materials derived from biogenic resources, such as oils, sucrose, lipids, resins, proteins, and amino acids, are also referred to as biopolymers (Reddy et al. 2013). Monodisperse spherical AgNPs with a diameter of 5–10 nm using pectin have been reported. Processing of pectin using alkali yields hydrolyzed pectin that contains galacturonic acid and some reducing sugars such as galactose and arabinose. The reducing sugars have the potential to convert Ag^+ into AgNPs and assist in stabilizing ultrafine NPs (Zahran, Ahmed, and El-Rafie 2014). Coseri et al. compared the characteristics of AgNPs formed at ambient temperature from pullulan and oxidized pullulan. A variety of forms and sizes of AgNPs can be made by altering the concentration of silver salt. According to the study, structure of the biopolymer determines the stability of AgNPs and is likely to be enhanced by more than six months using oxidized pullulan. A strong relationship between the zeta potential of NPs and their ability to fight against gram-positive bacteria was also discovered (Coseri et al. 2015). Cui et al. created Mo_2C NPs embellished with graphite carbon sheets (Mo_2C/GCSs) using a one-step solid-state synthesis method. In order to create Mo_2C/GCS hybrids, sodium alginate, a naturally occurring polymer generated from brown seaweed, is employed. These NPs were discovered to have excellent catalytic potential, and great electrochemical stability in acidic conditions (Cui et al. 2014). Other types of biopolymers that are frequently used in the fabrication of NPs include proteins and peptides. Shi et al. synthesized AuNPs by extracting intracellular proteins from the white rot fungus *Pycnoporus sanguineus*. Reductase enzymes help convert Au^{3+} into AuNPs, and *P. sanguineus* is thought to be able to create huge quantities of these enzymes. They also discovered an oppositional link between the size of the NPs generated and the concentration of the intracellular protein extract employed (Shi et al. 2015). Due to the fact that they are biocompatible and biodegradable, polysaccharides are thought to be the most promising substrates for making polysaccharide-based MNPs. Rao et al. reported AgNPs containing polyelectrolyte hydrogels (PEHs) from polysaccharides including chitosan and xanthan gum without using organic solvents or additional chemical reductants. On the NIH 3T3 fibroblast cell line, the AgNPs-embedded PEHs were evaluated for cell attachment, viability of cells, and antibacterial activity (Rao et al. 2016).

2.5.2 Metal Salt

NPs require a metal ion precursor and can be metallic, metal oxide-based, alloy-based, magnetic, etc.

2.5.2.1 Metallic NPs

Such NPs fall in nanoscopic dimensions (1–100 nm). In 1857, Faraday reported the existence of metallic NPs in solution (Faraday 1857), whereas Mie quantitatively described the color-changing behavior of NPs in solution form (Mie 1976). The significant features of metallic NPs include huge surface-area-to-volume ratio, great surface energies, explicit electronic structure due to their transition between molecular and metallic states, etc.

Silver, gold, platinum, copper, palladium, iron, and zinc NPs have easily been synthesized from their metal salts through green synthesis. Silver NPs are the most widely explored and synthesized because of their unique antibacterial property in combination with the facile reduction of ionic silver to its metallic form, and efficient plasmon excitation. AgNPs display a wide array of applications such as in medicine, textile, water purification, and catalysis. AuNPs are equally important, but not widely explored comparatively. NPs with other metals are widely reported in literature.

2.5.2.2 Bimetallic NPs

Bimetallic NPs consists of suitable amalgamation of two different metals with improved properties. As a consequence, these NPs have attracted a tremendous amount of attention from a technological as well as scientific viewpoint. Examples include Fe/Ni, Fe/Co, Pt/Pd bimetallic NPs.

2.5.2.3 Metal Oxide NPs

Synthesizing these NPs comprises establishing the metal ion and oxo (M-O-M) or hydroxo (M-OH-M) bonds, hence giving rise to polymers with these units in solution.

2.5.2.4 Alloy NPs

Linking different elements for producing alloys can significantly encompass the range of properties. Nano-alloys show properties very discrete from their bulk counterparts and can lead to improvement in some properties. Change in the composition, size of the clusters, and physicochemical properties can be simply tweaked by employing different metals.

2.5.2.5 Magnetic NPs

These NPs consist of two components, i.e., a magnetic component, which is iron, nickel, or cobalt, and a chemical constituent with a specific functional group. The magnetic component makes it possible to be manipulated magnetically.

2.6 ROLE OF PLANTS IN SYNTHESIS

Plants possess ability to soak up and detoxify inorganic metal ions when their absorption exceeds their removal. Numerous organic molecules present in plant tissues, like cysteine-rich peptides and metallothionein can chelate the metal ions and eliminate them through vacuolar sequestration, thus acting as efficient biological factories to reduce the contamination. Accumulation of the excess metal ions induce reactive oxygen species (ROS) in the cells which disrupts the cellular components causing some abnormalities in the plant body. The ROS homeostasis maintenance occurs through enzymatic antioxidant mechanisms and secondary metabolism of the plant through metabolites. Metal exposure such as copper, cadmium, lead, and nickel stimulate enzymes (phenylalanine ammonia-lyase and chalcone synthase) of the phenylpropanoid pathway. Similarly, copper stress induces the production of chlorogenic acid derivatives and procyanidin whereas maize with enhanced exposure to aluminum causes elevated levels of phenolic content. The enhanced levels of these compounds in response to the metal stress might be due to their capabilities for metal chelation. The flavonoids chelate through their–OH and C=O groups. Quercetin, a kind of flavonoid has multiple bidentate sites such as α-hydroxy-carbonyl, β-hydroxy-carbonyl or catechol with two –OH groups in ortho positions and forms stable complexes with various metal ions. High concentration of phenols disrupts the cells due to their reactive nature and at the same time, they induce denaturation of proteins, and hence are transported to extracellular spaces and vacuoles. Thus, the presence of phenolic compounds and metal ions in the same space boosts the development of MNPs. Numerous plants have been found to create and hoard NPs in their cells. *Medicago sativa* cultivated in $AuCl_4$ rich environment hoarded AuNPs (around 4 nm in size) in their tissues. Similarly, the same plants stored AgNPs to about 13.6% of their fresh weight when cultivated in $AgNO_3$-rich environment. *Phragmites australis* and *Iris pseudacorus* store CuNPs in and around their roots when grown in copper-rich medium.

2.7 MECHANISM OF GREEN SYNTHESIS OF NPs BY PLANT EXTRACTS

In-vitro synthesis of NPs using plant extract might proceed through the same mechanism as discussed above. When metal salt dissociates in solvent, the metal ions are surrounded to form hydroxyl complexes, growing crystallites of metal with oxygen-containing species. Heat plays an important role in supplying energy to the developing crystallite, which continues until the capping agents are activated and hinder further growth in the planes. This leads to the development of an explicit type of nanomaterial with distinct sizes and shapes. The reducing agents in the plant extract provide electrons to the metal ions converting them into NPs. These NPs possess a tendency to transform to low-surface energy which leads them to form aggregates. Aggregation can be prevented with elevated levels of reducing and stabilizing/capping agents. Proteins help to lock in metal ions, transform them into suitable nuclei for aggregation, and, then, form NPs. The–NH_2 groups found in proteins, –OH and –COOH of polyphenols/amino acids, –OH of the polysaccharides, and –COOH of organic acids for the plant extract bind the metal ions and crush the superoxide-driven Fenton reaction (a potential source of ROS), facilitating the development of metal-based NPs.

Protein-metal ion complexation is necessary for vacuolar sequestration when metal ions concentration goes high in plants, but the function of proteins in *in vitro* process is ambiguous. Plant extracts reduce metal ions to produce NPs even after boiling. Boiling changes the structures of proteins, however, the peptide bonds therein are still intact and hence are good at reducing. The proteins bind to the NPs via free amino groups or cysteine residues and stabilize them. Differences in concentration and conformations of the relevant phytochemicals in plant extracts and their resulting association with metal ions are believed to influence sizes and shapes of the resulting NPs.

2.8 SECONDARY METABOLITES IN PLANT EXTRACT-MEDIATED GREEN SYNTHESIS OF NPs

Diverse range of biomolecules ranging from proteins to species like terpenoids, alkaloids, amino acids, alcohols, polyphenols, glutathiones, polysaccharides, antioxidants, organic acids, etc., partake in the formation of NPs. The course of action of carbohydrates, terpenoids, phenols, alkaloids, phenolic and other organic acids, and proteins in reducing metal ions and imparting stability has also been proposed. Flavonoids have been widely assumed to take part in the green synthesis. Scientists have been using Fourier-transform infrared (FT-IR) spectroscopy studies to acquire hints and evidences on the molecules possibly involved in the process of reduction and capping. For example, study of *Suaeda monoica*-based AuNPs suggested that molecules with –COOH, $-NH_2$, and –OH groups participated in the process. Similarly, isolated biomolecules like flavonoids, terpenoids, and chlorogenic acid are effectively fruitful for nanomaterial synthesis which further endorses their capability and involvement. Monodispersed (18.24 nm) sphere-shaped gold NPs have been reported using kaempferol as reducing and stabilizing/capping agent. Terpenoids isolated from the leaves extract of *Andrographis paniculata* facilitated the synthesis of ZnO NPs which was endorsed by the presence of C=O group.

High-performance liquid chromatography (HPLC) of AgNPs synthesized using leaf extract of *Withania somnifera* showed that numerous phenols selectively trapped in the AgNPs were present in the extract. Their characteristic UV-vis spectra and co-elution with pure standards confirmed that these compounds were catechin, p-coumaric acid, and luteolin-7-glucoside. Jain and Mehata reported the synthesis of AgNPs with both leaf extract of *Ocimum sanctum* and quercetin present in the extract separately. Both AgNPs demonstrated comparable optical, morphological, and antibacterial characteristics, endorsing the role of quercetin in the process.

Flavonoids are natural polyphenols that include flavone, flavanone, flavonol, flavanonol, and isoflavone derivatives and respond to biotic and abiotic stresses. They consist of two phenyl rings, linked by an oxygenated heterocycle, with hydroxyl groups at various positions and possess tremendous chelating and antioxidant activity. The number of –OH groups and structure holds importance in determining its chelation properties.

Aglycones (chrysin, apigenin, and genistein) provide coordination sites to metallic cations in the space between their 4-carbonyl group and 5-hydroxyl group. Two chrysin ligands bind a divalent cation whereas genistein and biochanin bind Cu(II) and Fe(III) with a 1:2 (metal:ligand) ratio. Apigenin have been reported to coordinate with Al(III) and Fe(II) with ratio of 1:3 in a dioxan aqueous solution, through 4′-hydroxyl group.

Two prospective bidentate binding sites exist in luteolin (5-hydroxyl-4-carbonyl and 3′-4′-hydroxyl groups of ring B) that bind Al^{3+} ions in 2:1. The parting of the coordination pocket is crucial to enhance the chelating properties. Quercetin possesses three bidentate binding points and form stable complexes with ions, such as Fe^{2+}/Fe^{3+}, Cu^{2+}, and Al^{3+} using 3-hydroxyl and 4-carbonyl groups.

In brief, the alteration of free –OH groups in the flavonoids change their chelating properties. Free flavonoids are stronger chelators than the O-methylated ligands.

2.9 ROLE OF MICROBES IN NPs SYNTHESIS

The efforts to develop efficient, reliable, non-toxic, and environmentally friendly techniques for synthesizing NPs are essential. This goal can be accomplished by utilizing natural sources, including biological systems, microorganisms, enzymes, and biodegradable polymers. Microbes such as bacteria, fungi, actinobacteria, and viruses were studied by various researchers to investigate the synthesis of the desired NPs (Fariq, Khan, and Yasmin 2017).

Microbial cells exhibit inherent adaptability to various environments, making them fast-growing and easily manageable. Among these microorganisms, bacteria and microalgae stand out due to their unique ability to synthesize diverse nanomaterials, including exopolysaccharides, nanowires, and nanocellulose (Lateef, Darwesh, and Matter 2021).

Fabricating NPs from live sources and specifically microbes like yeast, actinobacteria, viruses, fungi, and algae, offers a safer alternative compared to physical and chemical methods (Salem and Fouda 2021). Among these living sources, bacteria are particularly advantageous due to their ease of cultivation in artificial conditions and their rapid growth rate. They have the ability to adapt to metal concentrations that are higher and also they are capable of reducing the inorganic materials to NPs both extracellularly or intracellularly (Fariq, Khan, and Yasmin 2017).

Microbes are remarkably adept at absorbing metal ions from their surroundings or medium and enzymatically converting them into their elemental forms (Li et al. 2011). This biologically driven approach to NP fabrication holds great promise for sustainable and eco-friendly synthesis methods, paving the way for diverse applications across various fields.

In the process of synthesizing biogenic NPs using extracellular fabrication by microbes, microbial cultures are cultivated in suitable media. Following this, the microbial cells are separated from the broth via centrifugation, resulting in a supernatant enriched with microbial enzymes which are incorporated into metal ions solutions for efficient reduction of the latter (Al-Dhabi, Mohammed Ghilan, and Arasu 2018). This approach harnesses microbial enzymes outside the cells to create NPs in a controlled and efficient manner, with potential applications in various fields (Li et al. 2011).

The combination of metal ions and microbial biomass is cultured under closely monitored incubation conditions until a distinct color change is observed. The emergence of distinct colors serves as a sign that NPs are forming. For example, a light yellow to pinkish hue indicates the production of AuNPs, while a pale yellow to brownish shade denotes the synthesis of AgNPs. Additionally, a whitish-yellow to yellow color signifies the synthesis of zinc and manganese NPs (Edison and Pradeep 2020).

The color-based method offers a visually straightforward way to monitor NP synthesis progress, and the use of appropriate microbial species can significantly influence the outcome, ensuring the successful and rapid production of stable NPs.

Genetic engineering is another intriguing aspect with its capability to engineer NPs with precise physical, chemical, and biological properties. This opens up fascinating possibilities for tailoring NPs to specific functions and applications (Kang et al. 2008).

Utilizing genetically engineered microorganisms to produce well-defined NPs with specific shapes and sizes represents an innovative approach. Recently, various methods have been explored to enhance the capabilities of wild-type microorganisms for NP synthesis, including the identification of genes responsible for the synthesis of NPs, the incorporation of foreign genes, and cellular reprogramming. By leveraging genetic engineering techniques and understanding microbial pathways, researchers can design and optimize microorganisms to precisely control NP characteristics, fostering exciting possibilities for tailored nanomaterials with diverse applications in various fields.

The biomineralization of magnetosomes and the mechanism underlying formation of magnetite in *Magnetospirillum gryphiswaldense* is intricately regulated by specific genes. In a groundbreaking study, researchers successfully introduced the *Magnetosome biomineralization* pathway (mam genes) into a new host, *Rhodospirillum rubrum*, resulting in the creation of 24 nm-sized magnetite NPs encased within a protein shell (Komeili 2012).

These studies showcase the exciting potential of exploring microbial systems for NP synthesis and offer valuable insights into tailoring NPs for various applications through genetic manipulation and control of environmental factors.

2.10 GENERAL CHEMICAL MECHANISM OF SYNTHESIS

A number of reducing agents found in plant and microbial extracts are capable of reducing, capping, and stabilizing the metal ions. This process has been used to remove valuable metals via hyper-accumulation from the soil by growing suitable plants, known as phytomining (Karman, Diah, and Gebeshuber 2015). Phytochemical screening is usually performed to identify the presence or absence of polyphenols such as alkaloids, flavonoids, tannins, quinones, saponins, etc. (Bao et al. 2021) (Figure 2.5).

Plant-mediated synthesis of nanomaterial makes use of vital and abundant phytochemicals such as carbohydrates, amino acids, aldehydes, phenols, ketones, glucosides, ascorbic acid, alcohols, and polyols. The process includes three phases: activation phase, growth phase, and termination phase (Bouafia and Laouini 2021).

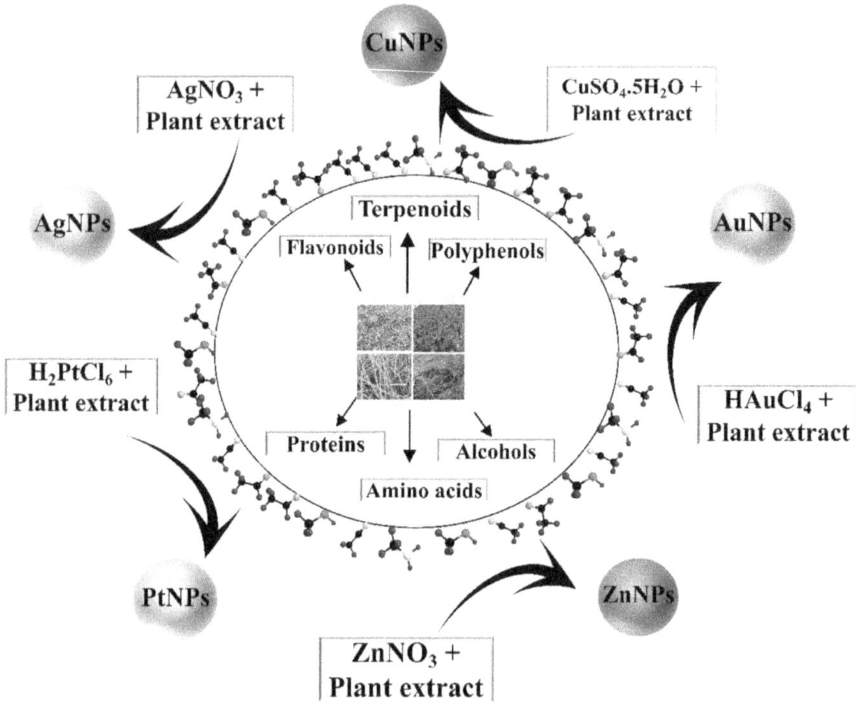

FIGURE 2.5 A representation of reducing abilities of different biomolecules in various plant extracts.

The first phase involves reduction of metal ions, utilizing the metabolites and biomolecules with reducing potential. During this phase, the mono- or divalent metal ions are transformed to zero-valent states and nucleation takes place. This nucleated specie then undergoes growth where other metal atoms join in, further reducing the metal ions and form particles of various morphologies such as cubes, rods, wires, etc. The growth imparts thermodynamic stability to the particles and the nucleation stretch helps in attaining different morphologies. This process is followed by termination phase where the morphology and size is finally defined. The plant metabolites cap the nucleated specie, thus stabilizing and preventing it from further growth and modification. The difference in conformation and concentration of these molecules are assumed to be key factors in defining the size and shape of the fabricated nanomaterial. Plants and microbes intrinsically contain different concentrations and composition of the active components which explains the morphological range and size of nanomaterials obtained with them (Dahl, Maddux, and Hutchison 2007). The composition, concentration, pH, temperature, and other physiological conditions of the reaction mixture also influence the size and shape of NPs. The pH, for instance, influences charge conversion even in the metabolites and affects their ability to chelate. This may also cause alteration in the morphology, dimensions of growth, and yield (Sarkar et al. 2023). Thus, the reduction and stabilization mechanisms need to be fully elucidated. The plant metabolites vary in their reducing potential and, hence, different phytochemical composition imparts different reducing properties to the extract. Tryptophan, tyrosine, and lysine perform exceptionally well in terms of reducing the metal ions rapidly (Wangoo et al. 2014).

FT-IR peaks into the functional groups responsible for reducing and stabilizing the nanomaterial through comparative studies with the biological extract (Song, Kwon, and Kim 2010). Some peaks found in the extract become less intense or disappear from the spectrum of the nanomaterial which indicates that the corresponding functional groups are responsible for reduction. Some peaks remain unaltered and hence suggest the role of the corresponding functional groups in stabilizing the nanomaterial (Mourdikoudis, Pallares, and Thanh 2018). For example, the FTIR data of PtNPs synthesized using *Maytenusroyleanus* show a decrease in peak intensities. The stretching O—H vibration of phenolic compounds, stretching C=O and N—H in amide shifted from 3310, 1608 and, 2938 cm^{-1} to 3412, 1615, and 2925 cm^{-1}, respectively, which establishes that the NPs are stabilized by various phytochemicals (Ullah et al. 2017). Nishanthi et al. attributed the bands around 1435 and 1280 cm^{-1} in the FTIR spectrum of PtNPs to secondary aromatic amines and polyols (Nishanthi, Malathi, and Palani 2019). Figure 2.6 shows the proposed mechanism of synthesis of ZnO NPs using *Nephelium lappaceum* peel extract. At 450°C, zinc complex with polyphenolic ellagic acid decompose and produce ZnONPs (Karnan and Selvakumar 2016).

Various microbes have the ability to produce NPs, however, the mechanisms are not well known due to the metabolic complexity and difficulty in detection of active components. The enzymes in microorganisms might be responsible for reduction of the metal ions (Ahmad et al. 2003). NADH-dependent reductase was suggested to be reducing Ag^{+} ions in *F. oxysporum*-based AgNPs synthesis, but the mechanism of the whole process is unknown.

Microbes are also equipped naturally with excellent potential to eliminate or reduce the metal ions for their survival. The microbial components involved in the process of synthesis might include reductases, naphthoquinones, flavonoids, and anthraquinones. Primarily, molecules like amino acids, peptides, proteins, and enzymes participate in reduction through electron shuttle or charge capping. It is important to estimate the threshold of the microbes to combat the toxicity of metal ions, followed by screening of the microbe if it really has the potential to reduce as all microbes are not necessarily capable of reduction.

Microbes-mediated synthesis is categorized into extracellular (Ahmad et al. 2003) and intracellular synthesis (Shi et al. 2015).

In extracellular synthesis, metal ions are trapped onto the surface of microbial cells and reduced in the presence of the various enzymes on the surface. In extracellular synthesis, generally, microorganisms are injected into a metal-rich medium, below its minimum inhibitory level, and incubated for one to two days at specific conditions. Color changes in the medium, and UV-visible, FT-IR, and Raman spectroscopy can suggest the formation of NPs. It has been hypothesized that electron shuttle NADH and NADH-dependent enzymes participate in the reduction of Ag and Au ions in *Rhodopseudomonas* (He et al. 2007) as demonstrated in Figure 2.7. The enzymes are believed to draw electrons from NADH to the metal ions. Hydrogenase enzymes have been widely predicted to be actively reducing U^{6+}, Se^{6+}, Tc^{7+}, and Cr^{6+} in *Micrococcus lactylilicus, Clostridium pasteurianum*, and some sulfate-reducing

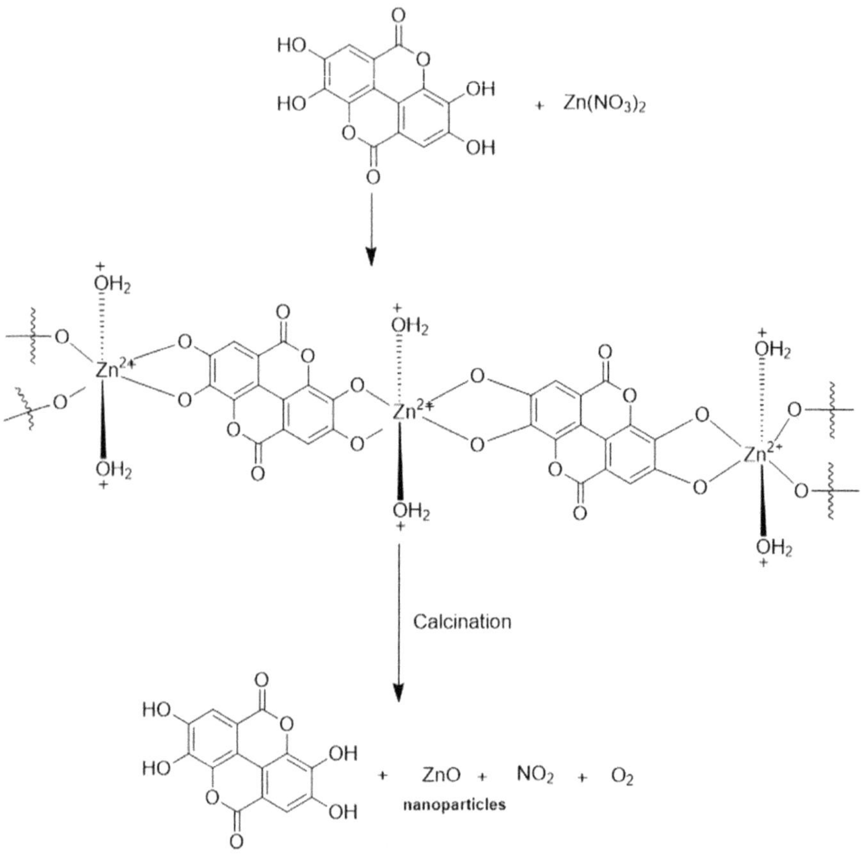

FIGURE 2.6 Proposed mechanism of synthesis of ZnO NPs by using *Nephelium lappaceum* peel extract. The hydroxyl groups of polyphenolic ellagic acid present in the extract bind with zinc ions to form a zinc-ellagate complex (pH 5–7) which yields ZnO NPs on calcination at 450°C (Karnan and Selvakumar 2016).

bacteria (Michel et al. 2001, Yanke, Bryant, and Laishley 1995, Woolfolk and Whiteley 1962). Extracellular synthesis is advantageous due to the facile harvest of NPs by centrifugation, with no other processing.

In intracellular synthesis, the transportation of metal ions into the microbial cell occurs followed by reduction through biomolecules. Microbes are introduced into metal ion solution and left to work. Synthesis of Ag-sulfide crystallites were reported using *Pseudomonas stutzeri* AG259 with high concentration of silver ions (Slawson, Trevors, and Lee 1992), demonstrated in Figure 2.8. The presence of **Ag2S** buildups at the poles and outside the cytoplasmic membrane shows the presence of H_2S which facilitates producing the Ag_2S, within 8 minutes. The Ag_2S deposits were found in vacuole-like bodies rooted in organic matrix. This method requires repeated centrifugation, further processing to disrupt the cells and recover NPs, and purification (Rana, Yadav, and Jagadevan 2020).

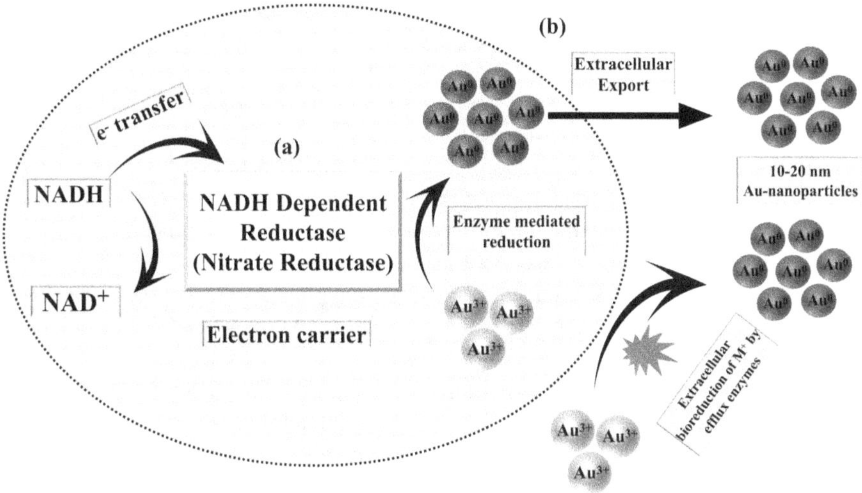

FIGURE 2.7 Proposed mechanism of extracellular synthesis of MNPs in bacteria (a) NADPH-dependent enzymatic metal reduction (b) bio-reduction by efflux enzymes.

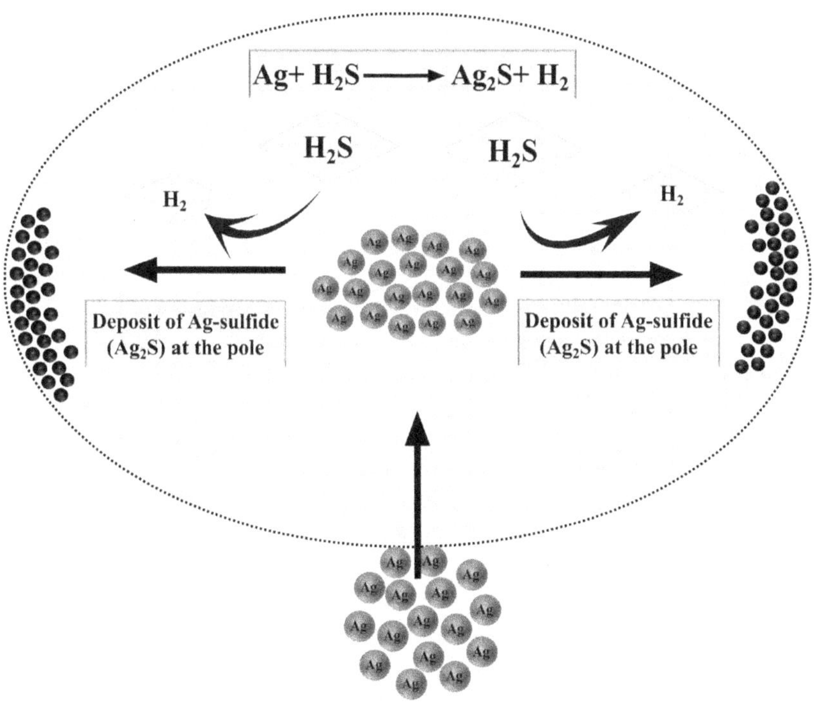

FIGURE 2.8 Mechanism of intracellular synthesis of NPs in bacteria.

2.11 FACTORS AFFECTING THE PRODUCTION OF MICROBE-MEDIATED NPs

The important factors that affect NP synthesis are enzymes, pH, temperature, and metal concentration (Kulkarni and Muddapur 2014).

Enzymes are crucial in the reduction and capping of metals in microorganisms through redox processes that take place in the intracellular or extracellular environment, frequently serving as nucleation sites. Several important enzymes that are vital in the reduction of metal ions are used in the manufacture of microbe-based NPs. Nitrate reductase, NADH-Dependent reductase, polyphenol oxidase (PPO), laccase, and other extracellular enzymes are a few prominent enzymes involved in this process (Ghosh et al. 2021).

In a study focused on silver NP synthesis using *Bacillus licheniformis*, it was observed that the biomass of this microorganism efficiently converted aqueous silver ions into stable silver NPs through the action of the enzyme nitrate reductase within just 24 hours of the reaction (Kalimuthu et al. 2008). Selecting appropriate microorganisms is crucial for enhancing the speed and efficiency of the NP synthesis process.

Another effective route for producing AgNPs involves NADH-dependent reductase. The bacterium *P. stutzeri* AG259 has been found to be involved in this process, wherein it releases electrons, leading to the oxidation of nicotinamide adenine dinucleotide (NAD) (Ahmad et al. 2003).

Depending on the microbial species and the metal ions used, various enzymes may participate in the synthesis of distinct NPs through different reduction pathways.

Numerous studies have investigated the effects of pH, temperature, and metal ion concentration on the synthesis of NPs. For instance, researchers found that at pH 7–8 and a temperature of 70°C, microorganisms like Arthrobacter sp. successfully synthesized face-centered cubic AgNPs in the size range of ~10–70 nm when using a lower concentration of $AgNO_3$ (1 mM). However, higher concentration of $AgNO_3$ (3 mM) yield clusters of AgNPs (Yumei et al. 2017). Interestingly, Yumei et al. noted that AgNP synthesis occurred only within the pH range of 7–8, with no NP formation observed at pH levels below 5 or above 8. Furthermore, they observed that at higher temperature (90°C), the fabrication time significantly reduces from 10 to 2 minutes. This study emphasizes the importance of carefully controlling these parameters to achieve NPs with specific properties. In another study, the mesophilic bacterium Shewanella algae was subjected to incubation in a medium with varying pH levels. The results of this study revealed that the pH of the solution influences both the size and distribution of the synthesized NPs (Konishi et al. 2004).

Active research is being constantly conducted to explore a wide range of aspects to optimize the generation of microbe-mediated NPs for various purposes. These include investigating novel enzymes, fine-tuning temperatures, optimizing the composition of nutrient media, and studying the impact of aerobic and anaerobic conditions, microbial growth phases, reaction times, pressure, light exposure, and radiation on NP synthesis. Through thorough analysis of these elements, researchers aim to enhance the structural and distinctive qualities of the synthesized NPs, paving the way for creative and effective application.

2.12 ADVANTAGES AND CONSTRAINTS OF GREEN SYNTHESIS

The green method of synthesis is simple, economical, and environment friendly and takes minimum time and less efforts. Protection of environment is need of the hour and hence mankind must adopt a greener approach, i.e., recycling and reusing. The feasibility of synthetic method should be decided keeping in mind the indigenous resources. A deeper understanding of some aspects of green synthesis is required as follows:

1. Microbes-mediated synthesis is slow and takes time from hours to days. Thus, optimization studies of certain factors such as type of microbes, pH, temperature, fermentation media, concentration of precursors, and reaction time are required. These factors also define the morphology, size, and monodispersity of NP.
2. It is of paramount importance to comprehensively investigate the involvement of each of the active metabolite, cellular components, mechanisms, and biochemical pathways including accurate identification and isolation of molecules responsible for reduction of metal ions. A clear understanding in this regard would carve a splendid trail for a sustainable evolution of green chemistry.
3. A comprehensive grasp of the structure of capping/stabilizing agents and functional groups is essential to acquire clarity on bio-friendliness and compatibility.
4. Although no robust equipment is required, large-scale production is still a challenge and needs to be prioritized. The toxicological evaluation of green nanomaterial, risk mitigation/safety profiling, handling, storage, and disposal calls for a clear assessment before being scaled up.
5. It is of great significance to investigate how physicochemical features of the NPs can be enhanced for specific applications. Genetically modified microbes with the knack to harvest a greater quantity of bioactive molecules could further boost the synthesis and stabilization of NPs. Similarly, an increased metal-trapping potential of these microbes could provide a futuristic approach to nanobiotechnology.
6. The isolation and purification of NPs from the reaction mixture require inquisitive exploration.

2.13 CONCLUSION, CHALLENGES AND FUTURE PROSPECTS

Such a tremendous success has triggered significant interest of researchers in academia and technology leading to the development of new metal-based nanomaterials with diverse applications in different fields. Though the vastness of metal-based nanomaterials has rendered immense research interest, these unique nanomaterials are only in their infancy.

Research helps in providing additional key resolutions to the major drawbacks. Although, a striking progress in the synthesis procedures of metal-based nanomaterials has been succeeded, yet numerous persisting challenges need to be resolved. The major limitations in the way of achieving a perfect synthesis technique for metal-based nanomaterials include high cost, size tuneability, inefficiency, and scalability. In addition, the assessment of specific synthesis mechanisms and production rate kinetics remains subtle. Besides, the production of noxious by-products and possible threats to the well-being are other major challenges. Moreover, further investigations are needed concerning the biocompatibility of metallic nanomaterials as they may possibly be toxic in high dosage. Therefore, newer, more viable, biocompatible, and eco-friendly synthetic techniques need to be prioritized for metal-based nanomaterials.

REFERENCES

Adeyemi, Jerry O, Ayodeji O Oriola, Damian C Onwudiwe, and Adebola O Oyedeji. 2022. "Plant extracts mediated metal-based nanoparticles: synthesis and biological applications." *Biomolecules* no. 12 (5):627.

Ahmad, Absar, Priyabrata Mukherjee, Satyajyoti Senapati, Deendayal Mandal, M Islam Khan, Rajiv Kumar, and Murali Sastry. 2003. "Extracellular biosynthesis of silver nanoparticles using the fungus *Fusarium oxysporum*." *Colloids and Surfaces B: Biointerfaces* no. 28 (4):313–318.

Akhter, Gulwaiz, Afroz Khan, Syed Ghazanfar Ali, Tabreiz Ahmad Khan, Khwaja Salahuddin Siddiqi, and Haris M Khan. 2020. "Antibacterial and nematicidal properties of biosynthesized Cu nanoparticles using extract of holoparasitic plant." *SN Applied Sciences* no. 2:1–6.

Al-Dhabi, Naif Abdullah, Abdul-Kareem Mohammed Ghilan, and Mariadhas Valan Arasu. 2018. "Characterization of silver nanomaterials derived from marine *Streptomyces* sp. Al-Dhabi-87 and its in vitro application against multidrug resistant and extended-spectrum beta-lactamase clinical pathogens." *Nanomaterials* no. 8 (5):279.

Amin, Rashid, Sieun Hwang, and Sung Ha Park. 2011. "Nanobiotechnology: an interface between nanotechnology and biotechnology." *Nano* no. 6 (02):101–111.

Azhar, Sundus, Khuram Shahzad Ahmad, Isaac Abrahams, Wang Lin, Ram K Gupta, Muhammad Mazhar, and Daoud Ali. 2021. "Phyto-inspired Cu/Bi oxide-based nanocomposites: synthesis, characterization, and energy relevant investigation." *RSC Advances* no. 11 (49):30510–30519.

Bao, Yunhui, Jian He, Ke Song, Jie Guo, Xianwu Zhou, and Shima Liu. 2021. "Plant-extract-mediated synthesis of metal nanoparticles." *Journal of Chemistry* no. 2021:1–14.

Berta, Lavinia, Năstaca-Alina Coman, Aura Rusu, and Corneliu Tanase. 2021. "A review on plant-mediated synthesis of bimetallic nanoparticles, characterisation and their biological applications." *Materials* no. 14 (24):7677.

Bouafia, Abderrhmane, and Salah E Laouini. 2021. "Plant-mediated synthesis of iron oxide nanoparticles and evaluation of the antimicrobial activity: a review." *Mini-Reviews in Organic Chemistry* no. 18 (6):725–734.

Chandrakala, Vinayagasundaram, Valmiki Aruna, and Gangadhara Angajala. 2022. "Review on metal nanoparticles as nanocarriers: current challenges and perspectives in drug delivery systems." *Emergent Materials* no. 5 (6):1593–1615.

Chokkalingam, Mohan, Priyanka Singh, Yue Huo, Veronika Soshnikova, Sungeun Ahn, Jongpyo Kang, Ramya Mathiyalagan, Yeon Ju Kim, and Deok Chun Yang. 2019. "Facile synthesis of Au and Ag nanoparticles using fruit extract of *Lycium chinense* and their anticancer activity." *Journal of Drug Delivery Science and Technology* no. 49:308–315.

Chopra, Hitesh, Shabana Bibi, Inderbir Singh, Mohammad Mehedi Hasan, Muhammad Saad Khan, Qudsia Yousafi, Atif Amin Baig, Md Mominur Rahman, Fahadul Islam, and

Talha Bin Emran. 2022. "Green metallic nanoparticles: biosynthesis to applications." *Frontiers in Bioengineering and Biotechnology* no. 10:548.

Coseri, Sergiu, Alina Spatareanu, Liviu Sacarescu, Cristina Rimbu, Daniela Suteu, Stefan Spirk, and Valeria Harabagiu. 2015. "Green synthesis of the silver nanoparticles mediated by pullulan and 6-carboxypullulan." *Carbohydrate Polymers* no. 116:9–17.

Cui, Wei, Ningyan Cheng, Qian Liu, Chenjiao Ge, Abdullah M Asiri, and Xuping Sun. 2014. "Mo_2C nanoparticles decorated graphitic carbon sheets: biopolymer-derived solid-state synthesis and application as an efficient electrocatalyst for hydrogen generation." *American Chemical Society Catalysis* no. 4 (8):2658–2661.

Dahl, Jennifer A, Bettye L S Maddux, and James E Hutchison. 2007. "Toward greener nanosynthesis." *Chemical Reviews* no. 107 (6):2228–2269.

Dahoumane, Si Amar, Claude Yéprémian, Chakib Djédiat, Alain Couté, Fernand Fiévet, Thibaud Coradin, and Roberta Brayner. 2014. "A global approach of the mechanism involved in the biosynthesis of gold colloids using micro-algae." *Journal of Nanoparticle Research* no. 16:1–12.

Devi, Lamabam Sophiya, and Shyam R Joshi. 2015. "Ultrastructures of silver nanoparticles biosynthesized using endophytic fungi." *Journal of Microscopy and Ultrastructure* no. 3 (1):29–37.

Dikshit, Pritam Kumar, Jatin Kumar, Amit K Das, Soumi Sadhu, Sunita Sharma, Swati Singh, Piyush Kumar Gupta, and Beom Soo Kim. 2021. "Green synthesis of metallic nanoparticles: applications and limitations." *Catalysts* no. 11 (8):902.

Dutt, Yogesh, Ramendra Pati Pandey, Mamta Dutt, Archana Gupta, Arpana Vibhuti, Jasmina Vidic, V Samuel Raj, Chung-Ming Chang, and Anjali Priyadarshini. 2023. "Therapeutic applications of nanobiotechnology." *Journal of Nanobiotechnology* no. 21 (1):1–32.

Edison, Lekshmi K, and Pradeep N Sukumaran. 2020. "Actinobacterial nanoparticles: green synthesis, evaluation and applications." In *Green Nanoparticles: Synthesis and Biomedical Applications*, 371–384. Springer Nature.

El-Kassas, Hala Yassin, and Mostafa M El-Sheekh. 2014. "Cytotoxic activity of biosynthesized gold nanoparticles with an extract of the red seaweed *Corallina officinalis* on the MCF-7 human breast cancer cell line." *Asian Pacific Journal of Cancer Prevention* no. 15 (10):4311–4317.

Faraday, Michael. 1857. "X. The Bakerian Lecture—Experimental relations of gold (and other metals) to light." *Philosophical Transactions of the Royal Society of London* no. 147:145–181.

Fariq, Anila, Tabeer Khan, and Azra Yasmin. 2017. "Microbial synthesis of nanoparticles and their potential applications in biomedicine." *Journal of Applied Biomedicine* no. 15 (4):241–248.

Fawcett, Derek, Jennifer J Verduin, Monaliben Shah, Shashi B Sharma, and Gérrard Eddy Jai Poinern. 2017. "A review of current research into the biogenic synthesis of metal and metal oxide nanoparticles via marine algae and seagrasses." *Journal of Nanoscience* no. 2017:1–15.

Gardea-Torresdey, Jorge L, Eduardo Gomez, Jose R Peralta-Videa, Jason G Parsons, Horacio Troiani, and Miguel Jose-Yacaman. 2003. "Alfalfa sprouts: a natural source for the synthesis of silver nanoparticles." *Langmuir* no. 19 (4):1357–1361.

Ghosh, Shubhrima, Razi Ahmad, Kamalika Banerjee, Mohamed Fahad AlAjmi, and Shakilur Rahman. 2021. "Mechanistic aspects of microbe-mediated nanoparticle synthesis." *Frontiers in Microbiology* no. 12:638068.

Giri, Alok Kumar, Biswajit Jena, Bhagyashree Biswal, Arun Kumar Pradhan, Manoranjan Arakha, Saumyaprava Acharya, and Laxmikanta Acharya. 2022. "Green synthesis and characterization of silver nanoparticles using *Eugenia roxburghii* DC. extract and activity against biofilm-producing bacteria." *Scientific Reports* no. 12 (1):8383.

Guilger-Casagrande, Mariana, and Renata de Lima. 2019. "Synthesis of silver nanoparticles mediated by fungi: a review." *Frontiers in Bioengineering and Biotechnology* no. 7:287.

Guleria, Anjali, Harshita Sachdeva, Kirti Saini, Komal Gupta, and Jaya Mathur. 2022. "Recent trends and advancements in synthesis and applications of plant-based green metal nanoparticles: a critical review." *Applied Organometallic Chemistry* no. 36 (9):e6778.

Hassan, Saad El-Din, Amr Fouda, Ahmed A Radwan, Salem S Salem, Mohammed G Barghoth, Mohamed A Awad, Abdullah M Abdo, and Mamdouh S El-Gamal. 2019. "Endophytic actinomycetes *Streptomyces* spp mediated biosynthesis of copper oxide nanoparticles as a promising tool for biotechnological applications." *JBIC Journal of Biological Inorganic Chemistry* no. 24:377–393.

He, Shiying, Zhirui Guo, Yu Zhang, Song Zhang, Jing Wang, and Ning Gu. 2007. "Biosynthesis of gold nanoparticles using the bacteria *Rhodopseudomonas capsulata*." *Materials Letters* no. 61 (18):3984–3987.

Jena, Jayashree, Nilotpala Pradhan, Rati Ranjan Nayak, Bishnu P Dash, Lala Behari Sukla, Prasanna K Panda, and Barada K Mishra. 2014. "Microalga *Scenedesmus* sp.: a potential low-cost green machine for silver nanoparticle synthesis." *Journal of Microbiology and Biotechnology* no. 24 (4):522–533.

Kalimuthu, Kalishwaralal, Ramkumarpandian Suresh Babu, Deepak Venkataraman, Mohd Bilal, and Sangiliyandi Gurunathan. 2008. "Biosynthesis of silver nanocrystals by *Bacillus licheniformis*." *Colloids and Surfaces B: Biointerfaces* no. 65 (1):150–153.

Kang, Seung Hyun, Krassimir N Bozhilov, Nosang V Myung, Ashok Mulchandani, and Wilfred Chen. 2008. "Microbial synthesis of CdS nanocrystals in genetically engineered *E. coli*." *Angewandte Chemie International Edition* no. 47 (28):5186–5189.

Karman, Salmah B, S Zaleha M Diah, and Ille C Gebeshuber. 2015. "Raw materials synthesis from heavy metal industry effluents with bioremediation and phytomining: a biomimetic resource management approach." *Advances in Materials Science and Engineering* no. 2015:185071.

Karnan, Thenmozhi, and Stanly Arul Samuel Selvakumar. 2016. "Biosynthesis of ZnO nanoparticles using rambutan (*Nephelium lappaceum* L.) peel extract and their photocatalytic activity on methyl orange dye." *Journal of Molecular Structure* no. 1125: 358–365.

Karthik, Kannan, Saranya Dhanuskodi, Chandrakasan Gobinath, Seetharaman Prabukumar, and Sivaperumal Sivaramakrishnan. 2018. "Multifunctional properties of microwave assisted CdO–NiO–ZnO mixed metal oxide nanocomposite: enhanced photocatalytic and antibacterial activities." *Journal of Materials Science: Materials in Electronics* no. 29:5459–5471.

Khan, Naeem, and Asghari Bano. 2016. "Modulation of phytoremediation and plant growth by the treatment with PGPR, Ag nanoparticle and untreated municipal wastewater." *International Journal of Phytoremediation* no. 18 (12):1258–1269.

Khan, Tabeer, Sidra Abbas, Anila Fariq, and Azra Yasmin. 2018. "Microbes: nature's cell factories of nanoparticles synthesis." *Exploring the Realms of Nature for Nanosynthesis*: 25–50.

Khan, Yousaf, Uzma Sharafat, Saima Gul, Muhammad Ibrahim Khan, Muhammad Ismail, Murad Ali Khan, Rafia Younus, and Sher Bahadar Khan. 2023. "Novel in situ synthesis of quaternary core–shell metallic sulfide nanocomposites for degradation of organic dyes and hydrogen production." *Green Processing and Synthesis* no. 12 (1):20228128.

Khandel, Pramila, and Sushil Kumar Shahi. 2018. "Mycogenic nanoparticles and their bio-prospective applications: current status and future challenges." *Journal of Nanostructure in Chemistry* no. 8:369–391.

Khanehzaei, Hajar, Mansor B Ahmad, Kamyar Shameli, and Zahra Ajdari. 2014. "Synthesis and characterization of Cu@ Cu2O core shell nanoparticles prepared in seaweed

Kappaphycus alvarezii Media." *International Journal of Electrochemical Science* no. 9 (12):8189–8198.

Kitching, Michael, Meghana Ramani, and Enrico Marsili. 2015. "Fungal biosynthesis of gold nanoparticles: mechanism and scale up." *Microbial Biotechnology* no. 8 (6): 904–917.

Komeili, Arash. 2012. "Molecular mechanisms of compartmentalization and biomineralization in magnetotactic bacteria." *Federation of European Microbiological Societies Microbiology Reviews* no. 36 (1):232–255.

Konishi, Yasuhiro, Toshiyuki Nomura, Takeshi Tsukiyama, and Norizoh Saitoh. 2004. "Microbial preparation of gold nanoparticles by anaerobic bacterium." *Transactions-Materials Research Society of Japan* no. 29:2341–2344.

Kulkarni, Narendra, and Uday Muddapur. 2014. "Biosynthesis of metal nanoparticles: a review." *Journal of Nanotechnology* no. 2014:510246.

Lahiri, Dibyajit, Moupriya Nag, Hassan I Sheikh, Tanmay Sarkar, Hisham Atan Edinur, Siddhartha Pati, and Rina Rani Ray. 2021. "Microbiologically-synthesized nanoparticles and their role in silencing the biofilm signaling cascade." *Frontiers in Microbiology* no. 12:636588.

Lateef, Agbaje, Osama M Darwesh, and Ibrahim A Matter. 2021. "Microbial nanobiotechnology: the melting pot of microbiology, microbial technology and nanotechnology." *Microbial Nanobiotechnology: Principles and Applications*:1–19.

Li, Hui, Rangrang Fan, Bingwen Zou, Jiazhen Yan, Qiwu Shi, and Gang Guo. 2023. "Roles of MXenes in biomedical applications: recent developments and prospects." *Journal of Nanobiotechnology* no. 21 (1):1–39.

Li, Xiangqian, Huizhong Xu, Zhe-Sheng Chen, and Guofang Chen. 2011. "Biosynthesis of nanoparticles by microorganisms and their applications." *Journal of Nanomaterials* no. 2011:1–16.

Michel, Caroline, Myriam Brugna, Corinne Aubert, Alain Bernadac, and Mireille Bruschi. 2001. "Enzymatic reduction of chromate: comparative studies using sulfate-reducing bacteria: key role of polyheme cytochromes c and hydrogenases." *Applied Microbiology and Biotechnology* no. 55:95–100.

Mie, Gustav. 1976. "Contributions to the optics of turbid media, particularly of colloidal metal solutions." *Contributions to the Optics of Turbid Media* no. 25 (3):377–445.

Mourdikoudis, Stefanos, Roger M Pallares, and Nguyen T K Thanh. 2018. "Characterization techniques for nanoparticles: comparison and complementarity upon studying nanoparticle properties." *Nanoscale* no. 10 (27):12871–12934.

Mustapha, Tijjani, Norashiqin Misni, Nur Raihana Ithnin, Abdullahi Muhammad Daskum, and Ngah Zasmy Unyah. 2022. "A review on plants and microorganisms mediated synthesis of silver nanoparticles, role of plants metabolites and applications." *International Journal of Environmental Research and Public Health* no. 19 (2):674.

Nishanthi, Ramasami, Selvaraj Malathi, and Perumal Palani. 2019. "Green synthesis and characterization of bioinspired silver, gold and platinum nanoparticles and evaluation of their synergistic antibacterial activity after combining with different classes of antibiotics." *Materials Science and Engineering: C* no. 96:693–707.

Pal, Kaushik. 2021. *Bio-Manufactured Nanomaterials*. Springer.

Patil, Maheshkumar Prakash, Min-jae Kang, Irvine Niyonizigiye, Alka Singh, Jong-Oh Kim, Yong Bae Seo, and Gun-Do Kim. 2019. "Extracellular synthesis of gold nanoparticles using the marine bacterium *Paracoccus haeundaensis* BC74171^T and evaluation of their antioxidant activity and antiproliferative effect on normal and cancer cell lines." *Colloids and Surfaces B: Biointerfaces* no. 183:110455.

Rafique, Muhammad, Iqra Sadaf, M Shahid Rafique, and M Bilal Tahir. 2017. "A review on green synthesis of silver nanoparticles and their applications." *Artificial Cells, Nanomedicine, and Biotechnology* no. 45 (7):1272–1291.

Rajathi, Fatimson Arockiya Aarthi, Chermapandi Parthiban, Vijayakumar Ganesh Kumar, and Perumal Anantharaman. 2012. "Biosynthesis of antibacterial gold nanoparticles using brown alga, *Stoechospermum marginatum* (kützing)." *Spectrochimica Acta Part A: Molecular and Biomolecular Spectroscopy* no. 99:166–173.

Rana, Anu, Krishna Yadav, and Sheeja Jagadevan. 2020. "A comprehensive review on green synthesis of nature-inspired metal nanoparticles: mechanism, application and toxicity." *Journal of Cleaner Production* no. 272:122880.

Rao, Kummara Madhusudana, Anuj Kumar, Adnan Haider, and Sung Soo Han. 2016. "Polysaccharides based antibacterial polyelectrolyte hydrogels with silver nanoparticles." *Materials Letters* no. 184:189–192.

Reddy, Murali M, Singaravelu Vivekanandhan, Manjusri Misra, Sujata K Bhatia, and Amar K Mohanty. 2013. "Biobased plastics and bionanocomposites: current status and future opportunities." *Progress in Polymer Science* no. 38 (10–11):1653–1689.

Rostamizadeh, Elham, Alireza Iranbakhsh, Ahmad Majd, Sedigheh Arbabian, and Iraj Mehregan. 2020. "Green synthesis of Fe_2O_3 nanoparticles using fruit extract of *Cornus mas* L. and its growth-promoting roles in Barley." *Journal of Nanostructure in Chemistry* no. 10:125–130.

Salem, Salem S, and Amr Fouda. 2021. "Green synthesis of metallic nanoparticles and their prospective biotechnological applications: an overview." *Biological Trace Element Research* no. 199:344–370.

Sanghi, Rashmi, and Preeti Verma. 2009. "Biomimetic synthesis and characterisation of protein capped silver nanoparticles." *Bioresource Technology* no. 100 (1):501–504.

Sarkar, Jit, Deepanjan Mridha, Mubarak Ali Davoodbasha, Jishnu Banerjee, Sumeddha Chanda, Kasturi Ray, Tarit Roychowdhury, Krishnendu Acharya, and Joy Sarkar. 2023. "A state-of-the-art systemic review on selenium nanoparticles: mechanisms and factors influencing biogenesis and its potential applications." *Biological Trace Element Research no.* 201(10):5000–5036..

Sasidharan, Sisini, Radhika Poojari, Dhirendra Bahadur, and Rohit Srivastava. 2018. "Embelin-mediated green synthesis of quasi-spherical and star-shaped plasmonic nanostructures for antibacterial activity, photothermal therapy, and computed tomographic imaging." *American Chemical Society Sustainable Chemistry & Engineering* no. 6 (8):10562–10577.

Sharma, Virender K, Jan Filip, Radek Zboril, and Rajender S Varma. 2015. "Natural inorganic nanoparticles–formation, fate, and toxicity in the environment." *Chemical Society Reviews* no. 44 (23):8410–8423.

Sharwani, Abdulrahman Ahmed, Kannan Badri Narayanan, Mohammad Ehtisham Khan, and Sung Soo Han. 2022. "Photocatalytic degradation activity of goji berry extract synthesized silver-loaded mesoporous zinc oxide (Ag@ ZnO) nanocomposites under simulated solar light irradiation." *Scientific Reports* no. 12 (1):10017.

Shi, Chaohong, Nengwu Zhu, Yanlan Cao, and Pingxiao Wu. 2015. "Biosynthesis of gold nanoparticles assisted by the intracellular protein extract of *Pycnoporus sanguineus* and its catalysis in degradation of 4-nitroaniline." *Nanoscale Research Letters* no. 10:1–8.

Show, Pau-Loke, Kit Wayne Chew, Wee-Jun Ong, Sunita Varjani, and Joon Ching Juan. 2023. *New Trends in Nanobiotechnology*. Beilstein-Institut.

Singh, Kshitij R B, Gunjan Nagpure, Jay Singh, and Ravindra Pratap Singh. 2023. "Introduction: utility of nanobiotechnology for bioremediation." In *Nanobiotechnology for Bioremediation*, 1–21. Elsevier.

Slawson, Robin M, Jack T Trevors, and Hung Lee. 1992. "Silver accumulation and resistance in *Pseudomonas stutzeri*." *Archives of Microbiology* no. 158:398–404.

Song, Jae Yong, Eun-Yeong Kwon, and Beom Soo Kim. 2010. "Biological synthesis of platinum nanoparticles using Diopyros kaki leaf extract." *Bioprocess and Biosystems Engineering* no. 33:159–164.

Sriramulu, Mohana, and Shanmugam Sumathi. 2018. "Biosynthesis of palladium nanoparticles using *Saccharomyces cerevisiae* extract and its photocatalytic degradation behaviour." *Advances in Natural Sciences: Nanoscience and Nanotechnology* no. 9 (2):025018.

Thakkar, Kaushik N, Snehit S Mhatre, and Rasesh Y Parikh. 2010. "Biological synthesis of metallic nanoparticles." *Nanomedicine: Nanotechnology, Biology and Medicine* no. 6 (2):257–262.

Ullah, Sadeeq, Aftab Ahmad, Aoke Wang, Muslim Raza, Amin Ullah Jan, Kamran Tahir, Aziz Ur Rahman, and Yuan Qipeng. 2017. "Bio-fabrication of catalytic platinum nanoparticles and their in vitro efficacy against lungs cancer cells line (A549)." *Journal of Photochemistry and Photobiology B: Biology* no. 173:368–375.

Vijayaram, Seerengaraj, Hary Razafindralambo, Yun-Zhang Sun, Seerangaraj Vasantharaj, Hamed Ghafarifarsani, Seyed Hossein Hoseinifar, and Mahdieh Raeeszadeh. 2023. "Applications of green synthesized metal nanoparticles—A review." *Biological Trace Element Research no.* 202 (1):1–27.

Wang, Zhiqiang, Cheng Fang, and Mallavarapu Megharaj. 2014. "Characterization of iron–polyphenol nanoparticles synthesized by three plant extracts and their fenton oxidation of azo dye." *American Chemical Society Sustainable Chemistry & Engineering* no. 2 (4):1022–1025.

Wangoo, Nishima, Sarabjit Kaur, Manish Bajaj, Dharamvir Vir Singh Jain, and Rohit K Sharma. 2014. "One pot, rapid and efficient synthesis of water dispersible gold nanoparticles using alpha-amino acids." *Nanotechnology* no. 25 (43):435608.

Woolfolk, C A, and H R Whiteley. 1962. "Reduction of inorganic compounds with molecular hydrogen by Micrococcus lactilyticus I. Stoichiometry with compounds of arsenic, selenium, tellurium, transition and other elements." *Journal of Bacteriology* no. 84 (4):647–658.

Yadav, Sapna, Nutan Rani, and Kalawati Saini. 2022. "A review on transition metal oxides based nanocomposites, their synthesis techniques, different morphologies and potential applications." *IOP Conference Series: Materials Science and Engineering* no. 1225(1):012004. https://iopscience.iop.org/article/10.1088/1757-899X/1225/1/012004/meta.

Yanke, L. Jay, Richard D. Bryant, and Edward J. Laishley. 1995. "Hydrogenase I of *Clostridium pasteurianum* functions as a novel selenite reductase." *Anaerobe* no. 1 (1):61–67.

Yumei, Li, Li Yamei, Li Qiang, and Bao Jie. 2017. "Rapid biosynthesis of silver nanoparticles based on flocculation and reduction of an exopolysaccharide from *Arthrobacter* sp. B4: its antimicrobial activity and phytotoxicity." *Journal of Nanomaterials* no. 2017:1–8.

Zahran, Magdy Kandil, Hanan Basioni Ahmed, and Muhammad H. El-Rafie. 2014. "Facile size-regulated synthesis of silver nanoparticles using pectin." *Carbohydrate Polymers* no. 111:971–978.

Zinoviadou, Kyriaki G, Paschalis Kastanas, Maria Gougouli, and Costas G Biliaderis. 2022. "Innovative bio-based materials for packaging sustainability." In *Innovation Strategies in the Food Industry*, 173–192. Elsevier.

3 Nanotechnology Applications in Cancer Nanotheranostics and Magnetic Particle Imaging

Nurcan Dogan and Muhammad Irfan

3.1 INTRODUCTION

Nanotechnology brought an unseen revolution in many aspects of life. Nano-size magnetic particles with core diameters between 1 and 100 nm are mainly preferred for biomedical applications [1]. Superparamagnetic iron oxide nanoparticles (SPIONs) are widely used for imaging modalities such as contrast agent (black hole) in magnetic resonance imaging (MRI) [2], tracer agent (hot spot) in magnetic particle imaging (MPI) [3], magnetic heat source in magnetic hyperthermia (MHT) [4], and drug carrier in therapeutic applications [1]. Uncoated magnetite (Fe_3O_4) and maghemite (γ-Fe_2O_3) are well-known nanoparticles; however, agglomeration and oxidation are the main challenges. Magnetic nanoparticles (MNPs) can also be composed with the doping of transition metal ions such as manganese (Mn), nickel (Ni), cobalt (Co), zinc (Zn), etc. Spinel ferrites introduce low coercivity and high saturation magnetization with additional legend functionalities to the basic nanoparticles.

The toxicity of bare MNPs is another challenge before the clinical trial. The non-magnetic hydrodynamic shell is essential to protect the magnetic core from agglomeration, oxidation, and toxicity. Functionalized coating layers bring homogenous dispersibility, colloidal stability, and additional target ligands. Nanotechnological magnetic particles with biocompatible and biodegradable characteristics ensure multi-model capabilities [1]. Organic coating materials such as citric acid (CA), ascorbic acid (AA), tartaric acid (TA), lauric acid (LA), oleic acid (OA), malic acid (MA), chitosan, cetyltrimethylammonium bromide (CTAB), synthetic polymer (polyethylene glycol (PEG), polyacrylic acid (PAA), polyethyleneimine (PEI)), natural (dextran, albumin), inorganic (gold) are widely used as compatible and biodegradable agents. The physicochemical, electromagnetic, and biological properties of the MNPs can be controlled during the synthesis process. SPIONs behave as tiny magnets in the presence of external magnetic fields and time during imaging and therapy operations.

DOI: 10.1201/9781003378563-3

MPI is a non-invasive imaging technique that utilizes S-shaped (non-linear) magnetic characteristics of the SPIONs. The MPI signals that are only generated from SPIONs support quantifiability (signal strength linearly depends on biodistribution).

Magnetic nanomaterials by resolving toxicity issues have become an essential part of various medical tools that enable therapeutic precision and diagnostic functionalities [5,6]. In addition to basic biocompatible coating, functional surface layers that carry a wide range of medicine, nucleic acids, and proteins [7] enhance the functionality of the MNPs as drug carriers. Drug carriers are intelligently driven toward maligned cells, tumor regions, and metastasis [8]. Functional groups such as thiol, imide, hydroxyl, azido, amino, sulfhydryl, and carboxyl are used to attach biomolecules and therapeutic agents [9–12].

The goal for nanomaterials is to develop smart nanoparticles such as nanotheranostics that are capable of diagnosis, drug carrier, therapy (e.g., focused MHT), and follow-up monitoring at the patient recovery stage [13,14]. Multi-modeling and multifunctional capabilities of the nanotheranostics agents help to achieve integrated medical tools.

3.2 MAGNETIC NANOPARTICLES

MNPs have a significant contribution to medical imaging modalities. The nanoparticles are classified based on their physical and chemical properties. Organic, inorganic, and carbon-origin nanoparticles are the main categories. Additionally, MNPs are further classified based on the type of magnetic core, and hydrodynamic coating (shell). A single-domain nanoparticle with possible functionalities adapted from reference [15] is pictorially represented in Figure 3.1. The thickness of the

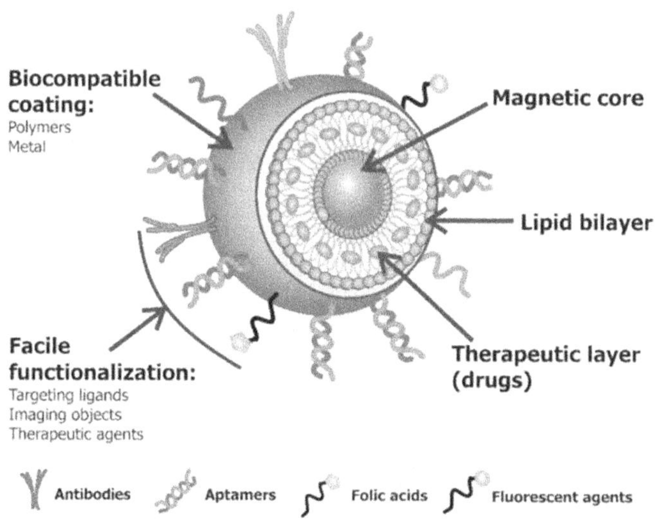

FIGURE 3.1 A generalized magnetic nanoparticle with different kinds of coatings, target ligands, and imaging agents. Therapeutic drugs are functionalized on the surface layers. A comprehensive representation of the single-domain nanoparticle was reused with permission from [15].

hydrodynamic coating layers ensures the minimum particle-to-particle interaction that prevents agglomeration phenomena.

3.2.1 Magnetic Response

The magnetic response of the nanoparticle is mainly based on the physicochemical characteristics such as magnetic core (material and size), and hydrodynamic coating (material and size). Imaging tools such as MPI and MRI need tracer and contrast agents that have an immediate response (minimum relaxation time) to the driving magnetic fields. The nanoparticles with a core diameter (<30 nm) are ideal candidates for imaging with minimum blurring issues. Hence superparamagnetic response of the contrast/tracer agent is inevitable. Superparamagnetic behavior is observed in magnetite (Fe_3O_4) and maghemite (γ-Fe_2O_4) which are the most known fundamental nanoparticles with magnetization similar to their bulk materials [16]. The magnetic response of SPIONs to direct (DC) applied field has a non-linear relationship commonly known as S-shaped behavior. The magnetization behavior changes abruptly at low applied magnetic fields and reaches saturation magnetization level at higher magnetic fields as shown in Figure 3.2.

As the applied field increases magnetization changes abruptly and becomes saturated after a certain point, so applying a higher field does not enhance magnetization. The relationship between the applied magnetic field and the magnetization response of the SPIONs can be modeled using the Langevin function [17,18]:

$$\mathcal{L}(\xi) := \begin{cases} \left(\coth(\xi) - \frac{1}{\xi}\right) & \xi \neq 0 \\ 0 & \xi = 0 \end{cases}. \tag{3.1}$$

Thermal stability is presumed for all nanoparticles under study. Thermal stability is ensured because the magnetizations of the SPIONs under investigation are studied at a constant temperature.

MNPs used for diagnosis, monitoring, and drug delivery have a nanoscale diameter. The overall magnetic response of the tracer/contrast agent is the result of the magnetic moment of individual nanoparticles. A magnetic moment of a single particle is represented by the symbol m. SPIONs maintain zero magnetic moment due to their random orientation at no magnetic field.

Let's assume ΔV as the volume with N^P MNPs. Overall, the magnetization (M) of the sample can easily be expressed as the summing moment of individual nanoparticles [17,18]:

$$M := \frac{1}{\Delta V} \sum_{i=0}^{N^P - 1} m_i. \tag{3.2}$$

As the external field is applied to SPIONs, they immediately start to align themselves in the direction of the driving field. SPIONs redirected and re-aligned in the

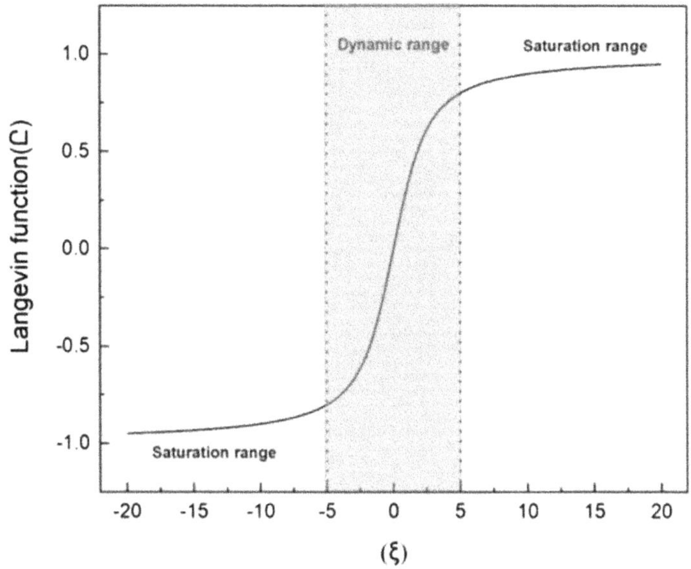

FIGURE 3.2 The Langevin function describes the response of magnetic nanoparticles under a direct magnetic field. It sharply changes from −5 to 5 (dynamic range). As the applied magnetic field increases, the Langevin function becomes saturated after a certain point [18].

presence of the external magnetic field. In the saturation mode nanoparticles are fully aligned in the direction of the applied field (H).

$$\boldsymbol{M}(\boldsymbol{H}) = M(H)e_H. \tag{3.3}$$

The magnetization response of the SPIONs resembles the Langevin function characteristics. So, equation $M(H)$ can be expressed as:

$$M(H) = cm\mathfrak{L}(\alpha H). \tag{3.4}$$

whereas $\alpha := \frac{\mu_0 m}{k_B T}$.

Here, T represents the temperature of the nanoparticles, μ_0 is the permeability of the free space, and k_B is the Boltzmann constant. In the meantime, c represents the particle concentration of the nanoparticles [17,18].

The magnetic moment of the nanoparticles can be described in terms of the magnetic core volume of the particle (V_c) and magnetization saturation (M_s). Here again, all nanoparticles are considered homogenous with single-domain characteristics.

$$m = V_c M_s. \tag{3.5}$$

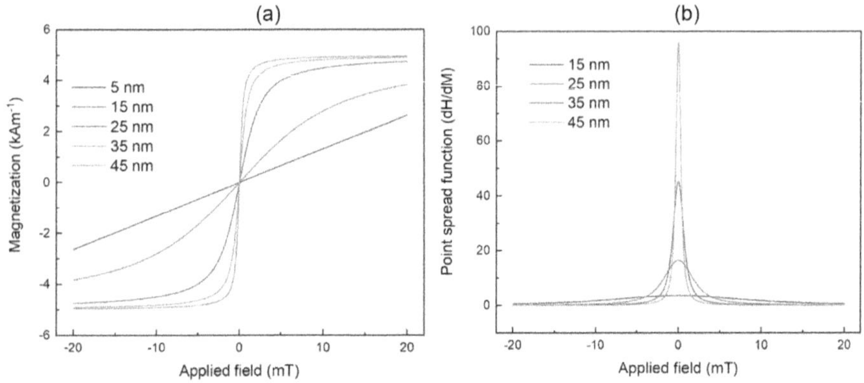

FIGURE 3.3 (a) The magnetization of magnetic nanoparticles with magnetic core diameters ranging from 5 nm to 45 nm [18]. A large diameter produces a sharp magnetization response, and (b) point spread functions also depend on magnetic core size, minimum full width at half maximum is achieved at a large-core diameter [18].

The core volume can further be elaborated with a core diameter (D_c) that helps to determine the volume of the spherical particle can be calculated as [15]:

$$V_c = \frac{1}{6}\pi D_c^3. \tag{3.6}$$

Practically, the iron concentration of the sample is more important as compared to particle concentration. If the iron concentration is kept constant, then particle concentration has an inverse relation with its particle volume. Saturation magnetization can be re-expressed as [17,18]:

$$M_s = cm. \tag{3.7}$$

The change of magnetization with the applied magnetic field is presented in Figure 3.3(a). The derivative of the magnetization response generates the point spread function (PSF). PSF helps to determine the spatial resolution of the MPI technique. As the magnetic core size of the nanoparticle increases the full width at half maximums (FWHM) reduces as shown in Figure 3.3(b).

Practically large nanoparticles are not preferred for imaging due to the high relaxation time that causes blurring. Image blurring and maximum operational frequency are the major challenges that evolved due to large particle sizes. Additionally, there are very specific regions of the human body where large nanoparticles do not pass through.

3.3 SYNTHESIS AND ANALYSIS OF MAGNETIC NANOPARTICLES FOR MAGNETIC PARTICLE IMAGING

Inorganic nanoparticles have the saturation magnetization approaches to the bulk materials. Metal oxide nanoparticles promise a wide range of biomedical applications;

however, the toxicity of the nanoparticles should be addressed. Hydrothermal, sol-gel, co-precipitation, and thermal cracking procedures, which are frequently used in the synthesis of MNPs, are given through the synthesis of various ferrite nanoparticles. In the presented study, nickel (Ni), manganese (Mn), and zinc (Zn) spinel ferrites are synthesized and coated with biocompatible and biodegradable materials to ensure the minimum toxicity effect with additional target ligands.

3.3.1 Synthesis of $NiFe_2O_4$ Spinel Ferrites

The most widely used hydrothermal synthesis process was preferred to develop nickel ferrites. It is a solution-reaction-based approach that is used for the formation of nanomaterials at a wide range of temperate scales. The following steps were followed in this synthesis method: initially, $FeCl_3.6H_2O$ and $NiCl_2$ salt were dissolved into distilled water. The mixture was stirred up at 500 rpm under the presence of nitrogen gas followed by the addition of NH_4OH aqueous solution. Finally, the mixture was placed into an autoclave and kept at 250°C for a specific time. After the heating operation, participants were washed with distilled water. The weak polyelectrolyte characteristics of PAA coating material have great potential to stabilize MNPs. Moreover, the excessive function groups of PAA make it attractive for MRI and MPI applications [19]. The CA hydrodynamic coating brings additional functionalities due to unfilled carboxylic acid groups as well. Additionally, CA coatings also prevent from clustering and oxidization of the magnetic cores [20,21].

3.3.1.1 Characteristics of $NiFe_2O_4$ Ferrites

Physicochemical properties of the nickel ferrites were evaluated by Fourier-transform infrared (FTIR) spectroscopy to explore chemical composition, X-ray diffraction (XRD) for the measurement of crystalline structure, scanning electron microscopy (SEM) for imaging of the sample, dynamic light scattering (DLS) to determine the hydrodynamic size, and thermogravimetric analysis (TGA) to figure out thermal stability. Nickel ferrites were also magnetically evaluated with electron-spin resonance (ESR) to measure metal complexes, and a physical properties measurement system (PPMS) for the evaluation of magnetic and electrical properties. The crystallite size of 9–12 nm was determined with XRD and confirmed by PPMS measurements. The evaluation of the physicochemical and magnetic properties of the $NiFe_2O_4$ ferrites proves the resemblance to the SPIONs.

MPI is a novel imaging technique that uses SPIONs for diagnostic and real-time imaging. A special device known as zero-dimensional MPI is proposed to evaluate potential tracer agents for biomedical applications [22].

3.3.1.2 Magnetic Particle Relaxometer (MPR) Evaluation

Instant evaluation of the MNPs just after the synthesis process could help to optimize and enhance the critical characteristics. Moreover, the transmit and receive signal setup of MPR and MPI tools are quite similar. The response of the MNPs is examined under very low frequency (VLF) magnetic fields. In this chapter, the MPR device at 9.9 kHz was utilized to generate sinusoidal magnetic fields with a peak amplitude of 15 mT. Operational frequency-based effective relaxation time,

TABLE 3.1
The Overall Results of the Nickel, Manganese, and Zinc Ferrite Nanoparticles Were Measured with MPR at 9.9 kHz

Sample Name	5th/3rd	FWHM (mT)	Relaxation Time (μs)
Perimag	0.62	5.70	4.5
VivoTrax	0.44	9.05	4.44
Fe_3O_4@OA	0.333	10.49	5.04
$NiFe_2O_4$	0.42	9.18	4.45
$NiFe_2O_4$@CA	0.49	7.98	3.08
$NiFe_2O_4$@PAA	0.49	7.75	3.10
$Mn_{0.5}Fe_{2.5}O_4$@OA	0.355	10.76	3.47
$MnFe_2O_4$@OA	0.330	11.73	2.24
$ZnFe_2O_4$	0.719	6.37	3.67
$ZnFe_2O_4$@PAA	0.636	4.89	2.68
$ZnFe_2O_4$@LA	0.603	5.39	2.68
$ZnFe_2O_4$@MA	0.588	5.68	2.46

FWHM of the point source nanoparticles, multiple harmonics that link with spatial resolution, and signal intensity based on iron concentration. The Neel and Brownian relaxation phenomena are quite important under time-dependent magnetic fields. The effective relaxation time of the nickel-based nanoparticles is experimentally calculated and presented in Table 3.1. This device inductively measures the non-linear magnetization response of the potential tracer agents. Specifically, multiple odd harmonics are associated with magnetic nanoparticle response as shown in Figure 3.4(a). The FWHM of the point-like sample helps to determine the spatial resolution of the nanoparticles in the MPI modality as shown in Figure 3.4(b).

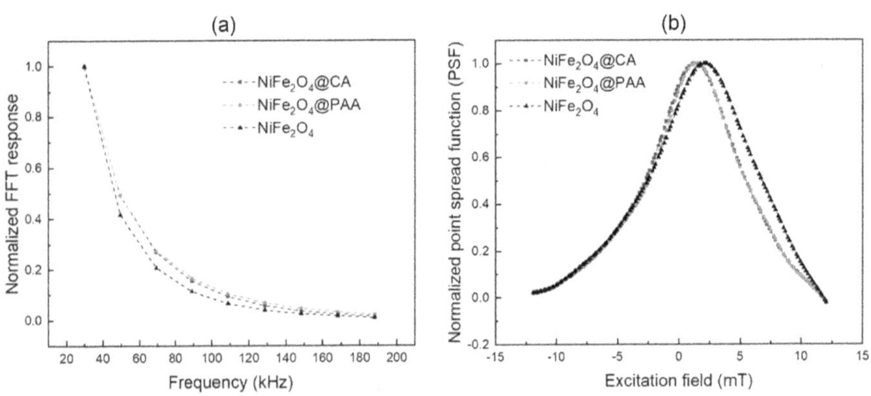

FIGURE 3.4 Magnetic particle relaxometer (MPR) evaluation of nickel ferrites at 9.9 kHz; (a) normalized odd harmonics response up to 20th harmonics and (b) point spread functions (PSFs) of the nickel ferrites for the determination of full width at half maximum (FWHM) in mT [23].

The time delay of the magnetization response to the applied sinusoidal drive field is commonly known as effective relaxation time. Magnetic core size, particle core size, particle size distribution, magnetic anisotropy, and dipole-dipole interaction affect the magnetic relaxation of the SPIONs. $NiFe_2O_4$@PAA sample has a quick response to the drive field as compared to the Perimag (Micromod GmbH, Germany) and VivoTrax (Magnetic Insight, USA) commercial nanoparticles. The relaxation behavior of the nanoparticles restricts the instantaneous response to the field free point (FFP) or field free line (FFL) scanning and becomes a major cause of blurring in MPI imaging. Therefore, low relaxation time is a highly desired characteristic of the potential tracer agents.

3.3.2 Synthesis of $Mn_xFe_{3-x}O_4$ Spinel Ferrites

A thermal decomposition synthesis process is commonly used to produce monodisperse and homogenous-shaped nanoparticles. The thermal decomposition method is utilized to develop MNPs with different concentrations (x = 0, 0.5, 1) of manganese (Mn). In the first stage, The Fe-oleate and Mn-oleate precursors were decomposed separately with a magnetic stirrer and slowly added methanol (CH_3OH) and sodium hydroxide (NaOH). In the second stage, two oleate complexes were decomposed using high boiling solvents (1-Octadecene) in the presence of a surfactant (OA). After the synthesis process, the nanoparticles were washed with ethanol and hexane two times and dried in an oven at 55°C. OA coating material improves the dispersion, high thermal stability, and crystallinity of the nanoparticles.

3.3.2.1 Characteristics of $Mn_xFe_{3-x}O_4$ Ferrites

Manganese-based nanoparticles were morphologically evaluated with XRD, FTIR, and thermal stability was analyzed by TGA. The physical and chemical properties are similar to superparamagnetic nanoparticles. $Mn_xFe_{3-x}O_4$ was also sample composition examined with ESR and electrical and magnetic properties were determined with PPMS. Additionally, the magnetic phase composition of the $Mn_xFe_{3-x}O_4$ nanoparticles coated with OA was determined with the Mössbauer spectra (MS) technique. The crystallite size of 11–18 nm is determined with XRD which is quite similar to the PPMS findings correlation Langevin curve fitting. Manganese ferrites coated with OA have great potential to become a tracer agent for MPI applications. MPR measurement tool at 9.9 kHz is exercised to evaluate relaxation time, signal intensity, and resolution of the newly synthesized manganese-based nanoparticles.

3.3.2.2 MPR Evaluation of Manganese Ferrites

Manganese nanoparticles were synthesized with a special MPR tool at 9.9 kHz. Fe_3O_4@OA, $Mn_{0.5}Fe_{2.5}O_4$@OA, and $MnFe_3O_4$@OA were studied under a sinusoidal magnetic field (peak value 15 mT). MPR measurement tool is considered a simpler (zero-dimension) variant of MPI scanners that do not use spatial encoding hardware components.

The fast Fourier transform (FFT) frequency spectrum of all samples is presented in Figure 3.5(a). Manganese with different concentration samples show a similar response because sample concentration does not affect the FFT response. The FFT

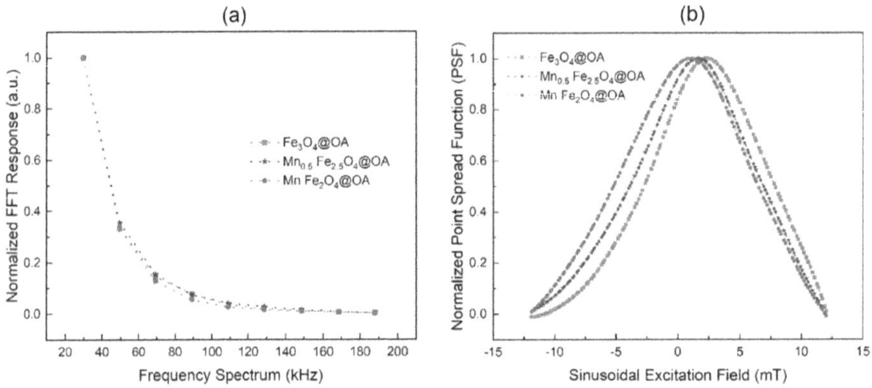

FIGURE 3.5 MPR characteristics of the manganese samples coated with OA at different concentrations at 9.9 kHz; (a) fast Fourier transform (FFT) frequency spectrum up to 20 odd harmonics and (b) point spread functions (PSFs) of the manganese ferrites for the determination of full width at half maximum (FWHM) in mT [24].

response is used in the system function reconstruction approach for the MPI scanner. All samples of manganese-doped MNPs show almost similar FWHM of 10.70 mT. However, the relaxation times of the samples are quite different from each other as presented in Table 3.1. The normalized PSF response of all samples is shown in Figure 3.5(b).

The FWHM of the PSFs plays a critical role in the spatial resolution of MPI. X-space reconstruction tool for MPI scanners uses direct time domain signal instead of FFT spectrum.

The spatial resolution Δx of the MPI scanners can be modeled as:

$$\Delta x = \frac{k_B T}{\mu_0 m G} \Delta\varepsilon_{FWHM}. \tag{3.8}$$

Here μ_0 is the magnetic permeability of the free space, m is the magnetic moment, $\Delta\varepsilon_{FWHM}$ is the FWHM of the PSF of the samples. G is the gradient strength of the static selection field.

The spatial resolution of the MPI imaging modalities mainly depends on the gradient field (T/m) and FWHM of the tracer obtained from PSFs. $MnFe_2O_4$@OA could produce 2.73 mm spatial resolution with a 4.3 T/m selection field MPI scanner. Moreover, $MnFe_2O_4$@OA has a minimum relaxation time (2.24 μs) outperformed other samples. MS also shows that manganese-based nanoparticles have enough potential to become tracer agents for MPI diagnosis and imaging.

3.3.3 Synthesis of $ZnFe_2O_4$ Spinel Ferrites

The zinc ferrite nanoparticles were synthesized by a combination of hydrothermal and co-precipitation methods processes. A magnetic core is developed in the hydrothermal process while coatings are applied in the co-precipitation process.

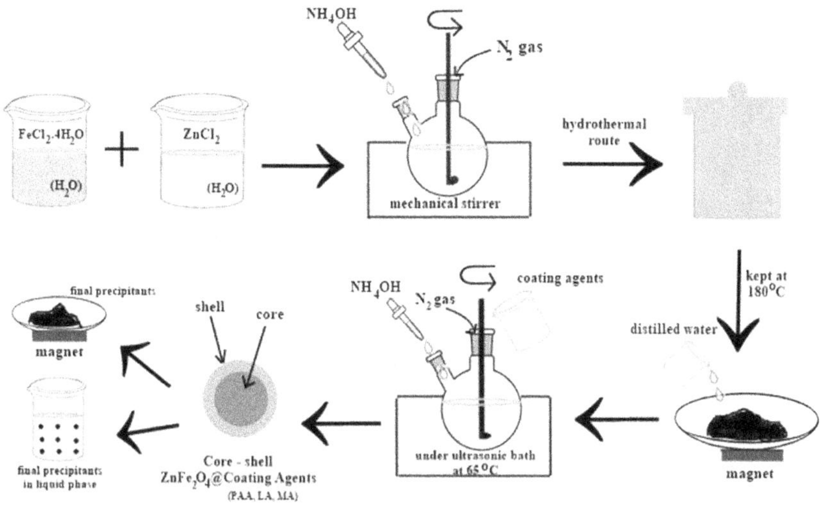

FIGURE 3.6 Zinc ferrites are developed by the combination of hydrothermal and co-precipitation techniques [25]. Magnetic cores were developed at the hydrothermal process 1st stage, while coating or shell structures were formed at the co-precipitation process 2nd stage.

$FeCl_3.6H_2O$ and $ZnCl_2$ salts were dissolved in distilled water. Under nitrogen gas, the mixture was stirred up with a mechanical stirrer at 500 rpm at room temperature (25°C) and NH_4OH was dropped into the solution. The solution was put into the Teflon-lined autoclave and kept at 180°C for 12 hours in the furnace. At the co-precipitation stage, the precipitants were dissolved into distilled water with the addition of capping agents (PAA, LA, and MA) in the ultrasonic bath at 65°C. The mixture was stirred mechanically under nitrogen gas to develop the core/shell structure of the MNPs. During this process, the stabilized flow of the nitrogen gas protects from the agglomeration of MNPs from the environmental conditions. The final participants were washed with distilled water. The flow chart of the synthesis phases of zinc ferrites is pictorially presented in Figure 3.6.

3.3.3.1 Characteristics of $ZnFe_2O_4$ Ferrites

Physical, chemical, electrical, and magnetic properties of zinc ferrites are evaluated with specific measurement tools. FTIR analysis ensured the presence of non-magnetic hydrodynamic coating material covering the magnetic cores. The crystallite size measured with the XRD technique of single-phase cubic spinel zinc ferrite nanoparticles was maintained within 14–18 nm with a stabilized synthesis process. Zinc ferrite nanoparticles coated with PAA, LA, and MA are also evaluated with PPMS. The core diameters determined with PPMS are in good agreement with the XRD technique. The capping agents PAA, LA, and MA played a significant role in the controllability of the nanoparticle size. MPR measurement tool at 9.9 kHz is utilized to evaluate relaxation time, signal intensity, and resolution of the $ZnFe_2O_4$ variants. The physicochemical and magnetic characterizations of the $ZnFe_2O_4$@ (PAA, LA, MA) nanoparticles proved that they have great potential for MPI applications.

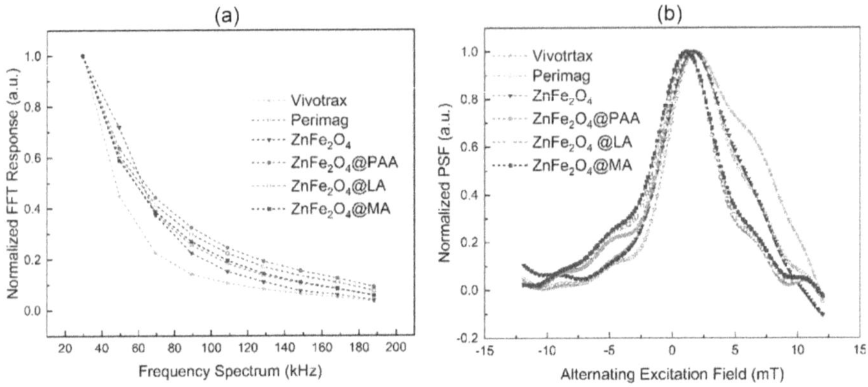

FIGURE 3.7 (a) Normalized FFT frequency spectrum of Perimag, VivoTrax, uncoated, and coated zinc-based nanoparticles and (b) normalized PSFs of the VivoTrax, Perimag, uncoated, and coated zinc-based nanoparticles [25].

3.3.3.2 MPR Evaluation of Zinc Ferrites

MPR tool provides a similar working environment to MPI scanners that help to virtually simulate the magnetic behavior of the zinc ferrites [22]. MPR determines the normalized signal intensity, spatial resolution (mT), effective relaxation time, and odd harmonics frequency spectrum. The results of MPR can be related to the physicochemical findings to re-synthesize the nanoparticles with optimized magnetic core and shell structure. MPR characterization of uncoated and coated (PAA, LA, MA) $ZnFe_2O_4$ nanoparticles was performed with a specially designed MPR [22] at 9.9 kHz with a 15 mT sinusoidal excitation field. The results obtained from MPR for uncoated and coated $ZnFe_2O_4$ nanoparticles were also compared with commercial nanoparticles such as Perimag (Micromod GmbH, Germany) and VivoTrax (Magnetic Insight, USA).

The slow decay of the $ZnFe_2O_4$@PAA sample enables it to produce higher spatial resolution as compared to other samples as shown in Figure 3.7(a). Moreover, the spatial resolution of the MPI scanners is classically defined by the FWHM (mm). However, MPR only determines the resolution in PSFs (FWHM, mT). The narrower half-peak width of the PSF means a higher resolution as shown in Figure 3.7(b).

A summary of key parameter comparisons of the MNPs discussed in this chapter is given in Table 3.1. Relaxation time and FWHM (mT) should be minimal for better performance in MPI environments. Among all samples, $ZnFe_2O_4$@PAA shows minimum FWHM (4.89 mT) and 2.68 μs relaxation time. In the meantime, this sample has a 0.636 (5th/3rd) that reflects the signal intensity. MPR tool has a great potential to discover many more properties such as viscosity and thermal mapping of the nanoparticles.

3.3.4 Cytotoxicity, Biodegradation, and Agglomeration

The magnetic characteristics and prospective applications of MNPs have transformed several biotechnological disciplines. However, the toxicity and biodegradation of

MNPs catch more attention on how they will end up in the environment and any potential negative impacts. The growing use of nanoparticles in biotechnology as a drug administrator and imaging agent requires a comprehensive study of toxicity and biodegradability.

The study of potential adverse effects of MNPs such as cytotoxicity, genotoxicity, and tissue-specific toxicity is known as toxicity evaluation. Iron oxide MNPs, known as SPIONs, exhibit superparamagnetic behavior and are considered reliable due to their biocompatibility and limited toxicity. Various magnetic cores, such as Fe, Zn, and Co, are incorporated to enhance the magnetic saturation of SPIONs and improve their functionality [23].

Adjusting the pH level below seven can prolong the lifetime of iron oxide MNPs [26]. Moreover, surface chemistry modifications provide MNPs with extra chemical and physical properties. Gold, polymers, carbons, and organic or inorganic molecules are considered safe coating materials for living systems. Furthermore, surface modifications that introduce a positive charge to MNPs enable faster uptake by the body [27].

The size and surface properties of MNPs influence their biodegradation rate. For example, positively charged PEI-coated drug-loaded nanoparticles enhance their uptake by cells, improve water solubility, and stabilize the system through electrostatic repulsion. MNP toxicity in vivo and in vitro can be controlled by adjusting the dosage [27], allowing them to be used as drug carrier systems.

Magnetic nanomaterials have numerous application areas including diagnostic bio-markers, bio-detection, bio-catalysis, scaffold engineering for bone repair, and molecule transport [28]. Besides drugs, MNPs can transport various cargo such as deoxyribonucleic acid (DNA), ribonucleic acid (RNA), oligonucleotides, proteins, or enzymes, providing a wide range of possibilities [29]. The quantity of healthy cells in a sample is known as cell viability, and the growth of cells is an essential indicator for recognizing the processes underlying the actions of certain genes, proteins, and pathways involved in the ability of cells to survive or die after being exposed to harmful substances.

3.4 CANCER THERANOSTIC APPLICATIONS OF MAGNETIC NANOPARTICLES

Nanotheranostics is the revolutionary concept of nanomedicine that integrates diagnostic and therapeutic agents in a single unit. It is a multidisciplinary research field that is a conjunction of nanotechnology, information, and biology. Diagnosis, drug delivery, and therapy are the basic operations that are simultaneously applied with theranostics agents. The processes linked to cancer theranostics are briefly illustrated in Figure 3.8. Tumor heterogenous effect and drug resistance are the main challenges for the fundamental cancer therapy techniques. Simultaneous target detection and drug distribution through theranostic agents could increase the effectiveness of the treatment. Hence, a customized treatment plan is necessary in the modern day because a single-line treatment is inapplicable to all patients.

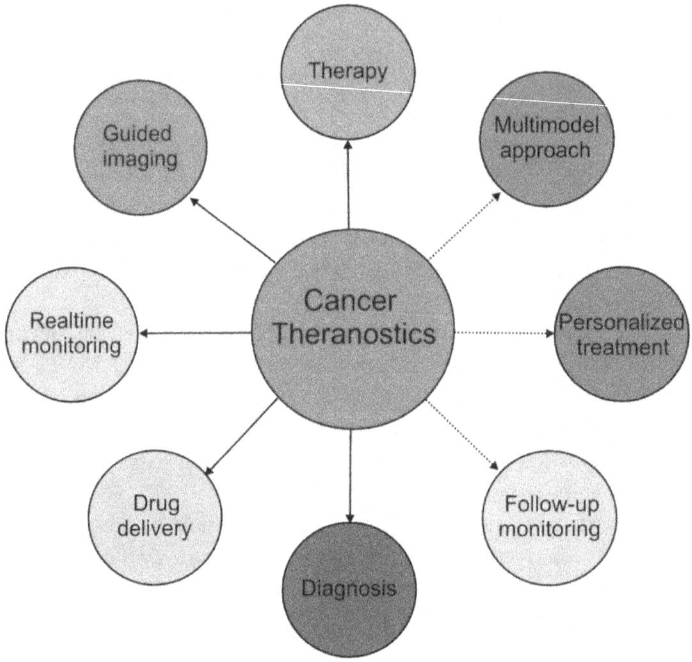

FIGURE 3.8 Cancer theranostics is a potential breakthrough in nanomedicine. Multimodal approach and personalized treatment are the ultimate objectives for nanomedicine and nanotechnology.

3.4.1 Diagnosis Tools

Multifunctional MNPs are used for scintigraphy, single photon emission computed tomography (SPECT), positron emission tomography (PET), MRI, and Raman spectroscopy. SPECT and PET are highly sensitive for biological examinations but have low spatial resolution [30]. MRI has excellent soft tissue contrast and morphological information, but it has low sensitivity [30]. MPI is another emerging diagnostic tool with higher sensitivity but a lack of structural information. Therefore, MRI is widely used clinically for the diagnosis of various diseases.

3.4.2 Magnetic Hyperthermia (MHT)

MHT is a non-invasive and externally controlled cancer therapy that uses biocompatible MNPs for heat generation in the presence of an external alternating magnetic field (AMF). MHT is based on the principle that the heat generated by the very small magnetic particles placed in the tumor tissue destroys the tumors. To generate heat, the particles must be exposed to an AMF produced by external electromagnetic coils. Theoretically, AMFs induce eddy currents in a conductive medium. The change of magnetization of iron against some resistive force releases heat into the environment. This method is used directly in the treatment of cancer or can act as a supportive

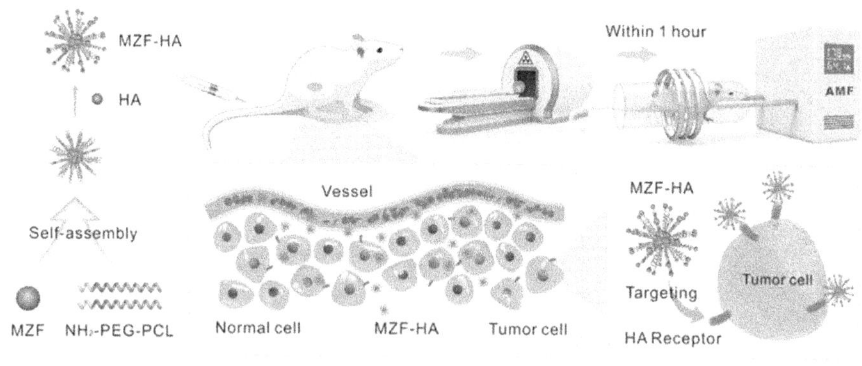

FIGURE 3.9 Targeted cancer treatment with manganese and zinc ferrite-based nanoparticles. Initially, radiotherapy is applied followed by a MHT process to enhance the effectiveness of the cancer treatment. Graphical illustration was reused with permission from [31].

treatment for chemotherapy, or radiotherapy. The heat released causes cell damage or aids drug delivery. SPIONs generate more heat than ferromagnets under the same conditions due to the high hysteresis losses of single-domain magnetic particles.

Generally, there are two ways to administer tracer agents for hyperthermia. The first is to inject nanoparticles directly into the tumor area. The second is to inject the nanoparticles into the vessels, which can provide a more homogeneous distribution in the circulatory system. When SPIONs are placed in a magnetic field, thermal energy is only released inside and very close to the tumor. This damage may affect only a few healthy cells around the tumor. The whole process raises the temperature inside the tumor to 42–45°C. The temperature range is sufficient to cause the death of cells. Such a high temperature activates multiple pathways that lead to cell destruction, allowing the treatment to be successful. This is important because the presence of cancer cells containing many somatic mutations tends to disappear with temperature increase (usually above 43°C) [31] as illustrated in Figure 3.9. Hyperthermia treatment can be repeated by immersing the patient in a magnetic field several times.

3.4.3 Personalized Medicine

Personalized treatment is a state-of-the-art concept of nanomedicine that gave hope in cancer therapy. Smart nanoparticles are used to figure out the therapeutic potency of the anticancer medicine. Pre-treatment cell viability test helps to establish an applicable cure plan. The role of nanotechnology in cancer therapy is vital for effective treatment protocols. A simplified cancer therapy protocol is presented in Figure 3.10. Personalized medicine is the best way to fight life-threatening cancer variants. Each patient responds differently to medication, so effective use of drugs is inevitable for life-saving.

Personalized medicine will be more attractive and effective with the greater use of nanotechnology. The history of the patient would also help to develop a disease-prevention approach. MNPs have rapidly progressed but still many challenges to

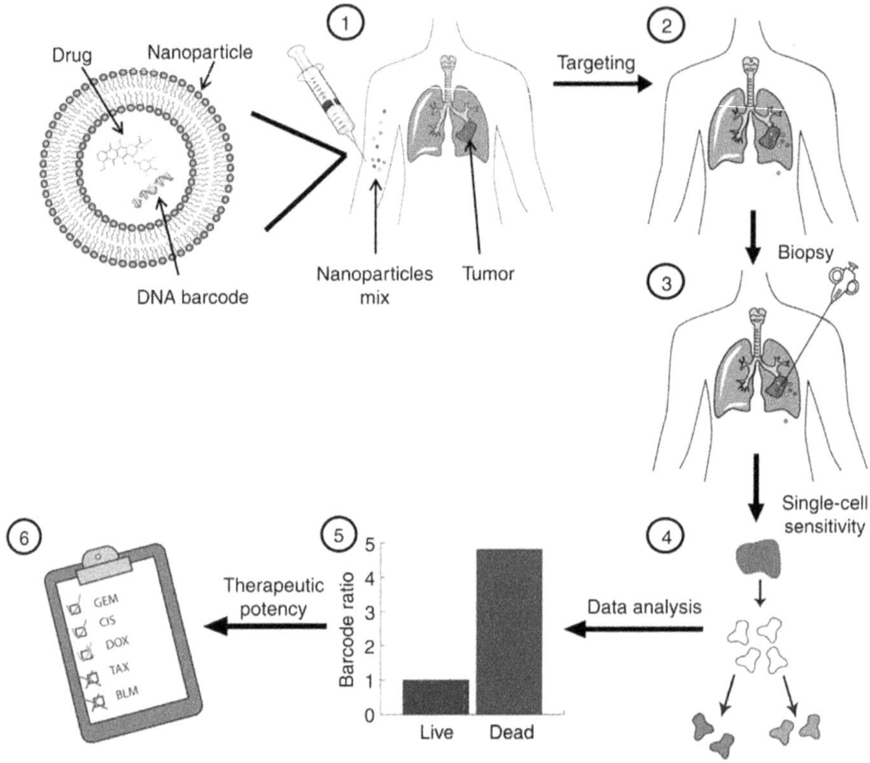

FIGURE 3.10 The use of nanotechnology in sensitivity determination of cancer to medicine. Custom-designed nanoparticles were utilized that carry patient-specific drugs on DNA barcodes. (1) Nanoparticles were administered through injection, (2) drug carriers reached the tumor cells in a certain time, (3) a biopsy operation was carried out for diagnosis, (4) the functionality of each cell was assessed, (5) the barcode ratio of live and dead cells was determined, and (6) based on the results personalized treatment plan was devised. A simplified diagnostic scheme was reused with permission from [32].

overcome for clinical translation. Multiple functionalities such as imaging, monitoring, and therapeutic characteristics make a single nanoparticle structure more complicated. This intradisciplinary field demands a great deal of coordination between scientists and engineers to develop effective nanotheranostic agents.

REFERENCES

1. Wahajuddin, and Sumit Arora. "Superparamagnetic Iron Oxide Nanoparticles: Magnetic Nanoplatforms as Drug Carriers." International Journal of Nanomedicine, 2012, 3445–71. https://doi.org/10.2147/ijn.s30320.
2. Lassenberger, Andrea, Andrea Scheberl, Andreas Stadlbauer, Alexander Stiglbauer, Thomas Helbich, and Erik Reimhult. "Individually Stabilized, Superparamagnetic Nanoparticles with Controlled Shell and Size Leading to Exceptional Stealth Properties and High Relaxivities." ACS Applied Materials & Interfaces 9, no. 4 (2017): 3343–53. https://doi.org/10.1021/acsami.6b12932.

3. Gleich, Bernhard, and Jürgen Weizenecker. "Tomographic Imaging Using the Nonlinear Response of Magnetic Particles." Nature 435, no. 7046 (2005): 1214–17. https://doi.org/10.1038/nature03808.
4. Cherukuri, Paul, Evan S. Glazer, and Steven A. Curley. "Targeted Hyperthermia Using Metal Nanoparticles." Advanced Drug Delivery Reviews 62, no. 3 (2010): 339–45. https://doi.org/10.1016/j.addr.2009.11.006.
5. Torchilin, Vladimir P. "Multifunctional, Stimuli-Sensitive Nanoparticulate Systems for Drug Delivery." Nature Reviews Drug Discovery 13, no. 11 (2014): 813–27. https://doi.org/10.1038/nrd4333.
6. Stark, Wendelin J. "Nanoparticles in Biological Systems." Angewandte Chemie International Edition 50, no. 6 (2011): 1242–58.https://doi.org/10.1002/anie.200906684.
7. Khandare, Jayant, Marcelo Calderón, Nilesh M. Dagia, and Rainer Haag. "Multifunctional Dendritic Polymers in Nanomedicine: Opportunities and Challenges." Chemical Society Reviews 41, no. 7 (2012): 2824–48. https://doi.org/10.1039/c1cs15242d.
8. Theek, Benjamin, Larissa Y. Rizzo, Josef Ehling, Fabian Kiessling, and Twan Lammers. "The Theranostic Path to Personalized Nanomedicine." Clinical and Translational Imaging 2, no. 1 (2014): 67–76. https://doi.org/10.1007/s40336-014-0051-5.
9. Shen, Zhiqiang, Mu-Ping Nieh, and Ying Li. "Decorating Nanoparticle Surface for Targeted Drug Delivery: Opportunities and Challenges." Polymers 8, no. 3 (2016): 83. https://doi.org/10.3390/polym8030083.
10. Goon, Ian Y., Leo M. Lai, May Lim, Paul Munroe, J. Justin Gooding, and Rose Amal. "Fabrication and Dispersion of Gold-Shell-Protected Magnetite Nanoparticles: Systematic Control Using Polyethyleneimine." Chemistry of Materials 21, no. 4 (2009): 673–81. https://doi.org/10.1021/cm8025329.
11. Lu, An-Hui, E. L. Salabas, and Ferdi Schüth. "Magnetic Nanoparticles: Synthesis, Protection, Functionalization, and Application." Angewandte Chemie International Edition 46, no. 8 (2007): 1222–44. https://doi.org/10.1002/anie.200602866.
12. Fattahi, Hassan, Sophie Laurent, Fujun Liu, Nasser Arsalani, Luce Vander Elst, and Robert N Muller. "Magnetoliposomes as Multimodal Contrast Agents for Molecular Imaging and Cancer Nanotheragnostics." Nanomedicine 6, no. 3 (2011): 529–44. https://pubmed.ncbi.nlm.nih.gov/21542690/.
13. Singh, Abhalaxmi, and Sanjeeb K. Sahoo. "Magnetic Nanoparticles: A Novel Platform for Cancer Theranostics." Drug Discovery Today 19, no. 4 (2014): 474–81. https://doi.org/10.1016/j.drudis.2013.10.005.
14. Yoo, Dongwon, Jae-Hyun Lee, Tae-Hyun Shin, and Jinwoo Cheon. "Theranostic Magnetic Nanoparticles." Accounts of Chemical Research 44, no. 10 (2011): 863–74. https://doi.org/10.1021/ar200085c.
15. Belyanina, Irina, Olga Kolovskaya, Sergey Zamay, Ana Gargaun, Tatiana Zamay, and Anna Kichkailo. "Targeted Magnetic Nanotheranostics of Cancer." Molecules 22, no. 6 (2017): 975. https://doi.org/10.3390/molecules22060975.
16. Cao, Derang, Hao Li, Lining Pan, Jianan Li, Xicheng Wang, Panpan Jing, Xiaohong Cheng, Wenjie Wang, Jianbo Wang, and Qingfang Liu. "High Saturation Magnetization of γ-Fe_2o_3 Nano-Particles by a Facile One-Step Synthesis Approach." Scientific Reports 6, no. 1 (2016). https://doi.org/10.1038/srep32360.
17. Knopp, Tobias, and Thorsten M. Buzug. Magnetic Particle Imaging: An Introduction to Imaging Principles and Scanner Instrumentation. Springer, Heidelberg New York Dordrecht London, 2012.
18. Irfan, Muhammad "Design and Implementation of Magnetic Particle Imaging (MPI) Scanner for Medical Applications. Ph.D. dissertation", Dept. Electron. Eng., Gebze Tech. Univ., Gebze, Turkey, 2021.
19. Sanchez, Laura M., Daniel A. Martin, Vera A. Alvarez, and Jimena S. Gonzalez. "Polyacrylic Acid-Coated Iron Oxide Magnetic Nanoparticles: The Polymer Molecular

Weight Influence." Colloids and Surfaces A: Physicochemical and Engineering Aspects 543 (2018): 28–37. https://doi.org/10.1016/j.colsurfa.2018.01.050.

20. Li, L., K.Y. Mak, C.W. Leung, K.Y. Chan, W.K. Chan, W. Zhong, and P.W.T. Pong. "Effect of Synthesis Conditions on the Properties of Citric-Acid Coated Iron Oxide Nanoparticles." Microelectronic Engineering 110 (2013): 329–34. https://doi.org/10.1016/j.mee.2013.02.045.
21. Riaz, S., S. Naseem, and X. Han. "Citric Acid Coated Iron Oxide Nanoparticles—Structural and Magnetic Properties." 2015 IEEE Magnetics Conference (INTERMAG), 2015. https://doi.org/10.1109/intmag.2015.7157349.
22. Irfan, Muhammad, Nurcan Dogan, Tarik Sapmaz, and Avhan Bingolbali. "Development of MPI Relaxometer for Characterization of Superparamagnetic Nanoparticles." Journal of Magnetism and Magnetic Materials 536 (2021): 168082. https://doi.org/10.1016/j.jmmm.2021.168082.
23. Irfan, Muhammad, Nuracan Dogan, Ayhan Bingolbali, and Fuad Aliew. "Synthesis and Characterization of $NiFe_2O_4$ Magnetic Nanoparticles with Different Coating Materials for Magnetic Particle Imaging (MPI)." Journal of Magnetism and Magnetic Materials 537 (2021): 168150. https://doi.org/10.1016/j.jmmm.2021.168150.
24. Dogan, Nuracan, Ozge Mercan Dogan, Muhammad Irfan, Fatmahan Ozel, A.S. Kamzin, V.G. Semenov, and I.V. Buryanenko. "Manganese Doped-Iron Oxide Nanoparticles and Their Potential as Tracer Agents for Magnetic Particle Imaging (MPI)." Journal of Magnetism and Magnetic Materials 561 (2022): 169654. https://doi.org/10.1016/j.jmmm.2022.169654.
25. Dogan, Nurcan, Gulsum Caliskan, and Muhammad Irfan. "Synthesis and Characterization of Biocompatible ZnFe2O4 Nanoparticles for Magnetic Particle Imaging (MPI)." Journal of Materials Science: Materials in Electronics 34, no. 5 (2023). https://doi.org/10.1007/s10854-022-09799-x.
26. Sun, Sheng-Nan, Chao Wei, Zan-Zan Zhu, Yang-Long Hou, Subbu S Venkatraman, and Zhi-Chuan Xu. "Magnetic Iron Oxide Nanoparticles: Synthesis and Surface Coating Techniques for Biomedical Applications." Chinese Physics B 23, no. 3 (2014): 037503. https://doi.org/10.1088/1674-1056/23/3/037503.
27. Liberman, Alina, Jason Fischer, and David Whitney. "Serial Dependence in the Perception of Faces." Current Biology 24, no. 21 (2014): 2569–74. https://doi.org/10.1016/j.cub.2014.09.025.
28. Argyo, Christian, Veronika Weiss, Christoph Bräuchle, and Thomas Bein. "Multifunctional Mesoporous Silica Nanoparticles as a Universal Platform for Drug Delivery." Chemistry of Materials 26, no. 1 (2013): 435–51. https://doi.org/10.1021/cm402592t.
29. Lancaster, Madeline A., Magdalena Renner, Carol-Anne Martin, Daniel Wenzel, Louise S. Bicknell, Matthew E. Hurles, Tessa Homfray, Josef M. Penninger, Andrew P. Jackson, and Juergen A. Knoblich. "Cerebral Organoids Model Human Brain Development and Microcephaly." Nature 501, no. 7467 (2013): 373–79. https://doi.org/10.1038/nature12517.
30. Roach, Mack, Jean-Louis Alberini, Alain Paul Pecking, Alessandro Testori, Francesco Verrecchia, Javier Soteldo, Ute Ganswindt, et al. "Diagnostic and Therapeutic Imaging for Cancer: Therapeutic Considerations and Future Directions." Journal of Surgical Oncology 103, no. 6 (2011): 587–601. https://doi.org/10.1002/jso.21805.
31. Wang, Yijue, Liqing Zou, Zhe Qiang, Jianhai Jiang, Zhengfei Zhu, and Jie Ren. "Enhancing Targeted Cancer Treatment by Combining Hyperthermia and Radiotherapy Using Mn–Zn Ferrite Magnetic Nanoparticles." ACS Biomaterials Science & Engineering 6, no. 6 (2020): 3550–62. https://doi.org/10.1021/acsbiomaterials.0c00287.
32. Yaari, Zvi, Dana da Silva, Assaf Zinger, Evgeniya Goldman, Ashima Kajal, Rafi Tshuva, Efrat Barak, et al. "Theranostic Barcoded Nanoparticles for Personalized Cancer Medicine." Nature Communications 7, no. 1 (2016). https://doi.org/10.1038/ncomms13325.

4 Biogenically Derived Metal Nanoparticles as a Sustainable Resource for Overcoming Bacterial Resistance

James Mordecai, Deepak Anand, Saleem Ullah, and Irshad Ahmad

4.1 INTRODUCTION

One of the most prevalent infectious diseases in the world is bacterial infection, which results from interaction with pathogenic bacterial strains. Although bacteria are ubiquitous, only a few are pathogenic [1]. A bacterial infection is more likely to occur when there is poor personal hygiene, sexual promiscuity, erroneous sewage disposal, and nosocomial contact. Incidence of typhoid fever, tuberculosis, pneumonia, meningitis, and gonorrhea continue to account for increased illnesses and deaths in many parts of the world. A report found that 33 bacterial pathogens were responsible for 7.7 million of the 13.7 million infection-related mortalities that took place globally in 2019 [2].

A good number of bacterial infections are readily treatable with the use of antibiotic medications. However, the rapid emergence of resistance to antibiotics makes treatment of these infections difficult and more expensive, as high doses of these drugs are needed to treat resistant bacterial strains. Antibiotic resistance is the state in which a bacterium develops multiple defenses against the disruptive effects of medications (antibiotics) used to treat an infection brought on by that bacterium [3]. Although the development of antibiotic resistance can occur spontaneously through natural selection or horizontal gene transfer, it is commonly caused by humans using medications and antibacterial drugs incorrectly and/or for a lengthy period when treating bacterial diseases [4].

Antibiotic resistance is thought to contribute to 700,000 deaths globally each year. By 2050, this figure is anticipated to increase to over 10 million deaths annually [5]. It is even a much bigger problem in older patients, immuno-compromised individuals, and recipients of organ transplants who developed complications due to bacterial infections [6]. Hence, it has become imperative to look for novel and effective ways to address and treat antibiotic-resistant infections.

DOI: 10.1201/9781003378563-4

For the alleviation and aversion of infectious diseases, innovative antimicrobial agents known as nanomaterials have emerged. These compounds have shown effectiveness against several strains of bacteria [7, 8]. Studies conducted in vitro show that nanomaterials have damaging impacts on different resistant bacterial strains, indicating the viability of using them in biological applications and drug delivery [9, 10]. One of the most promising types of nanomaterials for the amelioration of infections resistant to antibiotics is metal and metal oxide nanoparticles (NPs). NPs such as bismuth, copper, gold, selenium, and silver, and metal oxide NPs such as Al_2O_3, CaO, Fe_2O_3, MgO, TiO_2, or ZnO have been proposed and shown to be highly efficacious in combating bacterial resistance [11–16].

These NPs have antibacterial properties on their own and exhibit nonspecific activity, which inhibits the emergence of bacterial resistance [17]. The NPs can also be capped with antibiotics in a form of combination therapy of nano-based compounds and conventional antibiotics. The synergistic action of such combination therapy will allow for the use of very low doses of antibiotics with high antibacterial effects, such low doses will prevent any possible damage to human cells, and the NPs will promote specificity for only bacterial cells [18–20].

Despite the excitement around the application of NPs as antibacterial agents, there are drawbacks. NPs are usually produced either by chemical or physical methods. The physical method of making NPs uses sophisticated, expensive equipment, requires a considerable amount of energy, and yields little [21]. The chemical approach, however, uses and releases very harmful chemical by-products into the environment, including thiols, aromatic amines, and sodium borohydrides, despite being less expensive, simpler, and having a higher yield [22, 23].

Hence, the synthesis of biogenic NPs has been suggested as an alternate method for producing NPs. This method is reasonably easy, economical, and ecologically responsible [24]. The procedure uses cellular material from living entities like plants, fungus, algae, and bacteria as both a capping agent and a green reaction media. These biological extracts are non-toxic and non-volatile, and so are ideal mediums for NPs synthesis. Often time, biological extracts contain a blend of many active biomolecules, including carbohydrates, lipids, proteins, vitamins, polymers, and natural surfactants [25, 26].

This chapter discusses the fabrication of biogenic NPs, factors governing the synthesis of biogenic nanomaterials, effectiveness and potential of these green NPs, their proposed mechanisms of action, as well as challenges and prospects.

4.2 SYNTHESIS OF BIOGENIC NPs

4.2.1 Bacteria-Derived NPs

As the production of NPs using biogenic approaches continues to gain momentum, the use of bacteria for fabricating NPs has emerged as one of the most promising biogenic methods. Biogenic bacteria-derived NPs exhibit distinctive physiochemical characteristics and outstanding biological features, making them potential candidates for a variety of biomedical uses [27]. Bacteria act as nano-factories for the fabrication of NPs via extracellular or intracellular synthesis. Extracellular methods for generating NPs offer an edge over intracellular approach in terms of speed because it requires no additional steps to harvest NPs from the organisms [28]. In this method, metal ions are reduced in the presence of bacteria, which also serve as stabilizers, reducers, and shape-controlling agents. These cells have multiple forms of the enzyme reductase, which reduces metal ions into metal NPs [29]. Different bacterial species such as *Escherichia coli*, *Bacillus subtilis*, *Pseudomonas aeruginosa*, and *Lactobacillus* have been used for synthesizing NPs of several metals including platinum, silver, gold, and iron oxide [30]. One of the primary benefits of employing bacteria to synthesize NPs is that the procedure is straightforward, affordable, and doesn't require any hazardous chemicals or significant energy inputs. Additionally, the employment of bacteria enables the fabrication of NPs with exact size, shape, and surface chemistry, which are crucial factors that influence their biological activity [31].

Biogenic bacteria-derived NPs have shown promise in a variety of biological applications, including cancer therapy, medication delivery, antibacterial/antifungal activities, and imaging. For instance, gold NPs (AuNPs) synthesized using *Lactobacillus acidophilus* have been found to exhibit excellent anticancer activity against breast carcinoma [32]. Silver NPs (AgNPs) synthesized using *P. aeruginosa* have been shown to possess potent bactericidal activity against gram-positive and gram-negative bacteria [33]. A similar study showed that the resistant strains of *Salmonella typhi* and *Staphylococcus aureus* were successfully eliminated by AgNPs made with *Bacillus brevis* [34]. In addition, CuO NPs produced by *Halomonas elongata* exhibited efficient antimicrobial activity against *S. aureus* and *E. coli* [35]. Thus, the potential of biogenic bacteria-derived NPs as antibacterial agents and in antibiotic delivery needs to be further explored.

Apart from their biomedical applications, biogenic bacteria-derived NPs also have potential environmental applications such as water purification, soil remediation, and pollution control. For instance, iron oxide NPs (Fe_2O_3 NPs) synthesized using *B. subtilis* have been shown to efficiently chelate arsenic and lead ions from polluted water [36].

4.2.2 Fungi-Derived NPs

Fungi are now recognized as key mediators in the creation of NPs thanks to recent developments in nanotechnology. Like bacterial NP synthesis, myco-synthesis of NPs can be intracellular (within the mycelia) or extracellular (fungi extracts) [37, 38]. However, due to the abundance of several bioactive byproducts, high accrual, and improved production, fungi are more innovative than bacteria in the production of NPs [39, 40]. Additionally, fungi produce a more diverse range of enzymes than bacteria for the synthesis of NPs, and the mycelia network of fungi provides a large surface area for effective intracellular myco-synthesis of NPs [41, 42].

Biogenic synthesis of NPs using fungi involves chemically reducing metal ions in the presence of fungal cells or their metabolites. Quite a few proteins and enzymes produced by fungi serve as reducing mediators, stabilizers, and shape-controlling agents and are frequently employed in the fabrication of metallic NPs [43]. The most common myco-synthesized NPs are silver and gold.

A few studies have reported that *Fusarium oxysporum* secretes enzymes that aid in the fabrication of metal NPs [44]. For example, reductase enzymes in the cytosol of *F. oxysporum* rapidly reduce Ag^+ into Ag in the presence of $NADH^+$ [45]. Researchers studying *Coriolus versicolor* found that its mycelium contained hydroxyl groups which donated electrons to facilitate the reduction of silver ions to elemental silver for the fabrication of AgNPs [46]. Another study used *Rhizopus stolonifera* extracts to mediate the synthesis of monodispersed AgNPs [47], while a similar study highlighted the antimicrobial potential of AgNPs synthesized using *Candida glabrata* [48]. It has been demonstrated that AgNPs made with *Aspergillus flavus* have a potent antibacterial effect against a few bacteria [49].

AuNPs have also been reportedly synthesized with a range of fungal species. One study examining the ability of 29 thermophilic fungi to synthesize AuNPs via intracellular and extracellular approaches reported efficient formation of AuNPs of varying sizes [50]. A similar study using *Aureobasidium pullulans*, *F. oxysporum*, and *Fusarium* sp. indicated the role played by the vacuoles of these fungal species in the synthesis of AuNPs [42].

Besides synthesis of NPs, fungal metabolites have been used in producing capping agents and for post-production processing. For example, Sanghi and Verma [46] demonstrated that the amide bond in aliphatic and aromatic amines in certain proteins stabilizes fabricated NPs. The thiol (SH) group found in the proteins of some fungi extracts has also been shown to be involved in the capping of AgNPs [51]. Moreso, a nano conjugate of AuNPs, was recently made from the mycelia of *Rhizopus oryzae* [52].

Biogenic fungi-derived NPs really do have the potential for use in various biomedical and environmental applications; further research could unveil even more novel applications.

4.2.3 Yeast-Derived NPs

Yeasts are well-known single-celled (microscopic) fungi that have previously been used in the synthesis of NPs. Yeast cells secrete enzymes and other biomolecules that aid NP synthesis, including membrane-bound oxidoreductases. Several yeast species, including *Saccharomyces cerevisiae*, *Candida utilis*, and *Pichia pastoris*, have been utilized in the synthesis of metallic NPs. For instance, a recent study using *P. pastoris* synthesized gold and palladium NPs, the NPs were reportedly uniform in size with a diameter of 10–30 nm when analyzed with electron microscopy and XRD size analyzer [53]. To fabricate AgNPs with a diameter of 20 nm, another study employed a yeast strain extracted from an acid mine drainage system [54]. A similar study produced AgNPs via extracellular synthesis by directly exposing *S. cerevisiae*, *Rhodotorula glutinis*, and *Geotrichum candidum* to $AgNO_3$ solution [55]. In addition, when enough light is present, the glutamic acid and aspartic acid on the surface of some yeast cells photo-reduce silver ions into silver metal [56].

Yeast has some benefits over other fungal species, such as quick growth and controllable mass production of yeasts utilizing a straightforward nutrient culture media. Yeast cells also possess an easy encapsulation mechanism for metallic NPs, as the cells' membrane acts as a microcapsule during the fabrication and polymerization of NPs [57]. Overall, yeast cells are efficient, little factories for the synthesis of NPs.

4.2.4 Plant-Derived NPs

The green synthesis of NPs by plants is more efficient than all the microbial methods that have been discussed so far. This is because plants are larger and consequently contain more metabolites and phytochemicals such as terpenoids, phenols, and flavonoids, which are necessary for the production of NPs. Plant-assisted synthesis of NPs can either be extracellular or intracellular. The extracellular technique is used when plant extract is used as the primary component of NP production, while in the intracellular approach, phytochemicals and cytosolic enzymes are used to synthesize NPs within the cellular structure of a plant material. The synthesized NPs are retrieved by breaking apart the cell walls of the plant tissue [58].

A good number of plant species, including *Aloe vera*, *Azadirachta indica*, and *Ocimum sanctum*, have been employed in the production of NPs. For example, *A. vera* extract has been utilized as a reducing and stabilizing agent to produce AuNPs. The synthesized NPs were stable and exhibited antibacterial activity against both *E. coli* and *S. aureus* [59]. In another report, spherical AgNPs with efficient bactericidal activity were fabricated using leaf extract of *A. indica* (neem). Post-production analysis showed that phytochemical components of the leaf extract acted as a reducing and stabilizing agent during the formation of NPs [60]. A similar study used the root extract of *Morinda citrifolia* to fabricate AuNPs shown to possess antimicrobial activity [61]. More recently, stable copper NPs (CuNPs) were made and characterized with the aid of *O. sanctum* and *Piper nigrum* extracts [62]. An overview of the synthesis of NPs using various plant sources is depicted in Figure 4.1.

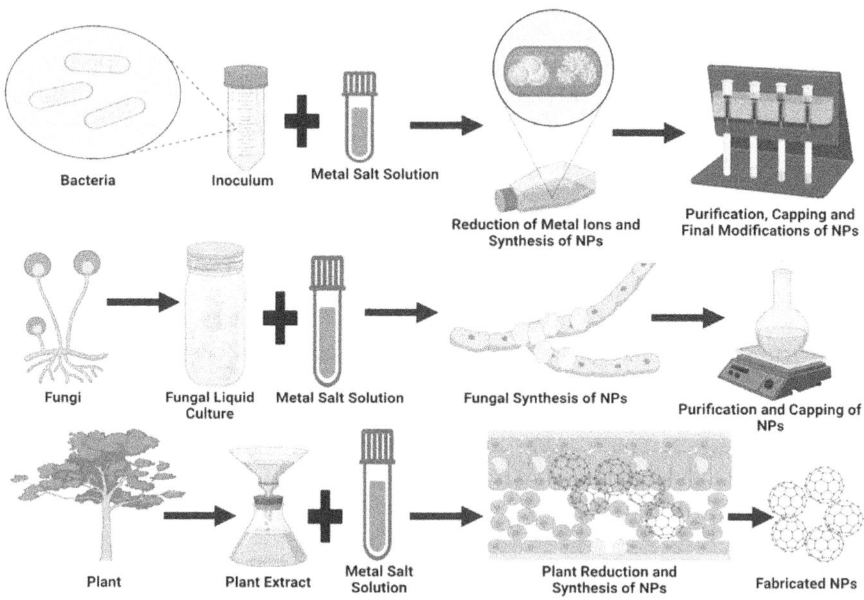

FIGURE 4.1 An overview of the synthesis of biogenic NPs. This image was designed with the aid of Biorender.com (accessed on 8 June 2023).

Plant-derived NPs have several benefits over other synthesis methods. It is ecologically benign, economical and has a greater yield. Given that plant extract has a greater number of phytochemicals that can maintain the stability of NPs and convert metal ions into metal NPs, synthesizing NPs from plants is comparably more efficient and yields better results [63].

4.3 FACTORS INDUCING THE SYNTHESIS OF BIOGENIC NPs

In order to seamlessly synthesize biogenic NPs with desirable qualities for various applications, several factors influencing the process of NP synthesis need to be fully understood and optimized. The following are some of the most important factors to consider.

4.3.1 Temperature

Temperature is an essential variable in the production of diverse NPs; it influences the speed of reactions, the fabrication of stable NPs, and the morphology of NPs. Temperature influences reaction turnover directly by regulating the activity of enzymes and proteins involved in green synthesis of NPs and enhancing the generation of nucleation centers [64]. In addition, the morphology of fabricated NPs is closely related to the prevailing temperature. For example, when AuNPs were synthesis from leaf extracts at varying temperatures, diverse shapes of NPs were formed. At 20°C, most of the NPs were triangular, and at 30°C to 40°C, the shape was octahedral, while at 50°C to 60°C, the shape changed to spheres [64].

In general, the effect of temperature on NP synthesis relies on the type of NP and the method of synthesis used. When the process of synthesis of metallic NPs and semiconductor NPs is compared, temperature seems to have a completely opposing effect. For semiconductor NPs, low temperatures are typically preferred. On the other hand, due to the enhanced rate of nucleation and growth in the case of metallic NPs, higher temperatures typically encourage the synthesis of larger particles. As an example, when precise sizes of colloidal AuNPs are required for a specific application, change of temperature ranges was used to achieve precise NP diameters. The sizes typically required are synthesized at a range of 60°C–100°C, higher temperatures produced larger particles with a wider size distribution [65].

Furthermore, biogenic synthesis of NPs is greatly influenced by temperature since it has a direct impact on the metabolic activity of the microorganisms involved. Higher temperatures can increase the metabolic rate of the microorganisms and thus enhance the rate of NP synthesis. However, too much heat can potentially harm microorganisms, resulting in a lower output of NP or instability [58].

4.3.2 The Concentration of Metal Ions

Another important factor that affects how different NPs are synthesized is the concentration of metal ions. Optimum metal ions concentration is required for the formation of the best size, shape, and stability of the NPs.

The sizes of NPs generated during green synthesis are greatly influenced by the amount of metal ions present. The generation of NPs is hampered at low concentrations, and the NPs that do form are typically of small sizes. The size of the NPs generated can, however, considerably become larger at high concentrations. For instance, during the production of AuNPs, increasing the amount of gold ions present resulted in an upsurge of the size of the NPs formed [66].

Concentration of ions also influences the shape of the NPs formed. The NPs that are produced at low concentrations are typically spherical or almost spherical. At high concentrations, however, the NPs can have various forms, including rods, triangles, and hexagons. AgNPs synthesized with high concentration of silver ions are known to form triangular and hexagonal NPs [67].

The stability of the NPs formed is also impacted by the degree of concentration of metal ions. The NPs that are generated at low concentrations are typically unstable and have a propensity to agglomerate or clump. However, NPs fabricated at high concentrations are often stable and well defined [68].

4.3.3 pH

The formation of NPs is critically dependent on the pH of the reaction medium. The pH affects the size, shape, and stability of the NPs formed; this is because the fabrication of metallic NPs is accelerated as the pH rises due to a surge in the quantity of nucleation centers.

At low pH values, the NPs formed are usually smaller in size, spherical, and unstable. While at high pH values, the NPs formed are larger in size, can have diverse shapes, and are stable and clearly defined [69].

4.3.4 Reaction Time

In addition to all the factors so far discussed, reaction time is an essential factor that can affect how different NPs are synthesized. It is a parameter whose value depends on the kind of procedure employed and the desired qualities of the NPs.

In general, longer reaction times tend to result in larger NPs. For instance, the Turkevich method of synthesizing AuNPs results in larger NPs being formed when reaction periods are prolonged [65]. The reaction time can also affect the shape of the NPs formed. Longer reaction times can lead to the fabrication of NPs with more complex shapes such as nanotubes and nanowires [70]. The stability, structural, and magnetic features of metallic NPs can also be influenced by the reaction time. Shorter reaction time tends to lead to aggregation of NPs, while longer reaction times form more stable NPs that are resistant to aggregation and possess unique magnetic properties [71].

4.4 ANTIMICROBIAL POTENTIAL OF BIOGENIC NPs

Metallic NPs have emerged as antimicrobial agents of great potential because their distinctive chemical and physical features have been harnessed and tailored toward the fight against resistant bacterial species. The antibacterial activity of metallic NPs is thought to be due to their capacity to interact and weaken bacterial cell walls, disrupt bacterial cell membranes, release reactive oxygen species (ROS), cause oxidative damage, and perturb several microbial metabolic pathways. These NPs have also been utilized as nanocarriers for drugs and antibiotics [72].

The most studied metallic NPs for antimicrobial efficacy include AgNPs, AuNPs, CuNPs, zinc oxide NPs (ZnONPs), and titanium dioxide NPs. Although these NPs do possess significant antimicrobial potential, biogenic NPs, synthesized using various biological sources, exhibit superior antibacterial activity compared to conventionally synthesized NPs [73].

4.4.1 Biogenic Silver NPs (AgNPs)

AgNPs are the most potent antibacterial nano-agents. The antimicrobial activity of biogenic AgNPs against a variety of pathogens, including bacteria, viruses, and fungus, has been documented in numerous investigations. One such study compared the antimicrobial activity of chemically synthesized AgNPs to biogenic-derived NPs and found that the biogenic AgNPs produced from *A. indica* leaf extract showed more bactericidal effect against *E. coli* and *S. aureus* [73]. Bhat et al. [74] also investigated the antibacterial activity of AgNPs derived from *A. indica* leaves. They found that the NPs negatively affected the growth curve and total protein concentration of cells of *S. aureus* [74]. In the same vein, a different study demonstrated that *Acalypha indica* leaf extract-derived biogenic AgNPs had strong antibacterial effects against *E. coli* and *Vibrio cholerae* [75]. Another study reported the fabrication of several metallic NPs, including AgNPs using the fungus *Aspergillus kambarensis*. Disc diffusion assay results with growth inhibition zones of 13.1–44.2 mm demonstrated the NPs' outstanding antifungal and antibacterial efficacy [76]. Furthermore, AgNPs biosynthesized using leaf extract of *Cucumis prophetarum* indicated a dosage-dependent inhibition activity against *S. aureus* and *S. typhi* [77].

The antimicrobial activity of biogenic AgNPs has been attributed to their ability to disrupt the cell membrane of microorganisms, leading to leakage of cellular constituents and eventual cell death. Biogenic AgNPs also exhibit superior antimicrobial activity compared to conventionally synthesized AgNPs due to their smaller size and larger surface area, which enhances their interactions with microbial cells. In addition, biogenic AgNPs exhibit low toxicity toward mammalian cells, making them a promising candidate for various applications in biology and medicine. Overall, the antimicrobial activity of biogenic AgNPs makes them a potential alternative to conventional antimicrobial agents and an important tool in combating the growing threat of antimicrobial resistance.

4.4.2 Biogenic Gold NPs (AuNPs)

The antibacterial properties of AuNPs are well documented. For instance, AuNPs synthesized with the aid of the stem of *Tinospora cordifolia* were reported to disrupt the biofilm-forming ability of *P. aeruginosa* [78]. In a different report, biogenic AuNPs synthesized using the fungus *F. oxysporum* demonstrated significantly more antibacterial activity than the antibiotic tetracycline against both gram-positive and gram-negative bacteria [79]. Moreso, the antibacterial action of AuNPs, has been shown to be affected by size. Researchers hypothesized that smaller AuNPs will demonstrate a more effective bactericidal impact than larger ones [80].

AuNPs exact antibacterial action via several mechanisms. The multivalency of AuNPs allows them to easily interact with surface membranes; the positive charge of the AuNPs interacts with the negative charge of the microbial cell membrane, leading to disruption of the bacterial cell [81]. AuNPs are usually capped with other biomolecules such as chitosan or antibiotics such as vancomycin to aid its attachment to bacterial membranes and to enhance the disruption of cells of multidrug-resistance bacteria [82]. Biogenic AuNPs also produce ROS through a few different ways, such as the generation of superoxide radicals and hydrogen peroxide. These generated ROS severely damage bacterial cells. Furthermore, AuNPs disrupt bacterial metabolism by inhibiting the activity of ATP synthase [83].

4.4.3 Biogenic Zinc Oxide NPs (ZnONPs)

Due to its distinctive antibacterial and antifungal capabilities, ZnONPs are currently being extensively researched. The potential for using ZnONPs for the formulation of novel antimicrobial agents to treat infectious diseases is immense; this is due to its broad-spectrum antimicrobial activity and low toxicity. ZnONPs have been shown to be capable of inhibiting the growth of various bacterial and fungal strains, including multidrug-resistant strains [84].

The potency of the antimicrobial activity of biogenic ZnONPs has been documented in numerous studies. A recent study evaluated the antibacterial activity of ZnONPs against *E. coli, S. aureus, P. aeruginosa*, and the gram-positive model *B. subtilis.* ZnONPs were reported to exhibit strong bactericidal activity on these bacteria species via damage to the cytoplasmic membranes of 70% of the bacterial cells after a 15-minute exposure to the NPs [85]. ZnONPs are effective against both bacterial and fungal strains. ZnONPs synthesized using *Lavandula pubescens* shoot extract were examined for their ability to inhibit the growth of *P. aeruginosa, S. aureus, Aspergillus niger*, and *Aspergillus terreus.* At different concentrations, the NPs showed bactericidal and fungicidal activity with significant zones of inhibition [86].

ZnONPs have unique physicochemical characteristics that are responsible for their antibacterial activities. ZnONPs are known to be photocatalytic and photo-oxidizing; when exposed to light, the NPs give off ROS, which disrupt the bacterial cell membrane and prevent respiration, resulting in cell death [87].

The size, surface area, and shape of ZnONPs also influence its antibacterial activity. Small-sized NPs can easily penetrate bacterial cells, and NPs with large surface area generate a greater concentration of ROS at the surfaces of the cells. In terms of shape, studies have shown that flower-shaped NPs had a stronger antibacterial impact than rod-shaped NPs against *S. aureus* and *E. coli* [88, 89].

4.4.4 Biogenic Copper NPs (CuNPs)

Biogenic CuNP is another potential antimicrobial agent that has gained intense attention lately. CuNPs have been shown to effectively disrupt the proliferation of viruses, bacteria and fungi. This broad-spectrum antimicrobial efficacy is attributed to their high surface area to volume ratio, small size, and ability to damage the viral envelope of viruses and microbial cell wall and membrane of bacteria and fungi.

A study comparing the antibacterial properties of CuNPs to its antifungal activity found CuNPs to be more potent against bacterial species. *Micrococcus luteus*,

S. aureus, E. coli, Klebsiella pneumoniae, and *P. aeruginosa*, and the fungal species *A. flavus, A. niger*, and *Candida albicans* were exposed to synthesize CuNPs; the NPs demonstrated a greater zone of inhibition in *E. coli* than *C. albicans* and stronger inhibitory effect in bacteria than fungi [90]. A similar study using chemically synthesized CuNPs against *S. aureus, B. subtilis, P. aeruginosa, Salmonella choleraesuis*, and *C. albicans* showed that CuNPs influenced the growth rate of these microorganisms. The NPs greatly inhibited the growth of the organisms of interest [91]. A more recent study reported the production of biogenic CuNPs fabricated using *Bambusa arundinacea* leaves aqueous extract solution. The researchers demonstrated the potent antibacterial activity of these biogenic CuNPs against *E. coli, S. aureus, B. subtilis*, and *Proteus vulgaris* [92].

4.4.5 Biogenic Titanium Oxide NPs (TiO_2 NPs)

Due to its potential uses in a variety of industries, including photocatalysis, solar cells, and electronics, the biogenic synthesis of titanium oxide NPs (TiO_2NPs) has come to receive attention. However, its antimicrobial potential hasn't escaped notice.

Studies have shown that a variety of microbes, including bacteria, fungi, and algae, can be used to biogenically synthesize TiO_2NPs. Biogenic TiO_2NPs were examined for their antimicrobial efficacy against a variety of harmful bacteria and fungi. A range of bacterial species have been effectively inhibited by TiO_2NPs produced from bacterial sources. Biogenic TiO_2NPs made from fungus like *F. oxysporum* and *A. niger* also showed strong antibacterial action against bacteria and fungi.

For instance, TiO_2NPs fabricated from aqueous *Psidium guajava* leaf extract were examined for activity as an antibacterial agent. These NPs successfully suppressed the proliferation of *S. aureus* and *E. coli* [93]. In a similar study, TiO_2NPs were synthesized using *Trigonella foenum-graecum* leaf extract, and their antibacterial effects against *Yersinia enterocolitica, P. vulgaris, Enterococcus faecalis, P. aeruginosa, Streptococcus faecalis, S. aureus, B. subtilis, E. coli*, and the fungus *C. albicans* were investigated. The test bacteria were efficiently suppressed by TiO_2NPs [94]. Another noteworthy study describes the robust antibacterial efficacy of TiO_2NPs produced from *Prunus yedoensis* leaf extract against *E. coli* and *S. aureus* [95]. Table 4.1 provides various biogenic NPs and their antibacterial activities reported in literature.

4.5 MECHANISM OF ACTION OF BIOGENIC NPs

The mechanism of action of biogenic NPs as antimicrobial agents is currently not fully determined, but it hypothetically involves several different processes. ROS generation by NPs is one such process. When biogenic NPs interact with microbes, they may produce ROS such as hydrogen peroxide, superoxide anions, and hydroxyl radicals that may disrupt the microorganisms' cell membranes, DNA, and proteins. Another alleged method of action purportedly compromises the strength of the cell membrane. Biogenic NPs can interact with a microorganism's cell membrane and rupture the lipid bilayer, causing internal components to seep out and ultimately resulting in cell death. Due to their more fragile outer membrane, gram-negative

TABLE 4.1
Biogenic Metallic NPs and Their Antibacterial Activity

Metal NP	Biological Source	Component of the Source used for Synthesis	Size and Shape	Bacterial Strain Inhibited	Reference
Ag	*Bacillus endophyticus*	Bacteria	5.1 nm spherical	*Escherichia coli*, *Salmonella typhi*, and *Staphylococcus aureus*	[96]
	Bacillus brevis	Bacteria	41–68 nm spherical	*S. typhi* and *S. aureus*	[97]
	Streptomyces kasugaensis	Bacteria	13.2 nm spherical	*E. coli*, *S. aureus* and *Bacillus subtilis*	[98]
	Aspergillus kambarensis	Fungi	50 nm spherical	*E. coli*, *B. subtilis* and *S. aureus*	[76]
	Azadirachta indica	Leaf extract	19–43 nm spherical	*E. coli* and *S. aureus*	[73]
	A. indica	Leaf extract	7.5 nm spherical	*S. aureus*	[74]
	Acalypha indica	Leaf extract	19–43 nm spherical	*E. coli* and *Vibrio cholerae*	[75]
	Cucumis prophetarum	Leaf extract	30–50 nm spherical	*S. aureus* and *S. typhi*	[77]
Au	*Klebsiella pneumoniae*	Bacteria	10–15 nm spherical	*E. coli* and *S. aureus*	[99]
Au	*Aspergillus terreus*	Fungi	10–19 nm spherical	*E. coli*	[100]
	Fusarium oxysporum	Fungi	22–30 nm spherical or hexagonal	*Bacillus cereus*, *E. coli*, *Pseudomonas aeruginosa* and *S. aureus*	[79]
	Areca catechu	Plant extract	13.7 nm spherical	*E. coli*, *Klebsiella pneumoniae*, *P. aeruginosa*, and *S. aureus*	[101]
	Tinospora cordifolia	Stem extract	16.1 nm spherical	*P. aeruginosa*	[78]
ZnO	*Desertifilum* sp.	Cyanobacteria	88 nm rod shaped	*S. aureus*, *E. coli*, and *P. aeruginosa*	[102]
	Lavandula pubescens	Shoot extract	10.76–30.93 nm rod shaped	*P. aeruginosa* and *S. aureus*	[86]
Cu	Antarctic Bacterial Consortia	Bacteria	30 nm spherical	*E. coli* and *S. aureus*	[103]
	Bambusa arundinacea	Leaf extract	15–30 nm spherical	*E. coli*, *S. aureus*, *B. subtilis* and *Proteus vulgaris*	[92]
TiO_2	*Psidium guajava*	Leaf extract	32.58 nm spherical	*S. aureus* and *E. coli*	[93]
	Trigonella foenum-graecum	Leaf extract	20–90 nm spherical	*Yersinia enterocolitica*, *P. vulgaris*, *Enterococcus faecalis*, *P. aeruginosa*, *Streptococcus faecalis*, *S. aureus*, *B. subtilis* and *E. coli*	[94]

bacteria are more susceptible to this mechanism [104]. Biogenic NPs can also alter how genes are expressed and proteins are made in bacteria, which can disturb cellular functions and ultimately cause cell death.

4.5.1 Generation of Reactive Oxygen Species (ROS)

ROS are oxygen-containing molecules that are extremely reactive as a result of possessing an outer orbital with unpaired electrons [105]. ROS can be produced by NPs via a variety of mechanisms, such as the Fenton reaction, Haber-Weiss reaction, and electron transfer.

The Fenton reaction occurs when hydrogen peroxide (H_2O_2) reacts with a metal ion, usually iron (Fe^{2+} or Fe^{3+}), to produce hydroxyl radicals (•OH) and a hydroxide ion (OH–). With transition metals functioning as a catalyst, this reaction takes place at neutral to acidic pH levels [106]. It has been observed that biogenic NPs such as iron oxide (Fe_3O_4) and silver (Ag) can trigger the Fenton reaction, and generate ROS [107].

In the Haber-Weiss reaction, oxygen gas and hydroxyl radicals are produced when hydrogen peroxide reacts with a superoxide radical. Transition metals like iron and copper act as catalysts for this process, which takes place at neutral to alkaline pH levels [108]. Biogenic NPs such as copper oxide (CuO) and titanium dioxide (TiO_2) have been reported to induce the Haber-Weiss reaction and the production of ROS [107].

The electron transfer mechanism, which results in the release of radicals consisting of superoxide and hydrogen peroxide, involves the transfer of electrons from a metal NP to oxygen molecules in the presence of water. Biogenic NPs like ZnO and gold (Au) embody this process [107].

The production of ROS by biogenic NPs has significant antimicrobial effects. ROS can cause oxidative stress in cells, which may negatively impact cellular macromolecules such as lipids, proteins, and DNA, which ultimately result in damage to microbial cell membranes, disruption of cellular function, and cell death.

4.5.2 Aerobic and Anaerobic Bacteria Species

It has been demonstrated that the kind of bacterium, either aerobic or anaerobic, influences the mechanism by which the NPs exhibit bactericidal activity.

The primary mechanism by which NPs display their antibacterial activity against aerobic bacteria is through ROS. The ROS produced by NPs can cause oxidative stress in bacterial cells, which can damage biological elements like DNA, proteins, and lipids [109].

For anaerobic bacteria, the mechanism of bactericidal efficacy is thought to be different from that of aerobic bacteria. Biogenic NPs exert their antibacterial activity against anaerobic bacteria through the disruption of bacterial membrane potential. This disruption can lead to the inhibition of essential metabolic pathways and enzymes. For example, the disruption of the activity of cytochrome oxidase results in a decrease in ATP production and bacterial cell death, and the disruption of bacterial electron transport chain leads to a decrease in cellular respiration and subsequent bacterial death [110].

4.5.3 Metal and its Oxide Biogenic NPs

Metal oxides (NPs) have a complex mode of action that alters based on the NP and the microorganism being attacked.

A possible mechanism of action is the emission of metal ions from the NPs, which can damage the bacterial cell wall and membrane. As a result, intracellular components seep out, ultimately resulting in cell death [107].

Metals and metal oxide NPs can also disrupt bacterial quorum sensing, which is how bacteria communicate with one another to coordinate group behavior. By obstructing this communication, NPs prevent the growth and development of microbial biofilms that are notoriously hard to treat with traditional antibiotics [111]. A detailed proposed mechanism of action of biogenic metallic NPs microorganisms is depicted in Figure 4.2.

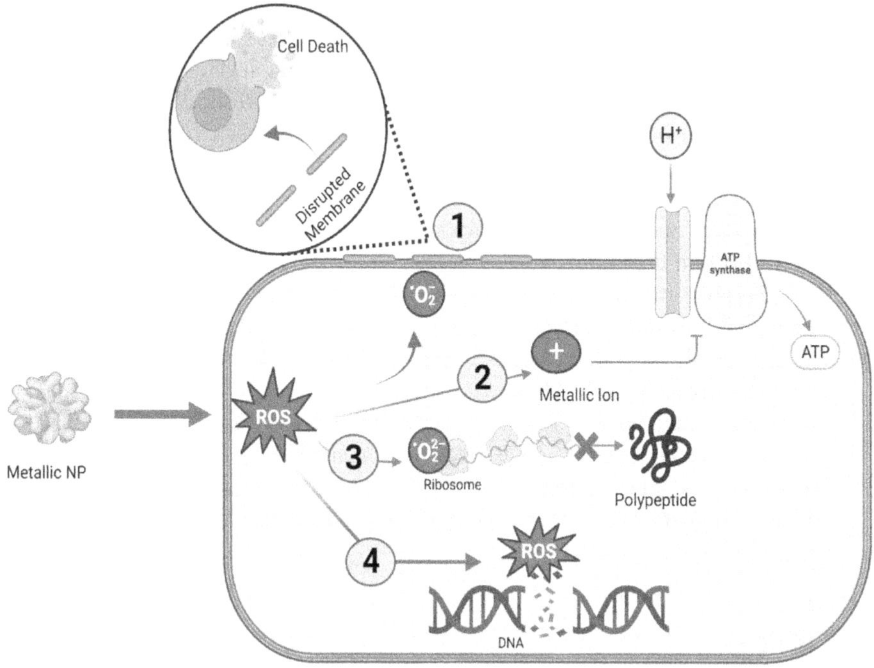

FIGURE 4.2 A synopsis of the various antibacterial mechanism of action of biogenic NPs. (1) Biogenic NPs generate ROS which disrupts bacterial cell membrane, causes cellular seepages, and eventually leads to cell death. (2) Biogenic metallic oxide NPs can produce metallic ions that interact with microbial membrane pumps and pathways resulting in the dysfunction of membrane potential and ATP synthesis. (3) NPs produce ROS that can inhibit the synthesis of cellular proteins or damage synthesized polypeptides. (4) ROS generated by NPs also causes bacterial DNA damage. This illustrative piece was created using the resources available on Biorender.com (accessed on 9 June 2023).

4.6 CONCLUSION AND FUTURE PERSPECTIVES

As drug resistance by microorganisms continues to be a major health challenge, future research could focus on exploring the use of nanotechnology-based interventions, particularly biogenic NPs synthesized through green synthesis. Here are some potential research perspectives:

i. **Mechanistic aspects of resistance:** A deeper comprehension of the processes behind the evolution of resistance to medications in microbes could be attained through further research. This could involve studying the genetic and molecular basis of resistance, as well as the environmental factors that influence its emergence.
ii. **Development of novel nanomaterials:** Researchers could explore the use of different biogenic NPs for the amelioration of drug-resistant illnesses. This could involve synthesizing NPs using different biomolecules or using different metal precursors to generate NPs with different physicochemical properties.
iii. **Optimization of synthesis protocols:** Researchers could focus on optimizing the green synthesis protocols to produce biogenic NPs. This could involve identifying the best conditions for synthesis, such as the optimum physicochemical parameters to produce NPs with the desired properties.
iv. **In vitro and in vivo testing:** Researchers could conduct further in vitro and in vivo testing of biogenic NPs to evaluate their efficacy against drug-resistant microorganisms. This could involve testing NPs on different bacterial strains and assessing their toxicity and biocompatibility.
v. **Combination therapies:** Researchers could investigate the use of combination therapies that incorporate biogenic NPs with other antibacterial agents. This could involve testing the synergistic effects of NPs and antibiotics or other natural compounds, to enhance their antibacterial activity and reduce the risk of resistance development.
vi. **Clinical trials:** Finally, future research could involve conducting clinical trials to assess the safety and efficacy of biogenic NPs as a treatment option for drug-resistant infections. This could involve testing the NPs on different patient populations, as well as comparing their efficacy to existing antibiotic treatments.

REFERENCES

1. Doron, S., Gorbach, S. L. Bacterial Infections: Overview. *International Encyclopedia of Public Health*. 2008, 273–282.
2. GBD 2019 Antimicrobial Resistance Collaborators. Global mortality associated with 33 bacterial pathogens in 2019: a systematic analysis for the Global Burden of Disease Study 2019. *Lancet (London, England)*. 2022, *400*(10369), 2221–2248.
3. Tanwar, J., Das, S., Fatima, Z., Hameed, S. Multidrug resistance: an emerging crisis. *Interdisciplinary Perspectives on Infectious Diseases*. 2014, 541340.
4. Saha, M., Sarkar, A. Review on multiple facets of drug resistance: a rising challenge in the 21st century. *Journal of Xenobiotics*. 2021, *11*(4), 197–214.

5. Matthiessen, L., Bergström, R., Dustdar, S., Meulien, P., Draghia-Akli, R. Increased momentum in antimicrobial resistance research. *Lancet (London, England)*. 2016, *388*(10047), 865.
6. Haak, B. W., Wiersinga, W. J. Uncovering hidden antimicrobial resistance patterns within the hospital microbiome. *Nature Medicine*. 2020, *26*(6), 826–828.
7. Wang L., Hu C., Shao L. The antimicrobial activity of nanoparticles: present situation and prospects for the future. *International Journal of Nanomedicine*. 2017, 12:1227–49.
8. Han C., Romero N., Fischer S., Dookran J., Berger A., Doiron A.L. Recent developments in the use of nanoparticles for treatment of biofilms. *Nanotechnology Reviews*. 2017, 6(5):383–404.
9. Vimbela, G.V., Ngo, S.M., Fraze, C., Yang, L., Stout, D.A. Antibacterial properties and toxicity from metallic nanomaterials. *International Journal of Nanomedicine*. 2017, 12, 3941–3965.
10. Nocito, G., Sciuto, E.L., Franco, D., Nastasi, F., Pulvirenti, L., Petralia, S., Spinella, C., Calabrese, G., Guglielmino, S., Conoci, S. Physicochemical characterization and antibacterial properties of carbon dots from two mediterranean olive solid waste cultivars. *Nanomaterials*. 2022, 12, 885.
11. Beyth N., Houri-Haddad Y., Domb A., Khan W., Hazan R. Alternative antimicrobial approach: nano-antimicrobial materials. *Evidence Complementary Alternative Med*. 2015, 2015:1–16.
12. Liao S., Zhang Y., Pan X., Zhu F., Jiang C., Liu Q., et al. Antibacterial activity and mechanism of silver nanoparticles against multidrug-resistant *Pseudomonas aeruginosa*. *International Journal of Nanotechnology*. 2019, 14:1469–1487.
13. Singh A., Ahmed A., Prasad K.N., Khanduja S., Singh S.K., Srivastava J.K. Antibiofilm and membrane-damaging potential of cuprous oxide nanoparticles against *Staphylococcus aureus* with reduced susceptibility to vancomycin. *Antimicrobial Agents Chemotherapy*. 2015, 59(11):6882–90.
14. Huang X.Q., Chen X., Chen Q.C., Yu Q.Q., Sun D.D., Liu J. Investigation of functional selenium nanoparticles as potent antimicrobial agents against superbugs. *Acta Biomaterialia*. 2016, 30:397–407.
15. Maji J., Pandey S., Basu S. Synthesis and evaluation of antibacterial properties of magnesium oxide nanoparticles. *Bulletin of Material Science*. 2020, 43(1):1–10.
16. Ijaz U., Bhatti I.A., Mirza S., Ashar A. Characterization and evaluation of antibacterial activity of plant mediated calcium oxide (CaO) nanoparticles by employing *Mentha piperita* extract. *Materials Research Express*. 2017, 4:105402.
17. Pant B., Pokharel P., Tiwari A.P., Saud P.S., Park M., Ghouri Z.K., et al. Characterization and antibacterial properties of aminophenol grafted and Ag NPs decorated graphene nanocomposites. *Ceramics International*. 2015, 41(4):5656–62.
18. Kumar N., Das S., Jyoti A., Kaushik S. Synergistic effect of silver nanoparticles with doxycycline against *Klebsiella pneumoniae*. *International Journal of Pharmacy and Pharmaceutical Sciences*. 2016, 8(7):183–6.
19. Panáček A., Smékalová M., Kilianová M., Prucek R., Bogdanová K., Věčerová R., et al. Strong and nonspecific synergistic antibacterial efficiency of antibiotics combined with silver nanoparticles at very low concentrations showing no cytotoxic effect. *Molecules*. 2016, 21(1):1–17.
20. Lopez-Carrizales M., Velasco K., Castillo C., Flores A., Magaña M., Martinez-Castanon G., et al. In vitro synergism of silver nanoparticles with antibiotics as an alternative treatment in multiresistant uropathogens. *Antibiotics*. 2018, 7(2):50.
21. Rajput, N. Methods of preparation of nanoparticles – a review. *International Journal of Advanced Engineering and Technology*. 2015, 7(6) 1806.
22. Schmidt, H. Nanoparticles by chemical synthesis, processing to materials and innovative applications. *Applied Organometallic Chemistry*. 2001, 15(5) 331–343.

23. Fabiano, B., Reverberi, A.P., Varbanov, P.S. Safety opportunities for the synthesis of metal nanoparticles and short-cut approach to workplace risk evaluation. *Journal of Cleaner Production*. 2019, (209) 297–308.
24. Ahmed, S., Chaudhry, S.A., Ikram, S. A review on biogenic synthesis of ZnO nanoparticles using plant extracts and microbes: a prospect towards green chemistry. *Journal of Photochemistry and Photobiology B: Biology*. 2017, (166) 272–284.
25. Akhtar, M.S., Panwar, J., Yun, Y-S. Biogenic synthesis of metallic nanoparticles by plant extracts. *ACS Sustainable Chemistry & Engineering*. 2013, 1(6) 591–602.
26. Sharma, D., Kanchi, S., Bisetty, K. Biogenic synthesis of nanoparticles: a review. *Arabian Journal of Chemistry*. 2015, 11(2).
27. Kumari, S., Li, Z. F., Anwar, M.N. Bacterial cellular mechanisms for synthesis of green nanostructured compounds. (Ed) *In Micro and Nano Technologies, Biogenic Sustainable Nanotechnology, Elsevier*. 2022, Pages 5976.
28. Singh, P., Kim, Y.-J., Zhang, D., Yang, D.-C. Biological synthesis of nanoparticles from plants and microorganisms. *Trends in Biotechnology*. 2016, 34(7) 588–599.
29. Lahiri, D., Nag, M., Sheikh, H. I., Sarkar, T., Edinur, H. A., Pati, S., Ray, R. R. Microbiologically-synthesized nanoparticles and their role in silencing the biofilm signaling cascade. *Frontiers in Microbiology*. 2021, 12, 636588.
30. Ghosh, S., Ahmad, R., Zeyaullah, M., Khare, S. K. Microbial nano-factories: synthesis and biomedical applications. *Frontiers in Chemistry*. 2021, 9, 626834.
31. Ramrakhiani, L., Ghosh, S. Metallic nanoparticle synthesised by biological route: safer candidate for diverse applications. *IET Nanobiotechnology*. 2018, 12(4), 392–404.
32. Casley, R.E., Jose, C.A., Kevin, D.M., Salvacion, E.M., Myla, S. Cytotoxic potential on human breast and lung cancer cells of the biosynthesized gold nanoparticles from the reduction of chloroauric acid by lactic acid isolated from *Lactobacillus acidophilus*. *Frontiers in Materials*. 2022, 9(9), 1–10.
33. Bruna, T., Maldonado-Bravo, F., Jara, P., Caro, N. Silver nanoparticles and their antibacterial applications. *International Journal of Molecular Sciences*. 2021, 22(13), 7202.
34. Saravanan, M., Barik, S.K., Mubarak Ali, D., Prakash, P., Pugazhendhi, A. Synthesis of silver nanoparticles from *Bacillus brevis* (NCIM 2533) and their antibacterial activity against pathogenic bacteria. *Microbial Pathogenesis*. 2018, 116, 221–226.
35. Rad, M., Taran, M., Alavi, M. Effect of incubation time, $CuSO_4$ and glucose concentrations on biosynthesis of copper oxide (CuO) nanoparticles with rectangular shape and antibacterial activity: Taguchi method approach. *Nano Biomedicine and Engineering*. 2018, 10, 25–33.
36. Casentini, B., Gallo, M., Baldi, F. Arsenate and arsenite removal from contaminated water by iron oxides nanoparticles formed inside a bacterial exopolysaccharide. *Journal of Environmental Chemical Engineering*. 2019, 7(1), 102908.
37. Molnár, Z., Bódai, V., Szakacs, G., Erdélyi, B., Fogarassy, Z., Sáfrán, G., Varga, T., Kónya, Z., Tóth-Szeles, E., Szucs, R. Green synthesis of gold nanoparticles by thermophilic filamentous fungi. *Scientific Reports*. 2018, 8, 3943.
38. Zhao, X., Zhou, L., Riaz Rajoka, M.S., Yan, L., Jiang, C., Shao, D., Zhu, J., Shi, J., Huang, Q., Yang, H. Fungal silver nanoparticles: Synthesis, application and challenges. *Critical Reviews in Biotechnology*. 2018, 38, 817–835.
39. Alghuthaymi, M.A., Almoammar, H., Rai, M., Said-Galiev, E., Abd-Elsalam, K.A. Myconanoparticles: Synthesis and their role in phytopathogens management. *Biotechnology & Biotechnological Equipment*. 2015, 29, 221–236.
40. Castro-Longoria, E., Vilchis-Nestor, A.R., Avalos-Borja, M. Biosynthesis of silver, gold and bimetallic nanoparticles using the filamentous fungus *Neurospora crassa*. *Colloids and Surfaces*. 2011, 83, 42–48.

41. Hu, X., Saravanakumar, K., Jin, T., Wang M.H. Mycosynthesis, characterization, anticancer and antibacterial activity of silver nanoparticles from endophytic fungus *Talaromyces purpureogenus*. *International Journal of Nanomedicine*. 2019, 14, 3427–3438.
42. Zhang, X., He, X., Wang, K., Yang, X. Different active biomolecules involved in biosynthesis of gold nanoparticles by three fungus species. *Journal of Biomedical Nanotechnology*. 2011, 7, 245–254.
43. Sonawane, H., Shelke, D., Chambhare, M., Dixit, N., Math, S., Sen, S., Narayan Borah, S., Islam, N.F., Joshi, S.J., Yousaf, B., Rinklebe, J., Sarma, H. Fungi-derived agriculturally important nanoparticles and their application in crop stress management – Prospects and environmental risks. *Environmental Research*. 2022, 212, 113543.
44. Riddin, T., Gericke, M., Whiteley, C. Analysis of the inter-and extracellular formation of platinum nanoparticles by *Fusarium oxysporum* f. sp. lycopersici using response surface methodology. *Nanotechnology*. 2006, 17(14) 3482.
45. Ahmad, A., Mukherjee, P., Senapati, S., Mandal, D., Khan, M.I., Kumar, R., Sastry, M. Extracellular biosynthesis of silver nanoparticles using the fungus Fusarium oxysporum. *Colloids and Surfaces B: Biointerfaces*. 2003, 28(4) 313–318.
46. Sanghi, R., Verma, P. Biomimetic synthesis and characterisation of protein capped silver nanoparticles. *Bioresource Technology*. 2009, 100(1) 501–504.
47. AbdelRahim, K., Mahmoud, S.Y., Ali, A.M., Almaary, K.S., Mustafa, A.E.-Z.M., Husseiny, S.M. Extracellular biosynthesis of silver nanoparticles using *Rhizopus stolonifer*. *Saudi Journal of Biological Sciences*. 2017, 24, 208–216.
48. Jalal, M., Ansari, M., Alzohairy, M., Ali, S., Khan, H., Almatroudi, A., Raees, K. Biosynthesis of silver nanoparticles from oropharyngeal *Candida glabrata* isolates and their antimicrobial activity against clinical strains of bacteria and fungi. *Nanomaterials* 2018, 8, 586.
49. Khleifat K, Qaralleh H, Al-Limoun M, et al. Antibacterial activity of silver nanoparticles synthesized by Aspergillus flavus and its synergistic effect with antibiotics. *Journal of Pure and Applied Microbiology*. 2022, 16(3):1722–1735.
50. Molnár, Z., Bódai, V., Szakacs, G., Erdélyi, B., Fogarassy, Z., Sáfrán, G., Varga, T., Kónya, Z., Tóth-Szeles, E., Szűcs, R., Lagzi, I. Green synthesis of gold nanoparticles by thermophilic filamentous fungi. *Scientific Reports*. 2018, 8(1), 3943.
51. Tan, Y., Wang, Y., Jiang, L., Zhu, D. Thiosalicylic acid-functionalized silver nanoparticles synthesized in one-phase system. *Journal of Colloid and Interface Science*. 2002, 249(2) 336–345.
52. Das, S.K., Das, A.R., Guha, A.K. Gold nanoparticles: microbial synthesis and application in water hygiene management. *Langmuir* 2009, 25(14) 8192–8199.
53. Elahian, F., Heidari, R., Charghan, V.R., Asadbeik, E. Mirzaei, S.A. Genetically modified *Pichia pastoris*, a powerful resistant factory for gold and palladium bioleaching and nanostructure heavy metal biosynthesis. *Artificial Cells, Nanomedicine, and Biotechnology*. 2020, 48(1), 259–265.
54. Mourato, A., Gadanho, M., Lino, A.R., Tenreiro, R. Biosynthesis of crystalline silver and gold nanoparticles by extremophilic yeasts. *Bioinorganic Chemistry and Applications*. 2011, 546074.
55. Zahran, M.K., Mohamed, A.A., Mohamed, F.M., El-Rafie, M.H. Optimization of biological synthesis of silver nanoparticles by some yeast fungi. *Egyptian Journal of Chemistry*. 2013, 56, 91–110.
56. Nam, K.T., Lee, Y.J., Krauland, E.M., Kottmann, S.T., Belcher, A.M. Peptide mediated reduction of silver ions on engineered biological scaffolds. *ACS Nano*. 2008, 2(7) 1480–1486.
57. Ma, G., Zhao, Z., Liu, H. Yeast cells encapsulating polymer nanoparticles as Trojan particles via in situ polymerization inside cells. *Macromolecules* 2016, 49, 1545–51.

58. Singh, A., Gautam, P. K., Verma, A., Singh, V., Shivapriya, P. M., Shivalkar, S., Sahoo, A. K., Samanta, S. K. Green synthesis of metallic nanoparticles as effective alternatives to treat antibiotics resistant bacterial infections: A review. *Biotechnology reports (Amsterdam, Netherlands).* 2020, 25, e00427.
59. Kamala Nalini, S.P., Vijayaraghavan, K. Green synthesis of silver and gold nanoparticles using *Aloe vera* gel and determining its antimicrobial properties on nanoparticle impregnated cotton fabric. *Journal of Nanotechnology Research.* 2020, 2, 042–050.
60. Ahmed, S., Saifullah, M., Ahmad, B.L., Swami, S. Green synthesis of silver nanoparticles using *Azadirachta indica* aqueous leaf extract. *Journal of Radiation Research and Applied Sciences.* 2016, 9(1), 1–7.
61. Suman, T., Rajasree, S.R., Ramkumar, R., Rajthilak, C., Perumal, P. The Green synthesis of gold nanoparticles using an aqueous root extract of *Morinda citrifolia* L. *Spectrochimica Acta, Part A: Molecular and Biomolecular Spectroscopy.* 2014, 118, 11–16.
62. Nirmala, M., Kavitha, B., Nasreen, S., Kiruthika, S., Divya B., Manochitra, E. Green Synthesis of Copper Nanoparticles Using *Ocimum sanctum* L. (Tulsi) and *Piper nigrum* L. (Pepper Seed) for Pollution Free Environment. *World Scientific News.* 2020, 150, 105–117.
63. Mohammadinejad, R., Shavandi, A., Raie, D.S., Sangeetha, J., Soleimani, M., Hajibehzad, S.S., Thangadurai, D., Hospet, R., Popoola, J.O., Arzani, A. Plant molecular farming: Production of metallic nanoparticles and therapeutic proteins using green factories. *Green Chemistry.* 2019, 21(8), 1845–1865.
64. Sneha, K., Sathishkumar, M., Kim, S., Yun, Y.-S. Counter ions and temperature incorporated tailoring of biogenic gold nanoparticles. *Process Biochemistry.* 2010, 45(9) 1450–1458.
65. Dong, J., Carpinone, P. L., Pyrgiotakis, G., Demokritou, P., Moudgil, B. M. Synthesis of precision gold nanoparticles using Turkevich method. *Kona: Powder Science and Technology in Japan.* 2020, 37, 224–232.
66. Rashid, A. G. Chapter 3 - Methods of nanostructured materials characterization. *(Ed): In Advanced Nanomaterials, Nanostructured Nonlinear Optical Materials, Elsevier.* 2018, Pages 79–116.
67. Iravani, S., Korbekandi, H., Mirmohammadi, S. V., Zolfaghari, B. Synthesis of silver nanoparticles: chemical, physical and biological methods. *Research in Pharmaceutical Sciences.* 2014, 9(6), 385–406.
68. Dheyab, M. A., Aziz, A. A., Oladzadabbasabadi, N., Alsaedi, A., Braim, F. S., Jameel, M. S., Ramizy, A., Alrosan, M., Almajwal, A. M. Comparative analysis of stable gold nanoparticles synthesized using sonochemical and reduction methods for antibacterial activity. *Molecules (Basel, Switzerland).* 2023, 28(9), 3931.
69. Okitsu, K., Sharyo, K., Nishimura, R. One-pot synthesis of gold nanorods by ultrasonic irradiation: the effect of pH on the shape of the gold nanorods and nanoparticles. *Langmuir.* 2009, 25(14), 7786–7790.
70. Khodashenas, B., Ghorbani, H.R. Synthesis of silver nanoparticles with different shapes. *Arabian Journal of Chemistry.* 2019, 12(8), 1823–1838.
71. Karade, V., Dongale, T., Sahoo, S.C., Kollu, P., Chougale, A., Patil, P., Patil, P. Effect of reaction time on structural and magnetic properties of green-synthesized magnetic nanoparticles. *Journal of Physics and Chemistry of Solids.* 2018, 120, 161–166.
72. Singh, R., Smitha, M., Singh, S.P. The role of nanotechnology in combating multidrug resistant bacteria. *Journal of Nanoscience and Nanotechnology.* 2014, 14(7) 4745–4756.
73. Khan, K., Javed, S. Silver nanoparticles synthesized using leaf extract of *Azadirachta indica* exhibit enhanced antimicrobial efficacy than the chemically synthesized nanoparticles: A comparative study. *Science Progress.* 2021, 104(2), 368504211012159.

74. Bhat, R. S., Almusallam, J., Al Daihan, S., Al-Dbass, A. Biosynthesis of silver nanoparticles using *Azadirachta indica* leaves: characterisation and impact on *Staphylococcus aureus* growth and glutathione-S-transferase activity. *IET Nanobiotechnology.* 2019, 13(5), 498–502.
75. Krishnaraj, C., Jagan, E. G., Rajasekar, S., Selvakumar, P., Kalaichelvan, P. T., Mohan, N. Synthesis of silver nanoparticles using *Acalypha indica* leaf extracts and its antibacterial activity against water borne pathogens. *Colloids and Surfaces. B, Biointerfaces.* 2010, 76(1), 50–56.
76. Gholami-Shabani, M., Sotoodehnejadnematalahi, F., Shams-Ghahfarokhi, M., Eslamifar, A., Razzaghi-Abyaneh, M. Physicochemical properties, anticancer and antimicrobial activities of metallic nanoparticles green synthesized by *Aspergillus kambarensis. IET Nanobiotechnology.* 2022, 16(1), 1–13.
77. Meena, P.R., Singh, A.P., Tejavath, K.K. Biosynthesis of silver nanoparticles using *Cucumis prophetarum* aqueous leaf extract and their antibacterial and antiproliferative activity against cancer cell lines. *ACS Omega*, 2020, 5(10), 5520–5528.
78. Ali, S. G., Ansari, M. A., Alzohairy, M. A., Alomary, M. N., AlYahya, S., Jalal, M., Khan, H. M., Asiri, S. M. M., Ahmad, W., Mahdi, A. A., El-Sherbeeny, A. M., El-Meligy, M. A. Biogenic gold nanoparticles as potent antibacterial and antibiofilm nano-antibiotics against *Pseudomonas aeruginosa. Antibiotics (Basel, Switzerland).* 2020, 9(3), 100.
79. Naimi-Shamel, N., Pourali, P., Dolatabadi, S. Green synthesis of gold nanoparticles using Fusarium oxysporum and antibacterial activity of its tetracycline conjugant. *Journal de Mycologie Médicale.* 2019, 29(1), 7–13.
80. Rajan, A., Vilas, V., Philip, D. Studies on catalytic, antioxidant, antibacterial and anticancer activities of biogenic gold nanoparticles. *Journal of Molecular Liquids.* 2015, 212, 331–339.
81. Guerrini, L., Alvarez-Puebla, R., Pazos-Perez, N. Surface modifications of nanoparticles for stability in biological fluids. *Materials.* 2018, 11(7), 1154.
82. Rajendran, N.K., Kumar, S.S.D., Houreld, N.N., Abrahamse, H. A review on nanoparticle based treatment for wound healing. *Journal of Drug Delivery Science and Technology.* 2018, 44, 421–430.
83. Cui, Y., Zhao, Y., Tian, Y., Zhang, W., Lü, X., Jiang, X. The molecular mechanism of action of bactericidal gold nanoparticles on *Escherichia coli. Biomaterials.* 2012, 33(7) 2327–2333.
84. Jones, N., Ray, B., Ranjit, K.T., Manna, A.C. Antibacterial activity of ZnO nanoparticle suspensions on a broad spectrum of microorganisms. *FEMS Microbiology Letters.* 2008, 279(1) 71–76.
85. Mendes, C.R., Dilarri, G., Forsan, C.F. et al. Antibacterial action and target mechanisms of zinc oxide nanoparticles against bacterial pathogens. *Scientific Reports.* 2022, 12, 2658.
86. Yagoub, A.E.A., Al-Shammari, G.M., Al-Harbi, L.N., Subash-Babu, P., Elsayim, R., Mohammed, M.A., Yahya, M.A., Fattiny, S.Z.A. Antimicrobial properties of zinc oxide nanoparticles synthesized from *Lavandula pubescens* shoot methanol extract. *Applied Sciences.* 2022, 12, 11613.
87. Sur, D.H., Mukhopadhyay, M. Role of zinc oxide nanoparticles for effluent treatment using Pseudomonas putida and Pseudomonas aureofaciens. *Bioprocess and Biosystems Engineering.* 2019, 42(2) 187–198.
88. Palanikumar, L., Ramasamy, S.N., Balachandran, C. Size-dependent antimicrobial response of zinc oxide nanoparticles. *IET Nanobiotechnology* 2014, 8(2) 111–117.
89. Talebian, N., Amininezhad, S.M., Doudi, M. Controllable synthesis of ZnO nanoparticles and their morphology-dependent antibacterial and optical properties. *Journal of Photochemistry and Photobiology B: Biology.* 2013, 120, 66–73.

90. Ramyadevi, J., Jeyasubramanian, K., Marikani, A., Rajakumar, G., Rahuman, A.A. Synthesis and antimicrobial activity of copper nanoparticles. *Materials Letters*, 2012, 71, 114–116.
91. Usman, M. S., El Zowalaty, M. E., Shameli, K., Zainuddin, N., Salama, M., Ibrahim, N. A. Synthesis, characterization, and antimicrobial properties of copper nanoparticles. *International Journal of Nanomedicine*. 2013, 8, 4467–4479.
92. Naradala, J., Allam, A., Tumu, V. R., Rajaboina, R.K. Antibacterial activity of copper nanoparticles synthesized by *Bambusa arundinacea* leaves extract. *Biointerface Research in Applied Chemistry*. 2022, 12(1), 1230–1236.
93. Santhoshkumar, T., Rahuman, A.A., Jayaseelan, C., Rajakumar, G., Marimuthu, S., Kirthi, A.V., Velayutham, K., Thomas, J., Venkatesan, J., Kim, S.-K. Green synthesis of titanium dioxide nanoparticles using Psidium guajava extract and its antibacterial and antioxidant properties. *Asian Pac. J. Trop. Med.* 2014, 7(12), 968–976.
94. Subhapriya, S., Gomathipriya, P. Green synthesis of titanium dioxide (TiO2) nanoparticles by Trigonella foenum-graecum extract and its antimicrobial properties. *Microbial Pathogenesis*. 2018, 116, 215–220.
95. Manikandan, V., Velmurugan, P., Jayanthi, P., Park, J.-H., Chang, W.-S., Park, Y.-J., Cho, M., Oh, B.-T. Biogenic synthesis from *Prunus yedoensis* leaf extract, characterization, and photocatalytic and antibacterial activity of TiO2 nanoparticles. *Research on Chemical Intermediates*. 2018, 44(4), 2489–2502.
96. Gan, L., Zhang, S., Zhang, Y., He, S., Tian, Y. Biosynthesis, characterization and antimicrobial activity of silver nanoparticles by a halotolerant *Bacillus endophyticus* SCU-L. *Preparative Biochemistry & Biotechnology*. 2018, 48(7), 582–588.
97. Saravanan, M., Barik, S. K., MubarakAli, D., Prakash, P., Pugazhendhi, A. Synthesis of silver nanoparticles from *Bacillus brevis* (NCIM 2533) and their antibacterial activity against pathogenic bacteria. *Microbial Pathogenesis*. 2018, 116, 221–226.
98. Wypij, M.; Golinska, P.; Dahm, H.; Rai, M. Actinobacterial-mediated synthesis of silver nanoparticles and their activity against pathogenic bacteria. *IET Nanobiotechnology*. 2016, 11, 336–342.
99. Prema, P.; Iniya, P.; Immanuel, G. Microbial mediated synthesis, characterization, antibacterial and synergistic effect of gold nanoparticles using *Klebsiella pneumoniae* (MTCC-4030). *RSC Advances*. 2016, 6, 4601–4607.
100. Priyadarshini, E., Pradhan, N., Sukla, L.B., Panda, P.K. Controlled synthesis of gold nanoparticles using *Aspergillus terreus* IF0 and its antibacterial potential against Gram negative pathogenic bacteria. *Journal of Nanotechnology* 2014, (2014).
101. Rajan, A., Vilas, V., Philip, D. Studies on catalytic, antioxidant, antibacterial and anticancer activities of biogenic gold nanoparticles. *Journal of Molecular Liquids*. 2015, 212, 331–339.
102. Ebadi, M., Zolfaghari, M. R., Aghaei, S. S., Zargar, M., Noghabi, K. A. Desertifilum sp. EAZ03 cell extract as a novel natural source for the biosynthesis of zinc oxide nanoparticles and antibacterial, anticancer and antibiofilm characteristics of synthesized zinc oxide nanoparticles. *Journal of Applied Microbiology*. 2022, 132(1), 221–236.
103. John, M. S., Nagoth, J. A., Zannotti, M., Giovannetti, R., Mancini, A., Ramasamy, K. P., Miceli, C., Pucciarelli, S. Biogenic synthesis of copper nanoparticles using bacterial strains isolated from an Antarctic consortium associated to a psychrophilic marine ciliate: characterization and potential application as antimicrobial agents. *Marine Drugs*. 2021, 19(5), 263.
104. Sintubin, L., Gusseme, B., Van der Meeren, P., Pycke, B.F., Verstraete, W., Boon, N. The antibacterial activity of biogenic silver and its mode of action. *Applied Microbiology and Biotechnology*. 2011, 91(1) 153–162.
105. Phaniendra, A., Jestadi, D. B., Periyasamy, L. Free radicals: properties, sources, targets, and their implication in various diseases. *Indian Journal of Clinical Biochemistry: IJCB*. 2015, 30(1), 11–26.

106. Fischbacher, A., von Sonntag, C., Schmidt, T. C. Hydroxyl radical yields in the Fenton process under various pH, ligand concentrations and hydrogen peroxide/Fe(II) ratios. *Chemosphere*. 2017, 182, 738–744.
107. Canaparo, R., Foglietta, F., Limongi, T., Serpe, L. Biomedical applications of reactive oxygen species generation by metal nanoparticles. *Materials (Basel, Switzerland)*. 2020, 14(1), 53.
108. Kehrer J. P. The Haber-Weiss reaction and mechanisms of toxicity. *Toxicology*. 2000, 149(1), 43–50.
109. Gabrielyan, L., Badalyan, H., Gevorgyan, V., Trchounian, A. Comparable antibacterial effects and action mechanisms of silver and iron oxide nanoparticles on *Escherichia coli* and *Salmonella typhimurium*. *Scientific Reports*. 2020, 10(1), 13145.
110. Hochvaldová, L., Večeřová, R., Kolář, M., Prucek, R., Kvítek, L., Lapčík, L., Panáček, A. Antibacterial nanomaterials: Upcoming hope to overcome antibiotic resistance crisis. *Nanotechnology Reviews*. 2022, 11(1), 1115–1142.
111. Gómez-Gómez, B., Arregui, L., Serrano, S., Santos, A., Pérez-Corona, T., Madrid, Y. Unravelling mechanisms of bacterial quorum sensing disruption by metal-based nanoparticles. *The Science of the Total Environment*. 2019, 696, 133869.

5 Microbial Synthesis of Metal Nanoparticles for Nanomedicinal and Catalytic Applications

Javed Iqbal, Shumaila Ijaz, Naila Ijaz, Banzeer Ahsan Abbasi, Tabassum Yaseen, Zakir Ullah, Rashid Iqbal, Ghulam Murtaza, Zaman Ashraf, Tariq Mahmood, Sobia Kanwal, Iftikhar Ali, Ikram Ullah, and Mohsin Kazi

5.1 INTRODUCTION

Recently, a variety of fields have shown a particular interest in nanoparticles (NPs) as a subject. The Greek word "Nanos," which means "dwarf," is where the word "nano" comes from and refers to a size measurement that is 100 millionths of a meter or less (Thakkar et al., 2010, Narayanan and Sakthivel, 2010). A strand of DNA has a diameter of 2.5 nm (Lyubchenko and Shlyakhtenko, 1997); most viruses are 100 nm wide. (Wagner et al., 1992), and a typical bacterium is 1–3 m wide (Katz et al., 2003). NPs are defined as particulate dispersions of solid particles with at least one dimension and a size range of 10–1000 nm (Mohanraj and Chen, 2006, Thakkar et al., 2010). The surface-to-volume ratio is the most crucial characteristic of NPs because it makes it simpler for them to engage in particle interactions (Thakkar et al., 2010, Narayanan and Sakthivel, 2010). Organisms have adapted by developing defenses against metals to survive in environments with high metal concentrations. These techniques might entail modifying the chemical makeup of the poisonous metal to render it nontoxic, leading to the development of target metal NPs. As a result, NP creation is a resistance mechanism's "byproduct" to a particular metal and can be employed as a different method of creating them. Compared to their bulk material equivalents, NPs have special magnetic, electrical, physical, optical, chemical, and thermal capabilities (Durán et al., 2007, Husseiny et al., 2007). Next-generation biosensors, catalysts, electronics, and antimicrobials can benefit from these properties (Durán et al., 2007, Narayanan and Sakthivel, 2010).

One significant and intensively researched class of materials that exhibits considerable diversity and has numerous applications is metallic NPs. The focus of this review

DOI: 10.1201/9781003378563-5

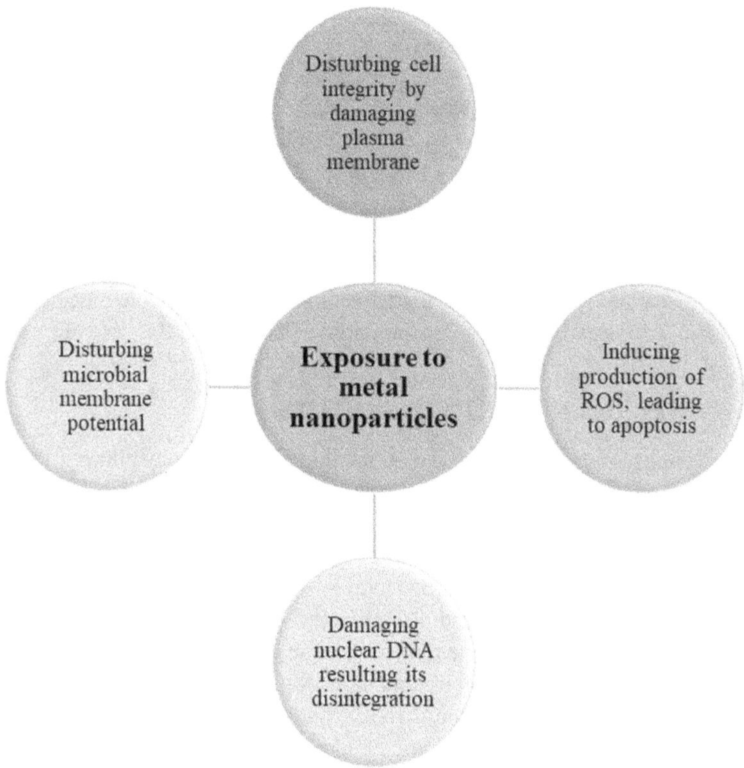

FIGURE 5.1 Consequences of a cell being exposed to metal nanoparticles.

will be on how biology and material science may work together to create a "green" way to produce metal NPs for a range of applications. There are important linkages between the production of NPs and what they might be used for. Numerous studies have revealed the antibacterial capabilities of silver nanoparticles (AgNPs) (Durán et al., 2007, Guzmán et al., 2009, Krishnaraj et al., 2010, Guzman et al., 2012, Iqbal et al., 2018). For example, it has been demonstrated that NPs such as silver and gold are effective at preventing the proliferation of gram-negative bacteria and gram-positive bacteria (Figure 5.1) (Guzmán et al., 2009, Lima et al., 2013). Due to the recent rise in antibiotic resistance, research has begun to focus on these antibacterial NPs as potential new medical instruments and the decrease in the discovery of new antibiotics.

Additionally, optical sensors made of AgNPs have been employed to create small molecule adsorbents (McFarland and Van Duyne, 2003). In contrast, it has been demonstrated that PtNP-based catalysts show much activity for formic acid electrooxidation (Waszczuk et al., 2002). Wet-chemical processes are often high-volume and low-cost and are also the most popular ways to create all of these NPs. However, their wide application in biological fields is constrained by the necessity for hazardous solvents and the pollution from chemicals employed in NP manufacturing (Li et al., 2011, Hameed et al., 2020, Uddin et al., 2021). To enable their employment in a wider range of industries, therefore, a "green" nontoxic technique for creating metallic NPs is needed. The use of biological techniques could be able to accomplish this. Metallic NP synthesis has been demonstrated in a wide variety of bacteria, fungi, and plants, each with unique benefits and drawbacks (Suresh et al., 2004, Bhainsa and D'Souza, 2006, Song and Kim 2009, Cai et al., 2022). Production factors for biological NPs include growth temperature, period of synthesis, ease of extraction, extracellular or intracellular synthesis, and the ratio of particles that were synthesized to those that were extracted from the sample. Choosing the best biological approach might be influenced by a variety of factors. Metal NPs that fall under examination are of utmost importance because, typically, only a small number of metals have been found to cause resistance in organisms, posing a potential constraint on the type of organism.

5.2 METALLIC NANOPARTICLE SYNTHESIS

However, a young scientific branch called synthetic biology is beginning to address these problems to create a more universal chassis that can synthesize multiple types of metallic NPs with the help of the same organism (Edmundson et al., 2014). There are two types of "natural" biogenic metallic NP production. The first is bioreduction, which involves physiologically reducing metal ions into chemical forms that are more stable. Numerous organisms can use dissimilatory metal reduction, which combines the reduction of a metal ion with the oxidation of an enzyme (Deplanche et al., 2010). As a result, metallic NPs become stable and inert, which may then be securely removed from a polluted material. Biosorption is the second classification. This does not need the use of energy and involves the attachment of metal ions to an organism, such as the cell wall, from an aqueous or soil sample. To generate stable complexes in the form of NPs, some fungi, bacteria, and plants produce peptides or have altered cell walls with metal ion-binding properties (Yong et al., 2002). Metallic NPs are gaining importance as a result of the variety of domains in which they may find use (Figure 5.2).

Therefore, it is essential to find a low-cost, ecologically benign technique to synthesize them. Numerous species have the capacity to synthesize NPs, making them potentially exploitable and amenable to modification to make them more effective for this use. The emphasis of the sections that follow will be on providing an overview and a description of the numerous biological processes used to create metallic NPs. Figure 5.3 shows a schematic illustration of various ways to synthesize metal NPs.

The extracellular and intracellular enzymes (Figure 5.4) required for the creation of NPs are likely to be found in microorganisms. Biogenic NPs are significantly more stable and have improved catalytic and specificity abilities. The bulk of microorganisms reside in the natural environment, which supports the microbial production of nanomaterials. As a result, creating and improving nanomaterials may be done much more easily. As nanotechnology develops, there is an increasing need to look for eco-friendly and economical methods of synthesizing nanomaterials. Investigating the fundamental processes of NPs' biogenesis, which can be thought of as powerful eco-friendly nanofactories, has become a priority. Nanotechnologists have observed calcium layers in S-layer bacteria, magnetite particles in diatoms, and calcium layers in siliceous materials. The improved knowledge of metal-microbe interactions is made possible by a number of biological applications, including biomineralization, bioleaching, and bioremediation. All biological techniques have the benefits of

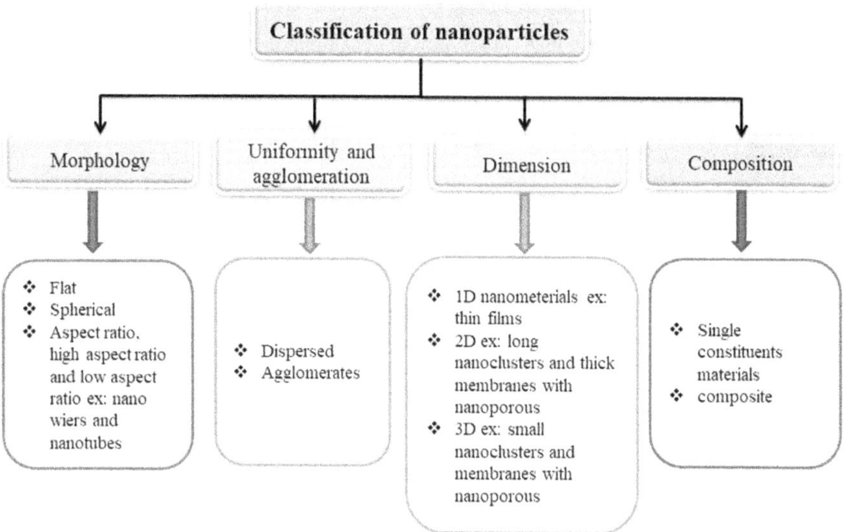

FIGURE 5.2 Classification of nanoparticles.

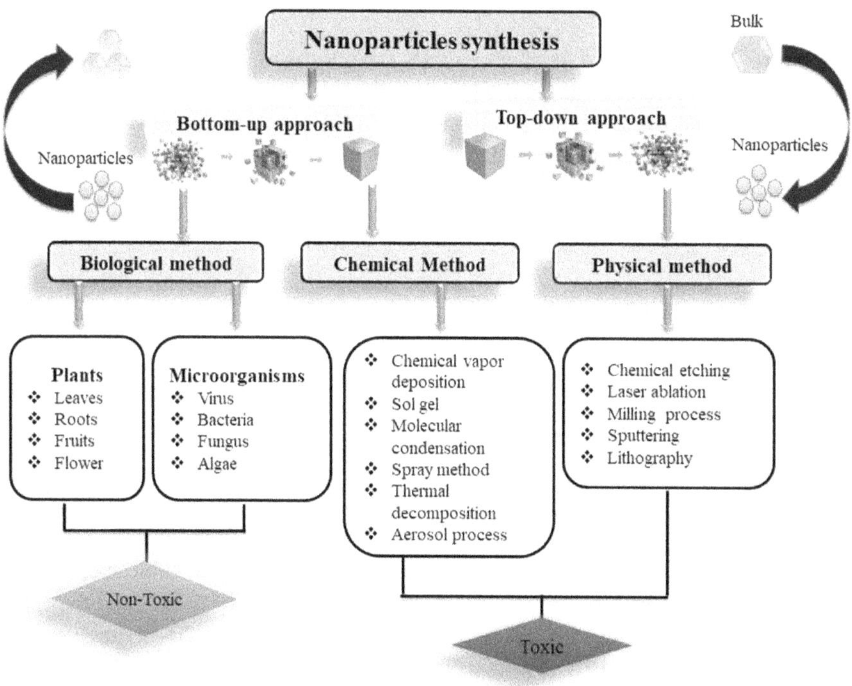

FIGURE 5.3 Schematic representation showing different approaches for the synthesis of metal nanoparticles.

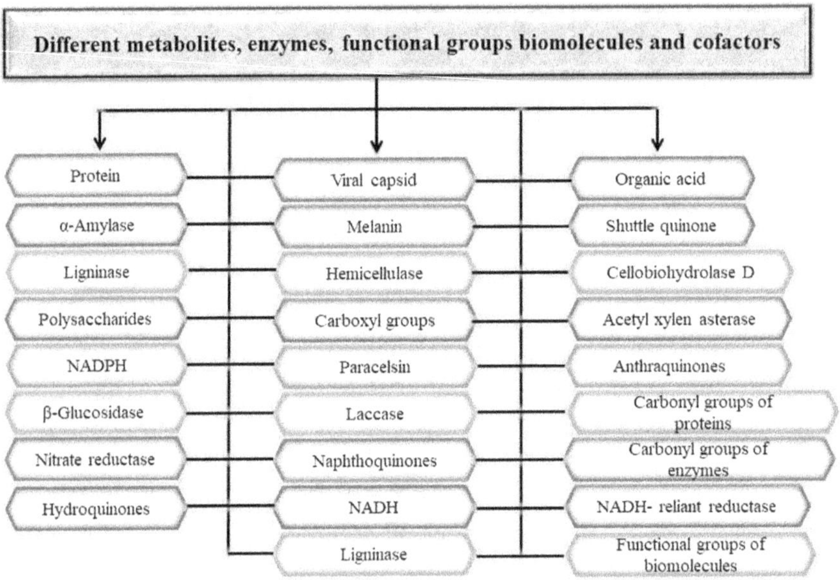

FIGURE 5.4 The microbial-mediated synthesis of nanoparticles involves a variety of enzymes, cofactors, metabolites, biomolecules, and functional groups.

being more dependable and environmentally friendly. Although biological methods of synthesis are currently widely used, it is still necessary to improve synthesis settings, comprehend the synthesis mechanism, and use effective downstream processing strategies (Mohammadian et al., 2007).

5.2.1 Chemical and Physical Methods of Metallic Nanoparticle Synthesis

There are numerous approaches to synthesize metallic NPs (Figure 5.5). A thorough grasp of the present chemical and physical procedures is required to research the biological ways of synthesizing NPs so that comparisons can be made and a foundation for improvement can be seen. Two alternative methods, namely, (i) the top-down approach and (ii) the bottom-up approach, can be used to synthesize NPs (Figure 5.6). In addition, three distinct approaches, physical, chemical, and biological, are used for the synthesis.

The literature has a considerable amount of analysis into the creation of metallic NPs. Several studies will be examined in this part to explain the "non-green" techniques used to create NP (Figure 5.7). AgNPs with tetrahedral and cubic shapes were made by Wiley et al. by bringing ethylene glycol and AgNO3 to 148°C, and they succeeded in doing this. Single crystal NPs with a diameter range of 20–80 nm were the end result. There was only a 10-minute reaction time, and the AgNPs were largely monodispersed (Wiley et al., 2004). However, since a relatively high temperature was required for the reaction to take place, producing commercial quantities would need a sizable amount of energy. AgNPs were successfully manufactured via a process created by Montazer et al. using polyamide textiles. Under SEM observation, it

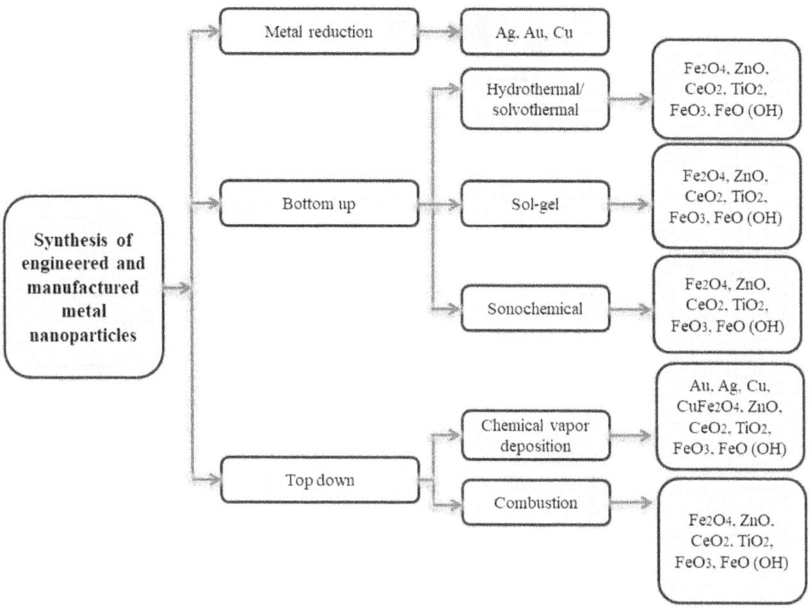

FIGURE 5.5 The production of manufactured and engineered metal-based nanoparticles using conventional methods.

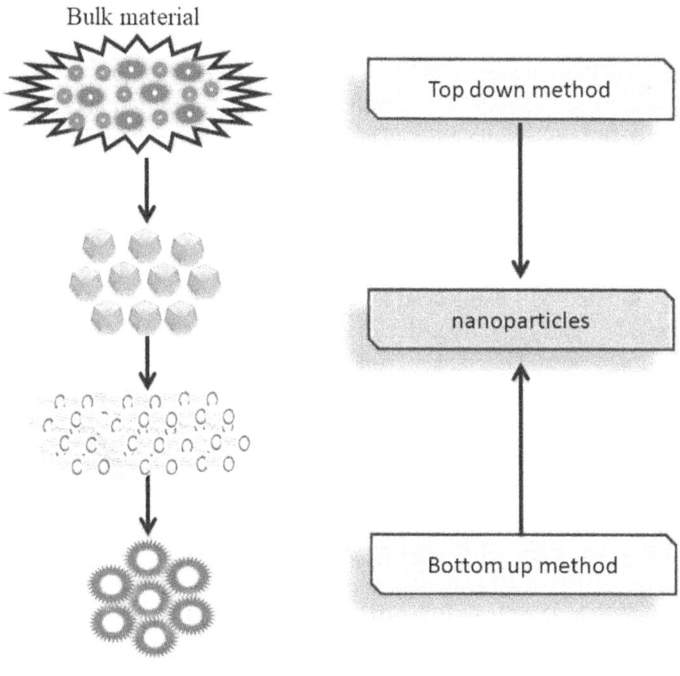

FIGURE 5.6 Two alternative methods, namely, (i) the top-down approach and (ii) the bottom-up approach, can be used to synthesize NPs.

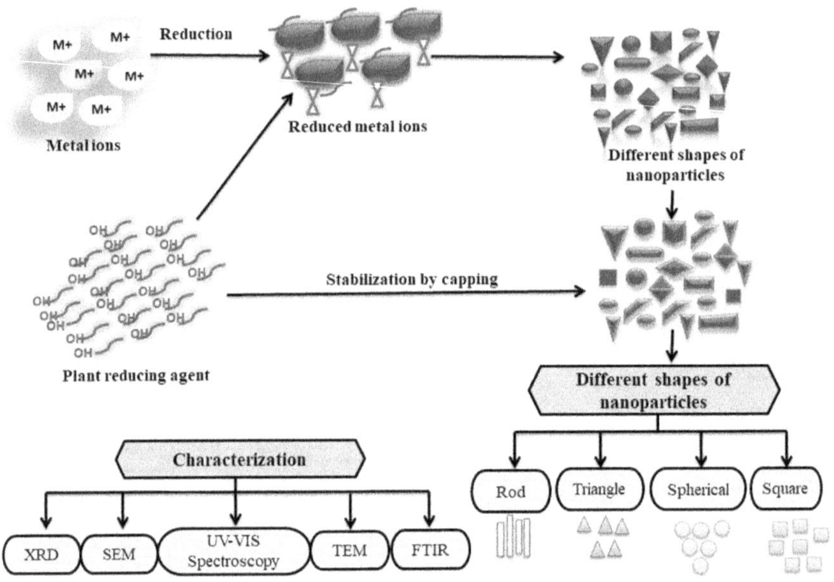

FIGURE 5.7 Metal nanoparticles are synthesized utilizing various methods, as shown in a schematic.

was discovered that the size of the NPs ranged from 20 to 150 nm, with an average size of 90 nm. A few aggregates were observed, and they were linked to the reaction's boiling point temperature.

By employing the polyamide fabric itself as the reducing agent for Ag^+, the synthesized NPs were stabilized. This significant finding rendered unnecessary the use of additional stabilizing medications. The fabric was then examined for its antibacterial action, and even after 30 washing cycles, it still had strong bactericidal properties (Montazer et al., 2014). A physical process without the use of reducing chemicals, which can contaminate the resulting NPs, has been used to produce stable platinum nanoparticles (PtNPs). By using a laser to heat up large amounts of material, Mafuné et al. created NPs using laser ablation. Mafuné and team used this approach in both pure water and SDS, and they discovered that stable PtNPs were generated in both situations. Next, TEM analysis of the PtNPs revealed that their size distribution ranged from 1 to 7 nm, which was too small for centrifugation to successfully separate those particles (Mafuné et al., 2003).

It has also been suggested that a variety of microbes, including bacteria, fungi, actinomycetes, and viruses, can produce different metal NPs. Numerous biological applications, including biomineralization, bioleaching, and bioremediation, were used in the past, and biocorrosion has made use of the interaction between metals and microbes (Klaus-Joerger et al., 2001). However, due to a few benefits over other approaches, microbial production of NPs has recently become recognized as a promising area of study. Depending on the type of microorganism, either intracellular or extracellular processes are used to create NPs (Mohamed et al., 2019). Table 5.1 summarizes earlier publications on the manufacture of metal NPs utilizing different microbes.

TABLE 5.1
Metal NPs Manufactured Utilizing Different Microbes

Microorganism Type	Nanoparticles	Microorganisms	Shape	Size (nm)	References
Yeast	Ag	*Rhodotorula* sp. ATL72	Spherical and oval	8–21	Soliman et al., 2018
	Ag	*R.* sp. ATL72	Spherical	2–20	Korbekandi et al., 2016
	Ag	*Cryptococcus laurentii*	Spherical	35–400	Fernández et al., 2016
	Ag	*C. laurentii*	Spherical	15–220	Cunha et al., 2018
	Ag	*Rhodotorula glutinis*	Spherical	15.5	Yang et al., 2017
	Au	*Saccharomyces cerevisiae*	Triangle, truncated triangle, and hexagon		Qu et al., 2018
	Au	*S. cerevisiae*	Spherical and pseudo-spherical	20–28	Sriramulu and Sumathi, 2018
	Se	*S. cerevisiae*	Hexagonal	32	Lian et al., 2019
	Se	*Magnusiomyces ingens LHF1*	Spherical and quasi-spherical	70–90	Lian et al., 2019
Bacteria	Se	*Lactobacillus acidophilus*	Spherical	2–15	Alam et al., 2020
	Se	*Bacillus subtilis*	Spherical	50–400	Wang et al., 2010
	Se	*Lysinibacillus* sp. *ZYM-1*	Cubic	100–200	Che et al., 2017
	Pd	*Shewanella oneidensis MR-1*	Spherical	10–100	Wang et al., 2019
	Pd	*Shewanella loihica*	Spherical	1–12	Wang et al., 2019
	Pt	*Jeotgalicoccus coquinae ZC15*	Spherical	5.74	Eramabadi et al., 2020
	Pt	*J. coquinae ZC15*	Spherical	2.83–61.03	Tuo et al., 2017
	Pt	*S. loihica*	Spherical	1–10	Ahmed et al., 2018a,b
	Au	*Mycobacterium* sp.	Spherical	5–55	Camas et al., 2018
	Au	*B. subtilis*	Spherical	20–25	Srinath et al., 2018

(*Continued*)

TABLE 5.1 (*Continued*)
Metal NPs Manufactured Utilizing Different Microbes

Microorganism Type	Nanoparticles	Microorganisms	Shape	Size (nm)	References
	Au	*Escherichia coli DH5α*	Spherical, triangles, and quasi-hexagons	25	Du et al., 2007
	Au	*Rhodopseudomonas capsulata*	Spherical	10–20	He et al., 2008
	Au	*Pseudomonas aeruginosa*	Spherical	15–30	Husseiny et al., 2007
	Ag	*Actinobacteria*	Spherical	13.2	Wypij et al., 2016
	Ag	*Pantoea ananatis*	Spherical	8.06–91.31	Monowar et al., 2018
	Ag	*Ochrobactrum anthropi*	Spherical	38–85	Thomas et al., 2014
	Ag	*Bacillus licheniformis*	Spherical	40	Kalishwaralal et al., 2008
	Ag	*Pseudomonas stutzeri*	Triangular	200	Klaus et al., 1999
	Ag	*B. subtilis*	Spherical	3–20	Alsamhary et al., 2020
Fungi	Ti	*Fusarium oxysporum*	Spherical	6–13	Bansal et al., 2005
	Si	*F. oxysporum*	Quasi-spherical	5–15	Bansal et al., 2005
	Pt	*F. oxysporum* f. sp. *Lycopersici*	Triangle, hexagons, square, and rectangles	10–50	Bansal et al., 2005
	Pt	*Penicillium chrysogenum*	Spherical	5–40	Riddin et al., 2006
	Au	*Rhizopus oryzae*	Spherical	5–40	Subramaniyan et al., 2018
	Au	*Aspergillus* sp.	Spherical	4–29	Kitching et al., 2016
	Au	*Pleurotus ostreatus*	Spherical	10–30	Shen et al., 2017
	Au	*Trichoderma harzianum*	Spherical	26–34	El Domany et al., 2018
	Au	*Cladosporium cladosporioides*	Spherical	60	Tripathi et al., 2018
	Ag	*Candida albicans ATCC 10231*	Spherical	10–20	Joshi et al., 2017
	Ag	*Ganoderma sessiliforme*	Spherical	45	Bonilla et al., 2017
	Ag	*Aspergillus terreus*	Spherical	7–23	Lotfy et al., 2021

	Ag	*F, oxysporum*	Spherical	21.3–37	Ahmed et al., 2018a,b
	Ag	*F. oxysporum*	Spherical	21.3–37	Elamawi et al., 2018
	Ag	*Trametes trogii*	Spherical and rod	5–65	Kobashigawa et al., 2019
	Ag	*Candida glabrata*	Spherical and rod	5–65	Jalal et al., 2018
	Ag	*Rhizopus stolonifer*	Spherical	2.86	AbdelRahim et al., 2017
Cyanobacteria	Ag	*Desertifilum*	Spherical	4.5–26	Hamida et al., 2020
	Ag	*Desertifilum*	Spherical	4.5–26	Hamida et al., 2020
	Ag	*Desertifilum*	Spherical	4.5–26	Hamida et al., 2020
	Ag	*Oscillatoria limnetica*	Quasi-spherical	3.30–17.97	Hamouda et al., 2019
	Ag	*O. limnetica*	Quasi-spherical	3.30–17.97	Hamouda et al., 2019

5.3 METAL NANOPARTICLES PRODUCED BY MICROORGANISMS

5.3.1 Virus-Mediated Nanoparticle Biosynthesis

The creation of semiconductor nanocrystals may use biomolecules such as amino acids and fatty acids as templates (Konishi et al., 2007). Biological templates such as protein cages (Jiang et al., 2005), DNA (Alivisatos, 2004), biolipid cylinders (Archibald and Mann, 1993), and bacterial rapidosomes (Douglas and Young, 1998) have been used to synthesize environmentally benign and inorganic nanostructures. Viruses that have been genetically modified and viral capsids that have self-assembled were used in the production of quantum dots and nanowires. In the presence of the tobacco mosaic virus, iron oxide NPs continued to be produced by oxidative hydrolysis. The fusion protein was produced by M13 bacteriophages and served as the nucleus for the growth of ZnS and cadmium sulfide (CdS) nanostructures (Mao et al., 2003).

5.3.2 Nanoparticle Production Using Yeast

It has also been reported that semiconductor NP generation occurs using yeast. Internally spherical and monodispersed quantum crystallites of CdS were created by *Candida glabrata* (Dameron et al., 1989) and metal-thiolate compounds with phytochelatins to detoxify metal ions (Kowshik et al., 2002). A decrease in gold ions in intracellular formulations of gold nanoparticles (GNPs) has also been observed (Kowshik et al., 2002). Due to nucleation on the cell boundary, the formation of triangular and hexagonal gold crystals was observed at a low pH of 2.0 (Gericke and Pinches 2006). Extracellular hexagonal SNPs were reported to be produced by the silver-tolerant yeast MKY3 strain (Kowshik et al., 2002). *Saccharomyces cerevisiae* was able to synthesize antimony oxide (Sb_2O_3) NPs with semiconductor-like properties under ambient circumstances. This may have happened as a result of the radial tautomerization of membrane-bound pH-dependent oxidoreductases (Agnihotri et al., 2009).

5.3.3 Synthesis of Nanoparticles through Actinomycetes

Actinomycetes were found to produce both external and intracellular NPs. Through *Thermomonospora* sp., monodispersed and spherical extracellular production of GNPs was discovered (Maheswari et al., 2012) a brand-new actinomycete with extremeophile For the intracellular metallic NPs produced on the inside of cell walls and cytoplasmic membranes, *Rhodococcus* sp. was examined. More particles were present in the cytoplasmic membrane than in the cell wall, which also suggests the existence of an enzyme (Gericke and Pinches, 2006). Actinomycetes were previously recognized as the main source of secondary metabolites, but these discoveries would lead to the selection of additional actinomycete strains for the creation of nanostructures (Agnihotri et al., 2009).

5.3.4 Algae-Mediated Synthesis of Nanoparticles

The utilization of tiny algae for metallic NPs has received relatively little attention. *Chlamydomonas reinhardtii* was recently employed in a study on the controlled biosynthesis of SNPs. Another experiment (Ahmad et al., 2003) investigated *Sargassum wightii*'s synthesis of metallic NPs using a methodical approach. This was the first account of the production of reasonably stable GNPs using marine algae. After reducing gold ions, extremely stable GNPs (8–12 nm) with high density were created. Certainly, under agitation conditions, 95% of the AuCl4 ions are bioreduced after 11–12 hours. The synthesis procedures have been expanded to generate platinum and palladium nanostructures by using their metallic chloride (Singaravelu et al., 2007).

5.3.5 Nanoparticle Production by Bacteria

Bacteriogenesis of NPs can occur both intracellularly and extracellularly. The ability of bacteria to detoxify chemicals as a result of ion efflux from specialized membrane proteins that act as proton anti-transporters is known as microbial resistance to heavy metals and ATPases in chemiosmotic cations (energy-dependent). Another cause of metal resistance is a change in solubility (Bharde et al., 2006). Accordingly, the transformation of poisonous, soluble inorganic ions into nontoxic, insoluble metallic nanoclusters leads to the detoxification of metal ions. Precipitation, biosorption, extracellular bioaccumulation, and extracellular complexation are further ways that bacteria might detoxify metals. The NPs produced extracellularly are quite significant in terms of commercial use. However, polydispersity continues to be a serious issue. The processes used in biological and microbial synthesis still need to be optimized to obtain monodispersity (Bruins et al., 2000).

Bioreactors can use biogenic NPs as catalysts and keep them on hand for ongoing use. Recent years have seen the use of microorganisms for the deposition process in mineral ores (Bao et al., 2007). In the Alaskan placer, Pedomicrobium has been observed to accumulate gold (Lengke and Southam, 2006). Octahedral-shaped GNPs with a 5–25 nm size range were created by reducing water-soluble Au^{3+} to Au^0. By reducing Fe(III), *Geosporobacter ferrireducens* precipitated gold in the periplasm (Figure 5.8) (Mann et al., 1984). Similarly, the mesophilic iron (III)-reducing strain of Shewanella algae formed GNPs in the periplasm (Konishi et al., 2007). To protect themselves from dangerous toxins, gram-negative bacteria use vesicles made of proteins, phospholipids, and lipopolysaccharides (LPS). Chloroauric acid's reduction to Au^0 serves as an illustration of *Escherichia coli* DH5-mediated synthesis (GNPs). Direct electrochemistry of blood proteins, including hemoglobin, can be understood using cell-bound NPs (Kashefi et al., 2001).

Surprisingly, the photosynthetic bacterium *Rhodobacter capsulatus* was recognized as a trivalent aurum-reducing agent in the log phase. The role of carotenoids embedded in cell membranes and NADH-dependent enzymes in bioreduction has been recognized (Du et al., 2007). Cell surfaces are the main focus of intracellular SNP production. Silver sulfide (Ag2S) crystallites, such as those of *Pseudomonas stutzeri* AG259, were found in periplasmic gaps and had equilateral, monoclinic, triangular, and hexagonal shapes (Feng et al., 2007). Silver-binding proteins offer a location that aided nucleation during the creation of SNPs (Joerger et al., 2000). Recent research has shown that a specific *Bacillus* species may convert metallic silver ions to atomic silver (Mouxing et al., 2006). According to Joerger et al. (2000),

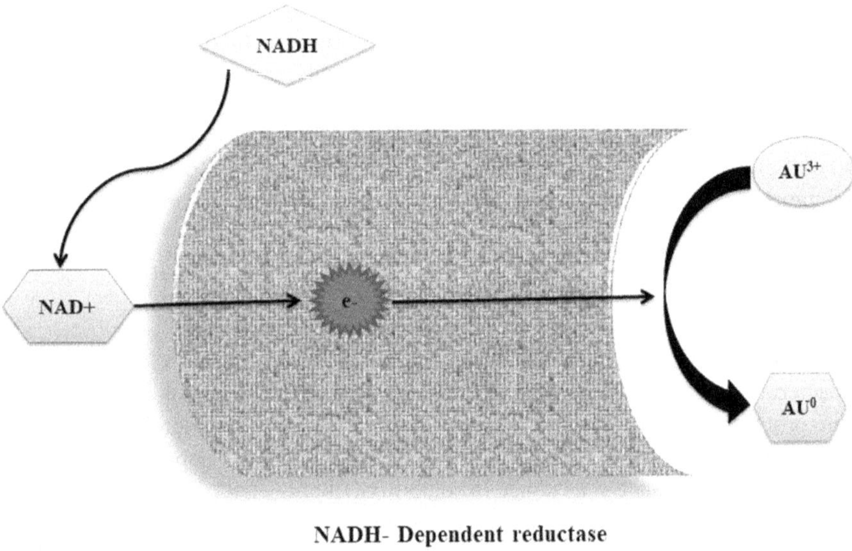

FIGURE 5.8 An idea for the process by which gold ions are converted into AuNPs. By using NADH-dependent reductase as an electron transporter to transfer electrons from NADH to produce NAD+, bioreduction is initiated. Gold ions are reduced to elemental AuNPs and receive electrons as a result.

this silver buildup on the cell wall is a result of a resistance process made possible by periplasmic proteins. When *P. stutzeri* AG259 came in contact with a saturated $AgNO_3$ solution, Klaus and colleagues (1999) discovered the arrangement of SNPs of different forms and diameters that are well defined inside the periplasm of the bacteria.

Similarly, *Corynebacterium* sp. ($SH0_9$) produced intracellular SNPs during a 72-hour incubation at 60°C with the diamine silver complex $(Ag(NH3)_2)^+$ and dried cells (Mouxing and colleagues 2006). Metallic NPs were found inside the bacterial cells of *Lactobacillus*, a common bacterium found in butter milk (Pugazhenthiran et al., 2009). It is interesting to note that alloy NPs were created when lactic acid bacteria from whey were subjected to mixed metallic ions (Zhang et al., 2005). It was discovered that *Shewanella oneidensis* MR-1 produces Pb nanocrystals inside the cell membrane and on the cell wall (Gupta et al., 1998). Additionally, *Desulfosporosinus* sp. transformed uranium's hexavalent U(VI) form into tetravalent U(IV) uraninite precipitates. In soils, soluble radionuclide pollutants may be suppressed, and soluble sediments may be converted to insoluble forms by uraninite precipitates (Nair and Pradeep, 2002). Magnetotactic bacteria may produce magnetic NPs that contain either iron sulfide, iron oxide, or both. These can be identified as bacterial magnetic particles (BacMPs) to differentiate them from other forms, such as synthetic magnetic particles (antimicrobial peptides [AMPs]).

These BacMPs are arranged in lines within the bacterial cells and operate as biological compass needles to aid bacterial migration in aquatic environments along gradients of oxygen using the Earth's geomagnetic field (Windt et al., 2005). BacMPs' organic membranes, which are composed of proteins and phospholipids, aid in the dissemination of magnetotactic bacteria in water environments. Marine and freshwater sediments are home to magnetotactic bacteria, which also produce greigite (Fe_3S_4), pyrrhotite, ferromagnetic iron sulfide, and pyrrhotite attached to internal membranes (Suzuki et al., 2002, Sweeney et al., 2004, and Matsunaga et al., 2005). A study showed that micro-electromagnets in bodily fluid can be used to control the assembly of magnetotactic bacteria. Numerous bacteria, including *Magnetospirillum magnetotacticum MS-1*, *Aquaspirillum magnetotacticum*, and *M. magnetotacticum*, were reported to demonstrate intracellular production of magnetic nanocrystals of magnetite (Fe_3O_4) (Farina, 1990).

Several bacteria, including *Klebsiella pneumonia* (Lee et al., 2004), *Clostridium thermoaceticum* (Philipse et al., 2002), and *E. coli* (Smith et al., 1998), produce CdS semiconductor nanocrystals intracellularly. It was discovered that the activity of cysteine desulfhydrase, which caused cysteine to become desulfhydrated, precipitated CdS nanocrystals. These CdS semiconductor NPs made by biotechnology have spherical and elliptical forms, as well as optical and photoactive characteristics. Desulfobacteriaceae microorganisms were used in another experiment to physiologically generate spherical ZnS NPs (Cunningham and Lundie, 1993). The NPs that were created inside of cells need extra procedures and downstream processing steps, such as washing with the right ultrasound or detergent treatment (Suzuki et al., 2002). Where reducing agents are located and how microbes create metallic NPs are related. After certain reductive enzymes involved in the reduction of metal ions are secreted, extracellular establishment of metallic NPs is feasible.

Extracellularly produced NPs have more uses than intracellular NPs in a variety of industries, including bioimaging, electronics, sensor technology, and optoelectronics. *Pseudomonas aeruginosa* (ATCC 90271) was employed in a study to produce extracellular GNPs with a particle size range of 5–40 nm (Reyes et al., 2009). *Aeromonas* sp. SH10 dried cells transformed silver nitrate into highly stable, uniform-sized SNPs (Labrenz et al., 2000). A quick synthesis of SNPs was observed when the cell filtrate of the *Enterobacteriaceae* family (*K. pneumonia*, *E. coli*, and *Enterobacter cloacae*) was given time to react with the silver nitrate solution. The direct involvement of the nitro reductase family in the reduction route was shown by the inhibition of the reduction process by the addition of piperitone, a nitrate reductase inhibitor. Similarly, *Bacillus licheniformis* was used to synthesize SNPs extracellularly (Husseiny et al., 2007). When *Acetobacter xylinum* cellulose membranes were exposed to biogenesis, homogenous SNPs were also discovered in a solution containing Ag^+ ions and triethanolamine (Ag^+-TAE) (Shahverdi et al., 2007). A type of Morganella that was resistant to silver obtained from an insect's gut was used to report on the biogenesis of SNPs (Bai et al., 2009). According to research by Bai et al. (Shahverdi et al., 2007), the photosynthetic bacteria *Rhodopseudomonas palustris* created spherical extracellular CdS nanocrystals. It was thought that *Klebsiella aerogenes* produced CdS crystallites that were spherical.

According to energy dispersive X-ray spectroscopy (EDS), the amounts of sulfur and cadmium were similar. Purple nonsulfur bacteria have been shown to produce CdS nanocrystals extracellularly and at room temperature (Shahverdi et al., 2007).

There have been descriptions of *R. palustris* (Parikh et al., 2008), *K. aerogenes*, *R. sphaeroides*, and *Gluconacetobacter xylinus*. Following precipitation, cellulose nanofibers from the bacteria were covered in *G. xylinus* CdS NPs. It is intriguing that bacteria have been shown to be capable of catalysis due to electrochemical energy, reduction potential, and electron shedding in areas with sufficient potency to cause magnetite to precipitate (Mouxing et al., 2006). *Actinobacteria* sp., an aerobic non-magnetotactic bacteria, has also been found to produce extracellular magnetite NPs (2005). Certain nonmagnetotactic bacteria, such as *Geobacter metallireducens*, have also created magnetite NPs outside of cells by reducing ferric oxide in an anaerobic setting and in the presence of acetate (Parikh et al., 2008).

5.3.6 Nanoparticles Mediated by Fungi

Since fungi have some benefits over bacteria in the manufacture of NPs, their usage in generating metallic NPs has received considerable attention. Important advantages to take into account are the simplicity of the economic feasibility, up scaling, and downstream processing, as well as the presence of mycelia, which offers a larger surface area. Due to their potential benefits, such as quick growth, simple scaling up, large biomass, and the capacity to reduce metallic ions for the manufacture of NPs both extracellularly and intracellularly, fungi have been used as a "nanofactory" for the production of MNPs (Boroumand Moghaddam et al., 2015). The intracellular method involves producing NPs inside the fungal cell by treating its biomass with a precursor salt solution after incubation for one day in a dark environment, in contrast to the extracellular processes, which involve treating the fungal filtrate or extracts with the appropriate precursor salts. It has been demonstrated that *Phoma* sp. and *Aspergillus flavus* both create AgNPs inside cells (Chen et al., 2003, Vala et al., 2014). Similar to *Verticillium luteoalbum*, *Penicillium chrysogenum*, and *Trichothecium* spp., *Fusarium oxysporum* and *Lycopersici* spp. have produced PtNPs (Riddin et al., 2006, Gericke and Pinches, 2006, Sheikhloo and Salouti, 2011). Extracellular approaches are an economically viable method for the biofabrication of AgNPs in *Aspergillus tamarii*, *Penicillium ochrochloron*, and *Aspergillus niger* since they do not require substantial downstream processing (Devi and Joshi, 2015). The intracellular approach for MNPs biosynthesis is expensive since it must deal with issues during downstream processing because it is carried out inside the cell.

Accordingly, numerous fungi strains have been described for the extracellular biofabrication of various NPs, such as Ti, CdS, zirconia, magnetite, Au, and Si (Bansal et al., 2004, 2005, Bharde et al., 2006, Kar et al., 2014, Chen et al., 2014, Castro-Longoria et al., 2011). *Penicillium brevicompactum* fungus cell filtrate was used to make AuNPs that were effective against mouse mayo blast carcinoma C2C12 cells cultured for 24, 48, and 72 hours. The cumulative incubation period and enhanced concentration of gold salt considerably amplified the percentage of cell death (Mishra et al., 2011). The extracellular AuNPs generated by the phytopathogenic fungus (*Helminthosporium solani*) ranged in size from 2 to 70 nm after incubation with aqueous chloroaurate ion solution. Human embryonic kidney cells efficiently absorbed NPs associated with an anticancer medication (doxorubicin). The importance of these conjugates as possible cancer-prevention medicines is demonstrated by this work (Kumar et al., 2008). Currently, AgNPs were created utilizing *Botryosphaeria rhodina*, an endophytic fungus, extracellular filtrate, which under in vitro conditions effectively scavenged free radicals and induced apoptosis in lung cancer cell lines (Akther et al., 2019).

In another work, biogenic AgNPs were created by a fungus (*Piriformospora indica*). AgNPs produced chemically or biologically were tested for in vitro cytotoxicity against normal cell lines. In comparison to chemically generated AgNPs, the results showed increased cytotoxic activity against MCF-7, HepG2, and HeLa cell lines (Aziz et al., 2019). Similarly, the synthesized MNPs and group were produced intracellularly, which again suggests that if they are to be used commercially, an additional step must be added to the purification process. Fungi also have a significant safety disadvantage, similar to bacteria. Numerous well-researched fungi, such as *F. oxysporum*, are pathogenic and may therefore be dangerous. Having been shown to be nonpathogenic, *Trichoderma reesei* and *Trichoderma asperellum* are both fungi that create AgNPs after exposure to silver salts, making them perfect for usage in industrial settings. *T. reesei* is already widely employed in industries such as food, animal feed, medicines, paper, and textiles.

The ability of the fungus to bioaccumulate metals and its tolerance to metals make it a promising candidate for the biological synthesis of metallic NPs. Extracellular sticking with metabolites may be mediated by a variety of processes, including physicochemical and biological processes, contributing to the complex process of metal buildup (metabolism-dependent accumulation). When *F. oxysporum* was utilized as the enzyme source, a quick extracellular synthesis of stabilized metal NPs was observed (Singaravelu et al., 2007). Extracellular synthesis, as opposed to intracellular synthesis, has additional benefits because of straightforward downstream processing processes (Riddin et al., 2006). Furthermore, the extraction of NPs from the biomass for intracellular production would require sophisticated equipment (Gade et al., 2008). The fungus *Phoma* may also produce SNPs extracellularly by combining cell extract with a silver nitrate aqueous solution (Sastry et al., 2003). By using nitrate reductase, which was created by SNPs, *F. oxysporum* may reduce the silver nitrate solution. A fungus' release of capping proteins gives the NPs stability and prevents them from aggregating (Rai et al., 2012). Mohammadian et al. (2007) described a quick photobiological approach for the production of SNPs (10–60 nm) utilizing different strains of *F. oxysporum*. The use of *F. semitectum* and *F. solani* in the synthesis of SNPs has been extensively studied (Mohanpuria et al., 2008). *T. asperellum* and *Cladosporium cladosporioides* (Ingle et al., 2009) are two examples. When protein capping ligands were utilized for consistency, extracellular spherical and polydispersed SNPs with diameters of 4–100 nm were produced (Mukherjee et al., 2008).

However, the amount of a cofactor, such as NADH, can be crucial for the creation of extremely stable alloy NPs of gold and silver with different compositions (Ahmad et al., 2005). Phytochelatin, a capping agent from the fungus *F. oxysporum*, and NADPH-dependent nitrate reductase were used in the in vitro creation of SNPs to clarify the method of synthesis for nanoscale materials (Senapati et al., 2005). As the responsible agent for the production process, enzymes were found to be notably active in the selective bioleaching of silicates from zirconia. After exposure to

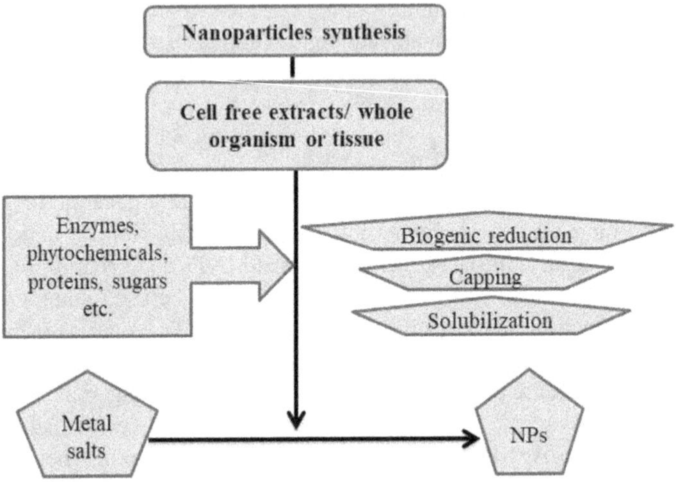

FIGURE 5.9 Schematic illustration of metal nanoparticle synthesis using cell-free extracts/ whole organisms or tissues involving enzymes, phytochemicals, proteins, sugars, etc.

F. oxysporum at room temperature, zirconia NPs are created when an aqueous solution of K_2ZrF_6 is hydrolyzed by extracellular proteins (Kumar et al., 2007). Through the use of statistical techniques, the impact of temperature, salt hexachloroplatinic acid concentration, and pH on the formation of NPs was understood using a response surface methodology (Riddin et al., 2006). A small number of fungal strains have also created different metal NPs of oxides and sulfides. When exposed to Cd^{2+} ions, *C. glabrata* formed intracellular CdS NPs. Similarly, intracellular production of PbS nanocrystals was reported in 2007 by Bansal et al. Research has shown that *S. cerevisiae* may create titanium oxide NPs. When cell-free extracts (Figure 5.9) were exposed one to one at room temperature to ferrous salts and ferric salts, *F. oxysporum* and *Verticillium* formed iron oxide NPs (Kowshik et al., 2002). According to these studies, myco MNPs may serve as a novel cancer therapeutic strategy.

5.4 NANOMEDICINAL AND CATALYTIC APPLICATIONS

Over the past ten years, medications made with nanotechnology have received much interest. The special characteristics of NPs, such as their small size and propensity to pass through delicate blood arteries and barriers, have made this field one of the most extensively studied and examined (Elsaesser and Howard, 2012). They offer significant benefits in terms of enhancing solubility, drug bioavailability, pharmacological activities, toxicity protection, distribution, and avoidance of physical and chemical deterioration as well as increased drug stability inside the body (Zoroddu et al., 2014). Nanomedicines have demonstrated a greater ability to bind with biomolecules and a decrease in tissue oxidative stress and inflammation. Over the years, thousands of unique nanomedicines have been developed; they have a variety of uses in treating various ailments. The use of nanomaterials in medicines means the employment

TABLE 5.2
Role of Nanotechnology and Its Various Applications in Different Ailments

Application	Role of Nanotechnology	Reference
Antimicrobial activity	Activity against bacteria active bacteriostasis	Seil and Webster, 2012
Cancer treatment	Finding carcinogenic and precancerous tumors Patients with solid tumors may undergo clinical testing therapeutic drug delivery to specified molecular targets	Sanna et al., 2014
Respiratory medicine	Aid in the treatment of pulmonary fibrosis, TB, and lung cancer Cystic fibrosis gene treatment using vectors Computed tomography (CT) and magnetic resonance imaging for the diagnosis of the lungs	Omlor et al., 2015
Gastroenterology	Treatment for colorectal and stomach cancer Inflammatory bowel disease treatment identification of various gastrointestinal disorders Organogenesis and surgical transplants	Brakmane et al., 2012
Reproductive medicine	For the treatment of genital infections, uterine fibroids, ectopic pregnancy, endometriosis, and more Transfer of genes and biological substances through sperm choosing particular gametes and embryos Gene therapy for problems of reproduction	Barkalina et al., 2014
Dermatology	Imaging and tailored treatments for skin cancer Immunomodulation and skin-based vaccine administration Antimicrobials and wound care	DeLouise, 2012

of nanotechnologies in very sophisticated molecular-level medicinal interventions. It gives researchers a place to start when looking for useful nanomaterials or nanomedicines. Numerous opportunities have been opened up in the field of medicine thanks to the development of nanomedicines, particularly in terms of drug delivery systems. They are an effective method to target specific areas and quickly penetrate the cell or sick regions due to their structural properties (Sebastian, 2017). Different kinds of NPs were identified based on therapeutic needs, depending on their use and place of origin. Liposomal, metal-based, iron oxide, and polymeric protein NPs have become the best. The application of metal NPs is the main topic of discussion in this review (Table 5.2).

5.4.1 Silver NPs (AgNPs)

Because of their special qualities, including catalytic activity and stability, AgNPs are regarded as excellent. Additionally, they have antibacterial, antiviral, and antifungal effects. AgNPs are used in antibacterial nanodevices due to their Ag^+ ion action. Because they have an antiproliferative effect and can result in cell death, they can be effectively employed as anticancer drugs (Venugopal et al., 2017). To reduce their toxicity and lengthen their biological retention duration, AgNPs may be loaded or coated, enabling the targeted killing of malignant cells. AgNPs from *Andrographis echioides*, which are frequently employed in human breast cancer cell lines, have been demonstrated to suppress the development of MCF-2 cells (Elangovan et al., 2015). The viability of tumor cells decreases when the concentration of AgNPs rises (Arivazhagan et al., 2004).

5.4.2 Gold NPs (AuNPs)

GNPs cause oxidative stress and have anticancer effects. They take up incident photons and transform them into heat that kills malignant cells. At some doses, cationic GNPs (2 nm in diameter) are hazardous (Adewale et al., 2019). GNPs can be found unoxidized. Comparatively speaking, smaller NPs exhibited a lower protein-to-protein ratio than larger NPs. According to reports, GNPs treated with HeLa cervical cancer showed that reactive oxygen species (ROS) are more prevalent, which caused lipids, proteins, and a number of other components to oxidize (Rao et al., 2016). When large NPs (those between 50 and 250 nm) were administered intravenously, AuNPs of 10 nm size were discovered to be widely dispersed throughout the liver, spleen, and circulation.

5.4.3 Iron Oxide NPs

The diameter of iron oxide NPs ranges from 1 to 100 nm. Magnetite (Fe_2O_3) and maghemite ($-Fe_2O_3$) are their two major forms. In isobutanol, which serves as a surfactant, ammonium hydroxide and sodium hydroxide were used to precipitate iron oxide NPs to create them (Farshchi et al., 2018). Without a surface coating, iron oxide NPs are unstable in aqueous conditions; in living organisms, they assemble and precipitate. Macrophages in the circulation sequester the aggregates created by unstable iron NPs. To reduce aggregation under specific circumstances, iron oxide NPs need to be coated with various moieties. Applications for iron oxide magnetic NPs in the so-called hyperthermia anticancer method are numerous; these materials heat-damage adjacent tissues. According to Kostyukova and Chung (2016), iron oxide NPs exhibit a variety of features, including good solubility, stability, dispersion, biocompatibility, and extended circulation period. Delivering high concentrations of NPs to the tumor bulk and tumor cells will boost the sensitivity of in vivo tumor imaging. According to a study, iron oxide NPs significantly alter the cellular morphology of MCF-7 cell lines by triggering apoptosis and necrosis (Peng et al., 2008).

Nevertheless, the use and applications of these nanomaterials (Figure 5.10) as medications must be limited. It is crucial to have a complete grasp of the possible risks prior to application and the toxicological effects of NPs on both the environment and human health. There is an urgent need to comprehend how NPs enter the body, how they operate at the cellular level, and how they affect public health. Understanding the surface functionality of nanomaterials inside biological systems is essential to comprehending their potential toxicological impacts. Before human usage of nanomaterial-based medications was approved, all these variables required more thorough research (Dikshit et al., 2021).

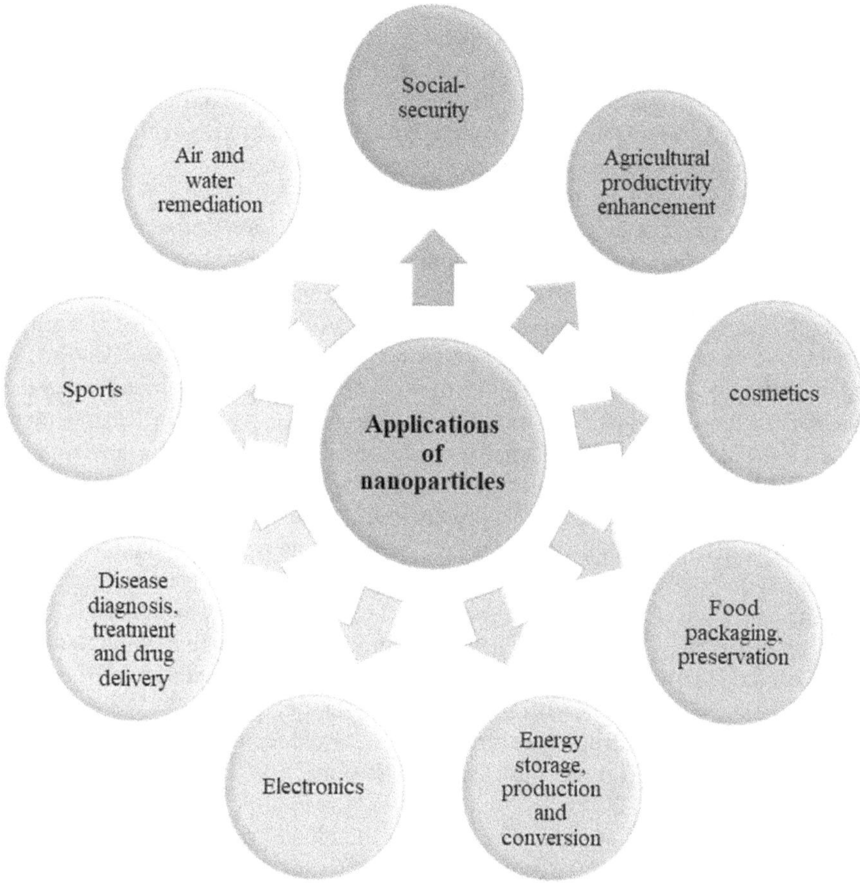

FIGURE 5.10 Different applications of NP are represented schematically.

5.4.4 Silver NPs (AgNPs): Cytotoxic Properties

As a result of their inherent cytotoxicity, AgNPs have been used extensively in the healthcare industry as antibacterial, anticancer, and biological agents. AgNP surface charges determine how harmful they are to cells (Suresh et al., 2012). AgNPs with a positive surface charge are more suited to remain on the luminal side of a blood artery or the surface of tissue for extended periods of time, which is a crucial mechanism for delivering anticancer medications (Schlinkert et al., 2015). Many cancer cell types have been treated using AgNPs due to their intrinsic cytotoxic properties, such as cervical carcinoma, hepatocellular carcinoma, lung cancer, and breast cancer (Jeyaraj et al., 2013, Gurunathan et al., 2013) (Vasanth et al., 2014). Smaller AgNPs were more efficient at generating ROS (Carlson et al., 2008). In addition to these cellular pathways, AgNPs have also shown antiangiogenic and antiproliferative properties (AshaRani et al., 2009; Gurunathan et al., 2009). The antiproliferative action of AgNPs in cancer cells is due to their capacity to harm DNA, break chromosomes, encourage genomic instability, and disrupt calcium (Ca^{2+}) homeostasis, which causes apoptosis and cytoskeletal instability. Damage to the cytoskeleton hinders cells from dividing and progressing through the cell cycle, which encourages cancer cells to become less proliferative. For instance, Lee et al. examined the intracellular trafficking of a single AgNP on a real-time single-NP level in an in vivo zebrafish embryo model system and its impact on early embryonic development.

Chorionic pore channels (CPCs) were shown to carry a single AgNP (5–46 nm) into and out of embryos, where it displayed Brownian diffusion (not active transport). The diffusion coefficient within the chorionic space (3×10^9 cm^2/s) was 26 times lower than that of egg water (7.7×10^8 cm^2/s). Glucose was employed by Thapa et al. as a reducing agent to incorporate graphene oxide into AgNPs (GO-AgNPs). Targeting of cancer cells expressing folate receptors was made possible by covalently fusing methotrexate (MTX) to Go-AgNPs through an amide bond, demonstrating the potential for the synergistic use of anticancer drugs and AgNPs in the treatment of cancer (Thapa et al., 2017). Another instance is the creation of a novel nanocomposite by Azizi and colleagues with the intention of creating AgNPs as a fresh anticancer drug that precisely targets tumor cells. A human breast cancer cell line called MDA-MB 231 was used to test the anticancerous effects of albumin-coated AgNPs. Apoptotic cell death was confirmed by DNA agarose gel electrophoresis after the cancer cells underwent morphological changes. AgNPs with a zeta potential of approximately 20 mV and a size of 90 nm could preferentially bind to tumor cells. Based on their LD50 against MDA-MB 231 (5 M), AgNPs appear to be a good candidate for use as a chemotherapeutic agent (Azizi et al., 2017). AgNPs conjugated with anti-seizure drugs (as a drug carrier) have been described by Ayaz and colleagues to treat central nervous system (CNS) infection in therapies other than oncology (Anwar et al., 2018).

Anti-seizure drugs that are known to cross the blood-brain barrier (BBB) were attached to the surface AgNPs as capping agents. Drugs such as diazepam, phenobarbitone, and phenytoin were conjugated to AgNPs, and these substances demonstrated overall anti-amoebic actions against both trophozoite and cyst phases. Furthermore, in both the trophozoite and cyst amoebic phases, fungicidal activity was significantly enhanced in comparison to that of the medicines alone. According to the researchers, the capacity of AgNP-based medications to attach to amoebae with receptors and ion channels on their cell membranes may be a plausible mechanism by which they can breach the BBB. AgNPs' cytotoxic properties have been widely applied to the food and healthcare sectors, including in the coating of medical devices, textiles, food storage, and environmental sensing (Sharma et al., 2009). Due to its particular toxicity to bacteria (Figure 5.11), silver is now used in a wide range of goods, such as antifouling surface coatings, packaging, and wound dressings. AgNP-coated bandages are another intriguing strategy since they can destroy dangerous microorganisms and promote improved healing in wounded tissue. Additionally, dental resin composites and coatings for medical equipment have been made with silver ions as an antibacterial agent (Sondi and Salopek-Sondi, 2004). Additionally, AgNPs have been used in food packaging to extend the shelf life of food without contaminating it.

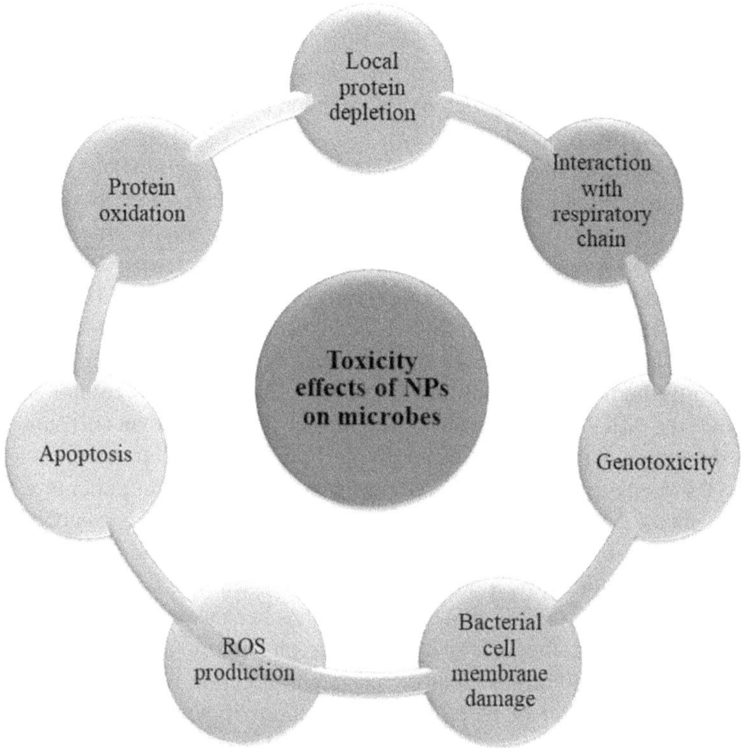

FIGURE 5.11 Toxicity effects of NPs on microbes.

5.4.5 PtNPs

Against a human breast cancer cell line, PtNPs made from *Streptomyces* sp. showed promising anticancer action (MCF-7). Detachment from the surface, significant cellular deformation, and decreased cell viability were all effects of PtNP treatment. Only 52.77% of the MCF-7 human cell lines were found to be viable in the presence of bacteriogenic PtNPs. The IC50 value was 31.2 g/mL. Bacteriogenic PtNP anticancer properties were discovered to be dose-dependent (Baskaran et al., 2017).

4-Nitrophenol (4-NP) and azo dyes were reduced catalytically using Pd-Pt alloy NP produced from *S. oneidensis* MR-1. The peak center of 4-NP was initially found at approximately 318 nm and shifted to 400 nm with the addition of NaBH4. The absorbance at 400 nm decreased and that at 300 nm increased when Pd-Pt alloy NPs were added, showing that 4-NP had been changed into 4-aminophenol (4-AP). In 4 minutes, the reduction reaction was complete. It was shown that Pd-PtNPs facilitated electron transfer from BH4-ions to the nitro group of 4-NP when they were present on the surface of the biomass. Azo dyes such as methyl orange (MO), methyl red (MR), and acid red 14 (AR 14) were used for the reduction reaction. The absorption peaks were located at 465, 431, and 506 nm for MO, MR, and AR 14, respectively. The absorbance of azo dyes by MO, MR, and AR 14 decreased in 2, 6, and 13 minutes, respectively, when Pt-Pd alloy NPs were introduced. When PtNPs were synthesized with AQDS present, the catalytic efficiency increased, and the Kapp value was. 0065 min^{-1}, which may have been caused by a decrease in particle size (Tuo et al., 2016, Xu et al., 2019).

Based on nanoscale metal products, Feridex is a superparamagnetic iron oxide particle aqueous colloidal solution (SPION). The surface of SPION, which has a hydrodynamic diameter of up to 150 nm and a diameter of approximately 5 nm, has been modified by dextran, resulting in an increase in particle size. However, ultra-SPION particles are much smaller than 50 nm. Feridex was given FDA approval in 1996 to be used as a contrast agent for magnetic resonance imaging (MRI). The cells of the MPS directly ingest iron oxide particle fragments. Malignant transformed cells have a limited capacity to absorb iron oxide. Because they were not altered by the use of liver, MRI and spleen lesions are easier to identify than surrounding tissues. Feridex manufacturing ceased in 2008. An FDA-approved medication called ferumoxytol, formerly known as feraheme, is used to treat adult patients with iron deficiency anemia. They allow macrophages in the MPS to release iron compounds. Recently, for MRI-based cancer and other cardiovascular disease diagnostics, ferumoxytol has been studied as a novel imaging agent. As a new superparamagnetic iron oxide colloidal blood pool contrast agent, it has been used (Weissig et al., 2014).

Numerous drug-protein nanoconjugates based on albumin are currently undergoing clinical testing. Albumin has attracted much attention recently as a potential drug carrier for a number of therapeutic medications important for enhancing the medications' pharmacokinetic profiles. The drug Abraxane is based on albumin, paclitaxel-conjugated albumin NPs measuring 130 nm in size. Targeted therapeutic medications include Kadcyla and Ontak, which are recombinant fusion proteins and immunoconjugates, respectively (Patravale et al., 2012).

Nanocrystals: a special class of pharmaceuticals called nanocrystals contains only drugs that are 100% insoluble in water and do not require any additional nanocarrier systems. The Noyes-Whitney equation caused the solid particles in the aqueous media to dissolve. This equation states that the surface area is a direct function of the dissolution velocity. As a result, micronization is well known as a formulation technique for substances with limited solubility. Due to their unique characteristics, such as a higher dissolving pressure, nanocrystals stand out in the scientific community (Weissig et al., 2014).

Metal NP-based medication development and commercialization have risen significantly in recent years. More than 150 businesses have created therapeutic items based on NPs. Nevertheless, 24 medicinal medicines based on NPs have received approval for clinical use. In this liposome, the two main classes that account for more than 80% of all NP-based products currently on the market, pharmaceuticals, are encapsulated medications and polymer-drug nanoconjugates. As the need for NP-based pharmaceutical products continues to rise, the use of NP-based products in biomedicine has recently received greater consideration. Despite the fact that the most recent developments need more consideration and research to fully comprehend, NP-based drug delivery systems still need further regulatory processes before receiving final therapeutic clearances (Patravale et al., 2012). Regardless of the NP type and intended use, this technique is very expensive, adaptable, and necessary. The traditional process for clinically approving medicinal substances as well as items made with NPs is a complicated improvement. Preclinical, clinical, postmarketing, and postclinical are the three periods that these can be generically separated. Preclinical is the first step. Drug discovery, formulation development, and animal research are all included in the preclinical phase. Trials in the clinical phase are further broken down into phase I, phase II, and phase III, while the postmarketing phase is known as phase IV (Baig et al., 2016).

These stages have rules that are exclusive to a given nation. Studying the effect, toxicity, safety, and extensive data production is a prerequisite for starting a clinical trial. Phase I clinical trials are carried out to establish protection and approve pharmacokinetic parameters such as dose and dosage actions. Phase II trials are carried out to show efficacy, safety, tolerance, and toxicity, and phase III clinical trials are often used to validate this efficacy (McCarthy et al., 2005, Zhang et al., 2008). A sizable patient population is necessary for the demonstration of a formulation's purported clinical benefit. The most expensive step in the medication development process is conducting clinical trials (Bawa, 2007, Rupp et al., 2007).

5.5 PROBLEMS, DIFFICULTIES, AND SAFETY ISSUES INVOLVING METAL NANOPARTICLES CAUSED BY MICROORGANISMS

Microorganisms serve as nanofactories and offer real opportunities for MNPs to be mass produced at the nanoscale. The potential of microbial nanotechnology has been revealed by early studies in the field. Prior to its use on an industrial and commercial level, there are some obstacles that must be overcome. Controlling the size and dispersion of MNPs is one such difficulty (Grasso et al., 2020). The relationship between the particle size and dispersity has a substantial impact on the behavior and characteristics of MNPs (size distribution). For commercial applications, the MNP monodisperse population is strongly advocated rather than polydisperse MNPs. Authors have examined the relationship between NP size and anticancer capability. For instance, monodisperse nanosilica-based drug nanoconjugates demonstrated that in terms of anticancer potential, the 50 nm-based silica NP-drug conjugate was superior to the 20 nm and 200 nm silica-based drug nanoconjugates (Tang et al., 2014).

According to other researchers, smaller AgNP sizes were more successful at inducing apoptosis than larger sizes. Monodispersed MNPs can be produced using a variety of optimization techniques, including adjusting the pH, temperature, precursor salt concentration, and culture growth conditions. Similarly, downstream processing, which includes MNP extraction and purification, must be included in the optimization processes. Centrifugation, freeze-thawing, osmotic shock, and other processes can change the physical characteristics of MNPs and lead to aggregation (Grasso et al., 2020). The purification and optimization of MNPs from microbes can benefit from knowing the capping agents. Due to the vast array of metabolites that are released by microbes, it can be difficult to precisely identify the capping agent (Ovais et al., 2018). Therefore, in-depth research must be performed to precisely identify the capping agents. To enhance the production of MNPs employing microorganisms, metabolic pathway knowledge and genetic engineering can be applied. Through genetic modification of the ATP metabolic pathway in *S. cerevisiae*, increased synthesis of CdSe QDs has recently been observed (Zhang et al., 2017). Enhancing the rate of decrease, which is often slow with microbes, is another significant difficulty. Due to their complicated requirements for sustaining culture and biosafety, microorganisms are slightly more expensive than alternative approaches and necessitate the use of additional resources, including the usage of plants for biofabrication. It is also suggested to use microorganisms that are not pathogenic to people or animals.

Despite the promising preclinical outcomes of inorganic NP in cancer theranostics, questions remain regarding their toxicity, biological destiny, and safety. The small size of MNPs gives them a special capacity to pass biological barriers and interact with fluids, which changes their pharmacokinetics (Kanneganti et al., 2006). Physiological factors, including exposure route, absorption, uptake, and excretion, are crucial. The production of oxidative stress by inorganic NPs is well established. In one study, possible neurotoxicity in rat brains was shown to be caused by a drop in antioxidant enzymes following nano-gold application. Additional research has shown increased neurotransmitter production and a chemical imbalance in the brain (Siddiqi et al., 2012).

Other organ toxicities, including those to the spleen, liver, and kidneys, have also been reported. NPs can produce oxidative stress by interacting with cellular machinery after building up in organs (Dick et al., 2003; Donaldson et al., 2004). The question of whether small NPs are more harmful than their bulkier counterparts or vice versa is contradictory. Numerous papers claim that when exposed to biological fluids, NPs aggregate, alter in surface charge, and change in physicochemical characteristics. With the available research, it is currently difficult to determine whether the noxiousness is brought on by the nanomaterials themselves (Choi et al., 2013). Due to their biomolecule caps, biogenic NPs are more biocompatible than chemogenic ones, according to numerous comparison studies (Chowdhury et al., 2014). One study comparing chemogenic AgNPs and biogenic AgNPs indicated that the biogenic AgNPs were superior to the chemogenic AgNPs in their ability to selectively kill cancer cells. Additionally, compared to chemogenic AgNPs, biological AgNPs have higher hemocompatibility.

Comparative research on chemogenic biogenic GNPs revealed similar results. In comparison to chemically manufactured biogenic gold and GNPs, NPs were proven to be an additional potent anticancer drug (Kummara et al., 2016, Virmani et al., 2019). In comparison to MNPs made using other methods, biogenic MNPs are highly recommended for use in nanomedicine due to their improved biocompatibility (Abdelghany et al., 2018, Rahman et al., 2019).

5.6 CONCLUSION

A novel, safe, and environmentally acceptable method for synthesizing metal NPs has been created by the interdisciplinary field of study known as nanobiotechnology. The need to create low-cost, environmentally acceptable technologies for creating nontoxic NPs has been developed by biological processes' increasing exposure to green synthesis (Ahmad et al., 2022, Khan et al., 2022). Alternative techniques can call for the use of dangerous substances. Thus, this is an environmentally friendly, cost-effective, clean, and sustainable method of microbial nanomaterial synthesis. The most effective source and potential biofactories for the environmentally friendly production of NPs are microorganisms. Researchers have demonstrated that a variety of fungi, bacteria, and yeast have a very high ability to synthesize unique metal and metal alloy NPs with various chemical compositions, forms, sizes, and controlled monodispersity over the past three decades. This shows that the detoxification route causes bacteria to create NPs. However, there is still uncertainty about the process of metal ion bioreduction. Nevertheless, there is still room for improvement in the efficiency of synthesis and downstream processing. To create monodisperse NPs with regulated sizes, it is also necessary to better investigate the mechanisms involved in the synthesis of nanomaterials. For the formation of NPs with numerous uses in the fields of agriculture, chemistry, and health, it is crucial to examine the effects of various environmental variables and combinatorial elements. To investigate the use of nanomaterials, it is important to correctly examine the surface chemistry of biogenic NPs. The use of fungi rather than bacteria to create organic "nanofactories" has the additional benefit of making downstream processing and management of the biomass considerably easier. Recently, it has been shown that many factors, such as metal ion concentration, solution pH, and reaction temperature, can regulate the structure and size of metal NPs produced by microorganisms. Microorganisms may be used as an alternative to chemical or physical procedures for the synthesis of metal NPs (Hameed et al., 2021, Malik et al., 2022).

REFERENCES

Abdelghany, T., Al-Rajhi, A.M., Al Abboud, M.A., Alawlaqi, M., Magdah, A.G., Helmy, E.A., Mabrouk, A.S. Recent advances in green synthesis of silver nanoparticles and their applications: about future directions, a review, BioNanoScience 8 (1) (2018) 5–16.

AbdelRahim, K., Mahmoud, S.Y., Ali, A.M., Almaary, K.S., Mustafa, A.E.Z.M., Husseiny, S.M. Extracellular biosynthesis of silver nanoparticles using *Rhizopus stolonifer*. Saudi J. Biol. Sci. 24 (2017) 208–216.

Adewale, O., Davids, H., Cairncross, L., Roux, S. Toxicological behavior of gold nanoparticles on various models: Influence of physicochemical properties and other factors. Int. J. Toxicol. 38 (2019) 357–384.

Agnihotri, M. Joshi, S., Kumar, A.R., Zinjarde, S. Kulkarni, S. Biosynthesis of gold nanoparticles by the tropical marine yeast Yarrowia lipolytica NCIM 3589. Mater. Letters. 63 (15) (2009) 1231–1234.

Ahmad, A., Senapati, S., Khan, M.I., Kumar, R., Sastry, M. Extra-/intracellular biosynthesis of gold nanoparticles by an alkalotolerant fungus, *Trichothecium* sp, J. Biomed. Nanotechnol. 1 (1) (2005) 47–53.

Ahmed, A.A., Hamzah, H., Maaroof, M. Analyzing formation of silver nanoparticles from the filamentous fungus *Fusarium oxysporum* and their antimicrobial activity. Turk. J. Biol. 42 (2018a) 54–62.

Ahmed, E., Kalathil, S., Shi, L., Alharbi, O., Wang, P. Synthesis of ultra-small platinum, palladium and gold nanoparticles by *Shewanella loihica* PV-4 electrochemically active biofilms and their enhanced catalytic activities. J. Saudi Chem. Soc. 22 (2018b) 919–929.

Ahmad, M., Ali, A., Ullah, Z., Sher, H., Dai, D. Q., Ali, M., Ali, I. Biosynthesized silver nanoparticles using *Polygonatum graminifolium* efficiently control fusarium wilt disease of tomato. Frontiers in Bioengineering and Biotechnology, 10 (2022) 988607.

Ahmad, A., Senapati, S., Khan, M.I., Kumar, R., Ramani, R., Srinivas, V., Sastry, M. Intracellular synthesis of gold nanoparticles by a novel alkalotolerant actinomycete, Rhodococcus species. Nanotechnology. 14 (7) (2003) 824.

Akther, T., Mathipi, V., Kumar, N.S., Davoodbasha, M., Srinivasan, H. Fungal-mediated synthesis of pharmaceutically active silver nanoparticles and anticancer property against A549 cells through apoptosis, Environ. Sci. Pollut. Res. – Int. 26 (13) (2019) 13649–13657.

Alam, H., Khatoon, N., Khan, M.A., Husain, S.A., Saravanan, M., Sardar, M. Synthesis of selenium nanoparticles using probiotic bacteria *Lactobacillus acidophilus* and their enhanced antimicrobial activity against resistant bacteria. J. Clust. Sci. 31 (2020) 1003–1011.

Alivisatos, P. The use of nanocrystals in biological detection. Nat. Biotechnoly. 22 (1) (2004) 47–52.

Alsamhary, K.I. Eco-friendly synthesis of silver nanoparticles by *Bacillus subtilis* and their antibacterial activity. Saudi J. Biol. Sci. 27 (2020) 2185–2191.

Anwar, A., Rajendran, K., Siddiqui, R., Raza Shah, M., Khan, N.A. Clinically approved drugs against CNS diseases as potential therapeutic agents to target brain-eating amoebae. ACS Chem. Neurosci. 10 (1) 2018 658–666.

Archibald, D.D., Mann, S. Template mineralization of self-assembled anisotropic lipid microstructures. Nat. 364 (6436), 430–433 (1993).

Arivazhagan, S., Velmurugan, B., Bhuvaneswari, V., Nagini, S. Effects of aqueous extracts of garlic (*Allium sativum*) and neem (*Azadirachta indica*) leaf on hepatic and blood oxidant-antioxidant status during experimental gastric carcinogenesis. J. Med. Food. 7 (2004) 334–339.

AshaRani, P., Hande, M.P., Valiyaveettil, S. Anti-proliferative activity of silver nanoparticles. BMC Cell Biol. 10 (2009) 65.

Aziz, N., Faraz, M., Sherwani, M.A., Fatma, T., Prasad, R. Illuminating the anticancerous efficacy of a new fungal chassis for silver nanoparticle synthesis. Front. Chem. 7 (2019) 65.

Azizi, M., Ghourchian, H., Yazdian, F., Bagherifam, S., Bekhradnia, S., Nyström, B. Anti-cancerous effect of albumin coated silver nanoparticles on MDA-MB 231 human breast cancer cell line. Sci. Rep. 7 (2017) 5178.

Baig, M.W., Ali, S.M., Beg, M.S. Nanotechnology companies in energy sector and its business purview in India. Int. J. Innov. Res. Sci. Eng. Technol. 2 (10) (2016) 71–77

Bai, H., Zhang, Z., Guo, Y., Yang, G. Biosynthesis of cadmium sulfide nanoparticles by photosynthetic bacteria Rhodopseudomonas palustris. Colloids Surf. B. 70 (1) (2009) 142–146.

Bansal, V., Rautaray, D., Ahmad, A., Sastry, M. Biosynthesis of zirconia nanoparticles using the fungus *Fusarium oxysporum*. J. Mater. Chem. 14 (22) (2004) 3303–3305.

Bansal, V., Rautaray, D., Bharde, A., Ahire, K., Sanyal, A., Ahmad, A., Sastry, M. Fungus-mediated biosynthesis of silica and titania particles. J. Mater. Chem. 15 (26) (2005) 2583–2589.

Bansal, V., Syed, A., Bhargava, S. K., Ahmad, A., Sastry, M. Zirconia enrichment in zircon sand by selective fungus-mediated bioleaching of silica. Langmuir. 23 (9) (2007) 4993–4998.

Barkalina, N., Charalambous, C., Jones, C., Coward, K., Nanotechnology in reproductive medicine: Emerging applications of nanomaterials. Nanomedicine: NBM, 10 (2014) e921–e938.

Baskaran, B., Muthukumarasamy, A., Chidambaram, S., Sugumaran, A., Ramachandran, K., and Rasu Manimuthu, T. Cytotoxic potentials of biologically fabricated platinum nanoparticles from *Streptomyces* sp. on MCF-7 breast cancer cells. IET Nanobiotechnol. 11 (3) (2017) 241–246. doi:10.1049/iet-nbt.2016.0040

Bawa, R. Patents and nanomedicine. Nanomed (Lond). 2 (2007) 351–374.

Bhainsa K.C., D'Souza S.F. Extracellular biosynthesis of silver nanoparticles using the fungus Aspergillus fumigatus. Colloids Surf. B Biointerfaces. 47 (2006) 160–164.

Bharde, Rautaray, D., Bansal, V., Ahmad, A., Sarkar, I., Yusuf, S.M., Sanyal, M., Sastry, M. Extracellular biosynthesis of magnetite using fungi. Small 2 (1) (2006) 135–141.

Bonilla, J.J.A., Guerrero, D.J.P., Sáez, R.G.T., Ishida, K., Fonseca, B.B., Rozental, S., López, C.C.O. Green synthesis of silver nanoparticles using maltose and cysteine and their effect on cell wall envelope shapes and microbial growth of *Candida* spp. J. Nanosci. Nanotechnol. 17 (2017) 1729–1739.

Boroumand Moghaddam, F. Namvar, M. Moniri, S. Azizi, R. Mohamad, Nanoparticles biosynthesized by fungi and yeast: a review of their preparation, properties, and medical applications. Molecules 20 (9) (2015) 16540–16565.

Brakmane, G., Winslet, M., Seifalian, A., Systematic review: The applications of nanotechnology in gastroenterology. Aliment. Pharmacol. Ther. 36 (2012) 213–221.

Bruins, M.R., Kapil, S., Oehme, F.W. Microbial resistance to metals in the environment. Ecotoxicol. Environ. Saf. 45 (3)(2000) 198–207.

Cai, F., Li, S., Huang, H., Iqbal, J., Wang, C., Jiang, X. Green synthesis of gold nanoparticles for immune response regulation: Mechanisms, applications, and perspectives. J. Biomed. Mater. Res. A., 110(2) (2022) 424–442.

Camas, M., Camas, A.S., Kyeremeh, K. Extracellular synthesis and characterization of gold nanoparticles using *Mycobacterium* sp. BRS2A-AR2 isolated from the aerial roots of the Ghanaian mangrove plant, *Rhizophora racemosa*. Indian J. Microbiol. 58 (2018) 214–221.

Carlson, C., Hussain, M., Schrand, A.M., Braydich-Stolle, L.K., Hess, K.L., Jones, R.L., Schlager, J.J. Unique cellular interaction of silver nanoparticles: Size-dependent generation of reactive oxygen species. J. Phys. Chem. B. 112 (2008) 13608–13619.

Castro-Longoria, E., Vilchis-Nestor, A.R., Avalos-Borja, M. Biosynthesis of silver, gold and bimetallic nanoparticles using the filamentous fungus Neurospora crassa. Colloids Surf. B Biointerfaces 83 (1) (2011) 42–48.

Che, L., Dong, Y., Wu, M.; Zhao, Y., Liu, L., Zhou, H. Characterization of selenite reduction by *Lysinibacillus* sp. ZYM-1 and photocatalytic performance of biogenic selenium nanospheres. ACS Sust. Chem. Eng. 5 (2017) 2535–2543.

Chen, J., Lin, Z., Ma, X. Evidence of the production of silver nanoparticles via pretreatment of *Phoma* sp. 3.2883 with silver nitrate. Lett. Appl. Microbiol. 37 (2) (2003) 105–108.

Chen, G., Yi, B., Zeng, G., Niu, Q., Yan, M., Chen, A., Du, J., Huang, J., Zhang, Q. Facile green extracellular biosynthesis of CdS quantum dots by white rot fungus Phanerochaete chrysosporium. Colloids Surf. B Biointerfaces 117 (2014) 199–205.

Choi, S.-J., Lee, J.K., Jeong, J., Choy, J.-H. Toxicity evaluation of inorganic nanoparticles: considerations and challenges. Mol. Cell. Toxicol. 9 (3) (2013) 205–210.

Chowdhury, S., Basu, A., Kundu, S. Green synthesis of protein capped silver nanoparticles from phytopathogenic fungus Macrophomina phaseolina (Tassi) Goid with antimicrobial properties against multidrug-resistant bacteria. Nanoscale Res. Lett. 9 (1) (2014) 1–11.

Cunha, F.A., da CSO Cunha, M., da Frota, S.M., Mallmann, E.J., Freire, T.M., Costa, L.S., Paula, A.J., Menezes, E.A., Fechine, P.B. Biogenic synthesis of multifunctional silver nanoparticles from *Rhodotorula glutinis* and *Rhodotorula mucilaginosa*: Antifungal, catalytic and cytotoxicity activities. World J. Microbiol. Biotechnol. 34 (2018) 1–15.

Cunningham, D.P. Lundie, L. *Precipitation of cadmium by Clostridium thermoaceticum*. Appl. Environ. Microbiol. 59 (1) (1993) 7.

Dameron, C.T., Reese, R.N., Mehra, R.K., Kortan, A.R., Carroll, P.J., Steigerwald, M.L., Brus, L.E., Winge, D.R. Biosynthesis of cadmium sulphide quantum semiconductor crystallites. Nature 338 (6216) (1989) 596–597.

DeLouise, L.A., Applications of nanotechnology in dermatology. J. Invest. Dermatol., 132 (2012) 964–975.

Deplanche K., Caldelari I., Mikheenko I.P., Sargent F., Macaskie L.E. Involvement of hydrogenases in the formation of highly catalytic Pd(0) nanoparticles by bioreduction of Pd(II) using *Escherichia coli* mutant strains. Microbiology 156 (2010) 2630–2640.

Devi, L.S., Joshi, S. Ultrastructures of silver nanoparticles biosynthesized using endophytic fungi. J. Microsc. Ultrastruct. 3 (1) (2015) 29–37.

De Windt, W., Aelterman, P., Verstraete, W. Bioreductive deposition of palladium (0) nanoparticles on Shewanella oneidensis with catalytic activity towards reductive dechlorination of polychlorinated biphenyls. Environ. Microbiol. 7 (3) (2005) 314–325.

Dick, C.A., Brown, D.M., Donaldson, K., Stone, V. The role of free radicals in the toxic and inflammatory effects of four different ultrafine particle types. Inhal. Toxicol. 15 (1) (2003) 39–52.

Dikshit, P.K., Kumar, J., Das, A.K., Sadhu, S., Sharma, S., Singh, S., … & Kim, B.S.. Green synthesis of metallic nanoparticles: Applications and limitations. Catalysts, 11 (8) (2021) 902.

Donaldson, K., Stone, V., Tran, C., Kreyling, W., Borm, P.J. Nanotoxicology, BMJ Publishing Group Ltd, 2004.

Douglas, T., Young, M. Host–guest encapsulation of materials by assembled virus protein cages. Nature 393 (6681) (1998) 152–155.

Du, L., Jiang, H., Liu, X., Wang, E. Biosynthesis of gold nanoparticles assisted by *Escherichia coli* DH5α and its application on direct electrochemistry of hemoglobin. Electrochem. Commun. 9 (2007) 1165–1170.

Durán, N., Marcato, P.D., De Souza, D.I.H., Alves, O.L., Esposito E. Antibacterial effect of silver nanoparticles produced by fungal process on textile fabrics and their effluent treatment. J. Biomed. Nanotechnol. 3 (2007) 203–208.

Edmundson, M.C., Capeness, M., Horsfall, L. Exploring the potential of metallic nanoparticles within synthetic biology. N Biotechnol. 31(2014) 572–578.

El Domany, E.B., Essam, T.M., Ahmed, A.E., Farghali, A.A. Biosynthesis physico-chemical optimization of gold nanoparticles as anti-cancer and synergetic antimicrobial activity using *Pleurotus ostreatus* fungus. J. Appl. Pharm. Sci. 8 (2018) 119–128.

Elamawi, R.M., Al-Harbi, R.E., Hendi, A.A. Biosynthesis and characterization of silver nanoparticles using Trichoderma longibrachiatum and their effect on phytopathogenic fungi. Egypt. J. Biol. Pest Control. 28 (2018) 1–11.

Elangovan, K., Elumalai, D., Anupriya, S., Shenbhagaraman, R., Kaleena, P., Murugesan, K. Phyto mediated biogenic synthesis of silver nanoparticles using leaf extract of *Andrographis echioides* and its bio-efficacy on anticancer and antibacterial activities. J. Photochem. Photobiol. B 151 (2015) 118–124.

Elsaesser, A., Howard, C. Toxicology of nanoparticles. Adv. Drug Deliver. Rev. 64, (2012) 129–137.

Eramabadi, P., Masoudi, M., Makhdoumi, A., Mashreghi, M. Microbial cell lysate supernatant (CLS) alteration impact on platinum nanoparticles fabrication, characterization, antioxidant and antibacterial activity. Mat. Sci. Eng. C. 117 (2020) 111292.

Farina, C.R. Conceiving due process. Yale JL & Feminism. 3 (1990) 189.

Farshchi, H., Azizi, M., Jaafari, M., Nemati, S., Fotovat, A. Green synthesis of iron nanoparticles by Rosemary extract and cytotoxicity effect evaluation on cancer cell lines. Biocatal. Agric. Biotechnol. 16 (2018) 54–62.

Feng, Y., Yu, Y., Wang, Y., Lin, X. Biosorption and bioreduction of trivalent aurum by photosynthetic bacteria Rhodobacter capsulatus. Curr. Microbio. 55 (2007) 402–408.

Fernández, J.G., Fernández-Baldo, M.A., Berni, E., Camí, G., Durán, N., Raba, J., Sanz, M.I. Production of silver nanoparticles using yeasts and evaluation of their antifungal activity against phytopathogenic fungi. Process Biochem. 51 (2016) 1306–1313.

Gade, A.K., Bonde, P., Ingle, A.P., Marcato, P.D., Duran, N., Rai, M.K. Exploitation of Aspergillus niger for synthesis of silver nanoparticles. J. Biobased. Mater. Bio. 2 (3) (2008) 243–247.

Gericke, M., Pinches, A. Microbial production of gold nanoparticles. Gold Bull. 39 (1) (2006) 22–28.

Grasso, G., Zane, D., Dragone, R. Microbial nanotechnology: challenges and prospects for green biocatalytic synthesis of nanoscale materials for sensoristic and biomedical applications. Nanomaterials 10 (1) (2020) 11.

Gupta, A., Maynes, M., Silver, S. Effects of halides on plasmid-mediated silver resistance in Escherichia coli. Appl. Environ. Microbiol. 64 (1998) 5042–5045.

Gurunathan, S., Han, J.W., Eppakayala, V., Jeyaraj, M., Kim, J.-H. Cytotoxicity of biologically synthesized silver nanoparticles in MDA-MB-231 human breast cancer cells. Biomed. Res. Int. 2013 (2013) 535796.

Gurunathan, S., Lee, K.-J., Kalishwaralal, K., Sheikpranbabu, S., Vaidyanathan, R., Eom, S.H. Antiangiogenic properties of silver nanoparticles. Biomaterials. 30 (2009), 6341–6350.

Guzman, M., Dille, J., Godet, S. Synthesis and antibacterial activity of silver nanoparticles against gram-positive and gram-negative bacteria. Nanomed.: Nanotechnol. Biol. Med. 8 (2012) 37–45.

Guzmán, M., Jean, G.D., Stephan, G. Synthesis of silver nanoparticles by chemical reduction method and their antibacterial activity. Int. J. Chem. Biomol. Eng. 2 (2009) 104–111.

Hameed, S., Ali Shah, S., Iqbal, J., Numan, M., Muhammad, W., Junaid, M., & Umer, F. Cannabis sativa-mediated synthesis of gold nanoparticles and their biomedical properties. Bioinspir. Biomim. Nan. 9 (2) (2020) 95–102.

Hameed, S., Khalil, A.T., Ali, M., Iqbal, J., Rahman, L., Numan, M., Shinwari, Z.K. Precursor effects on the physical, biological, and catalytic properties of *Fagonia indica* Burm. f. mediated zinc oxide nanoparticles. Microsc. Res. Tech. 84 (12) (2021) 3087–3103.

Hamida, R.S., Abdelmeguid, N.E., Ali, M.A., Bin-Meferij, M.M., Khalil, M.I. Synthesis of silver nanoparticles using a novel cyanobacteria *Desertifilum* sp. extract: their antibacterial and cytotoxicity effects. Int. J. Nanomedicine 15 (2020) 49.

Hamouda, R.A., Hussein, M.H., Abo-elmagd, R.A., Bawazir, S.S. Synthesis and biological characterization of silver nanoparticles derived from the cyanobacterium *Oscillatoria limnetica*. Sci. Rep. 9 (1) (2019) 1–17.

He, S., Zhang, Y., Guo, Z., Gu, N. Biological synthesis of gold nanowires using extract of *Rhodopseudomonas capsulata*. Biotechnol. Progr. 24 (2008) 476–480.

Husseiny, M.I., Abd El-Aziz, M., Badr, Y., Mahmoud, M.A. Biosynthesis of gold nanoparticles using *Pseudomonas aeruginosa*. Spectrochim. Acta A Mol. Biomol. Spectrosc. 67 (2007) 1003–1006.

Ingle, A., Rai, M., Gade, A., Bawaskar, M. Fusarium solani: a novel biological agent for the extracellular synthesis of silver nanoparticles. J. Nanopart. Res. 11 (2009) 2079–2085.

Iqbal, J., Abbasi, B.A., Khalil, A.T., Ali, B., Mahmood, T., Kanwal, S., Ali, W. Dietary isoflavones, the modulator of breast carcinogenesis: Current landscape and future perspectives. Asian Pac. J. Trop. Med. 11 (3) (2018) 186–193.

Jalal, M., Ansari, M.A., Alzohairy, M.A., Ali, S.G., Khan, H.M., Almatroudi, A., Raees, K. Biosynthesis of silver nanoparticles from oropharyngeal *Candida glabrata* isolates and their antimicrobial activity against clinical strains of bacteria and fungi. Nanomaterials. 8 (2018) 586.

Jeyaraj, M., Sathishkumar, G., Sivanandhan, G., MubarakAli, D., Rajesh, M., Arun, R., Kapildev, G., Manickavasagam, M., Thajuddin, N., Premkumar, K. et al. Biogenic silver nanoparticles for cancer treatment: An experimental report. Colloids Surf. B Biointerfaces. 106 (2013) 86–92.

Jiang, H., Moon, K.-S., Lu, J. Wong, C. Conductivity enhancement of nano silver-filled conductive adhesives by particle surface functionalization. J. Electron. Mater. 34 (2005) 1432–1439.

Joerger, R., Klaus, T., Granqvist, C. Biologically produced silver–carbon composite materials for optically functional thin-film coatings. Adv. Mater. 12 (6) (2000) 407–409.

Joshi, C.G., Danagoudar, A., Poyya, J., Kudva, A.K., Dhananjaya, B.L. Biogenic synthesis of gold nanoparticles by marine endophytic fungus—*Cladosporium cladosporioides* isolated from seaweed and evaluation of their antioxidant and antimicrobial properties. Process Biochem. 63 (2017) 137–144.

Kanneganti, T.D., Ozoren, N., Body-Malapel, M., Amer, A., Park, J.H., Franchi, L., Whitfield, J., Barchet, W., Colonna, M., Vandenabeele, P., Bertin, J., Coyle, A., Grant, E.P., Akira, S., Núnez, G., Bacterial RNA and small antiviral compounds activate caspase-1 through cryopyrin/Nalp3. Nature 440 (7081) (2006) 233–236.

Kar, P.K., Murmu, S., Saha, S., Tandon, V., Acharya, K, Anthelmintic efficacy of gold nanoparticles derived from a phytopathogenic fungus, Nigrospora oryzae. PLoS One 9 (1) (2014) e84693.

Kashefi, K., Tor, J.M., Nevin, K.P., Lovley, D. R. Reductive precipitation of gold by dissimilatory Fe (III)-reducing bacteria and archaea. Appl. Environ. Microbiol. 67 (7) (2001) 3275–3279.

Katz, A., Alimova, A., Min, X., Rudolph, E., Shah, M.K. et al. Bacteria size determination by elastic light scattering. IEEE J. Sel. Top. Quantum. Electron. 9 (2) (2003) 277–287.

Khan, S., Bibi, G., Dilbar, S., Iqbal, A., Ahmad, M., Ali, A., Ali, I. Biosynthesis and characterization of iron oxide nanoparticles from Mentha spicata and screening its combating potential against Phytophthora infestans. Front. Plant Sci. 13 (2022) 1001499.

Kitching, M., Choudhary, P., Inguva, S., Guo, Y., Ramani, M., Das, S.K., Marsili, E. Fungal surface protein mediated one-pot synthesis of stable and hemocompatible gold nanoparticles. Enzym. Microb. Technol. 95 (2016) 76–84.

Klaus-Joerger, T., Joerger, R., Olsson, E., Granqvist, C.G. Bacteria as workers in the living factory: Metal-accumulating bacteria and their potential for materials science. Trends Biotechnol. 19 (2001) 15–20.

Klaus, T., Joerger, R., Olsson, E., Granqvist, C.G. Silver-based crystalline nanoparticles, microbially fabricated. Proc. Natl. Acad. Sci. USA 96 (1999) 13611–13614.

Kalishwaralal, K., Deepak, V., Ramkumarpandian, S., Nellaiah, H., Sangiliyandi, G. Extracellular biosynthesis of silver nanoparticles by the culture supernatant of *Bacillus licheniformis*. Mater. Lett. 62 (2008) 4411–4413.

Kobashigawa, J.M., Robles, C.A., Ricci, M.L.M., Carmarán, C.C. Influence of strong bases on the synthesis of silver nanoparticles (AgNPs) using the ligninolytic fungi *Trametes trogii*. Saudi J. Biol. Sci. 26 (2019) 1331–1337.

Konishi, Y., Tsukiyama, T., Tachimi, T., Saitoh, N., Nomura, T., Nagamine, S. Microbial deposition of gold nanoparticles by the metal-reducing bacterium Shewanella algae. Electrochim. Acta. 53 (2007) 186–192.

Korbekandi, H., Mohseni, S., Mardani Jouneghani, R., Pourhossein, M., Iravani, S. Biosynthesis of silver nanoparticles using *Saccharomyces cerevisiae*. Artif. Cell Nanomed. Biotechnol. 44 (2016) 235–239.

Kostyukova, D., Chung, Y. Synthesis of iron oxide nanoparticles using isobutanol. J. Nanomater. 2016 (2016) 1–9.

Kowshik, M., Ashtaputre, S., Kharrazi, S., Vogel, W., Urban, J., Kulkarni, S.K., Paknikar, K.M. Extracellular synthesis of silver nanoparticles by a silver-tolerant yeast strain MKY3. Nanotechnology 14 (1) (2002) 95.

Kowshik, M., Vogel, W., Urban, J., Kulkarni, S.K., Paknikar, K.M., Microbial synthesis of semiconductor PbS nanocrystallites. Adv. Mater. 14 (11) (2002) 815–818.

Krishnaraj C., Jagan E.G., Rajasekar S., Selvakumar P., Kalaichelvan P.T. et al. Synthesis of silver nanoparticles using Acalypha indica leaf extracts and its antibacterial activity against water borne pathogens. Colloids Surf. B Biointerfaces 76 (2010) 50–56.

Kumar, S.A., Abyaneh, M.K., Gosavi, S., Kulkarni, S.K., Pasricha, R., Ahmad, A., Khan, M., Nitrate reductase-mediated synthesis of silver nanoparticles from $AgNO_3$. Biotechnol. Lett. 29 (2007) 439–445.

Kumar, S.A., Peter, Y.-A., Nadeau, J.L. Facile biosynthesis, separation and conjugation of gold nanoparticles to doxorubicin. Nanotechnology 19 (49) (2008), 495101.

Kummara, S., Patil, M.B., Uriah, T. Synthesis, characterization, biocompatible and anticancer activity of green and chemically synthesized silver nanoparticles–a comparative study. Biomed. Pharmacother. 84 (5) (2016) 10–21.

Labrenz, M., Druschel, G.K., Thomsen-Ebert, T., Gilbert, B., Welch, S.A., Kemner, K.M., Logan, G.A., Summons, R.E., De Stasio, G., Bond, P.L.. Formation of sphalerite (ZnS) deposits in natural biofilms of sulfate-reducing bacteria. Science 290 (5497) (2000) 1744–1747.

Lee, H., Purdon, A.M., Chu, V., and Westervelt, R.M. Controlled assembly of magnetic nanoparticles from magnetotactic bacteria using microelectromagnets arrays. Nano Lett. 4 (5) (2004) 995–998.

Lengke, M., Southam, G. Bioaccumulation of gold by sulfate-reducing bacteria cultured in the presence of gold (I)-thiosulfate complex. Geochim. Cosmochim. Acta. 70 (14) (2006) 3646–3661.

Lian, S., Diko, C.S., Yan, Y., Li, Z., Zhang, H., Ma, Q., Qu, Y. Characterization of biogenic selenium nanoparticles derived from cell-free extracts of a novel yeast *Magnusiomyces ingens*. 3 Biotech. 9 (2019) 1–8.

Lima E., Guerra R., Lara V., Guzmán A. Gold nanoparticles as efficient antimicrobial agents for *Escherichia coli* and Salmonella typhi. Chem. Cent. J. 7 (2013) 11.

Li X., Xu H., Chen Z.S., Chen G. Biosynthesis of nanoparticles by microorganisms and their applications. J. Nanomater. 2011 (2011) 1–16.

Lotfy, W.A., Alkersh, B.M., Sabry, S.A., Ghozlan, H.A. Biosynthesis of silver nanoparticles by *Aspergillus terreus*: Characterization, optimization, and biological activities. Front. Bioeng. Biotechnol. 9 (2021) 633468.

Lyubchenko Y.L., Shlyakhtenko L.S. Visualization of supercoiled DNA with atomic force microscopy in situ. Proc Natl Acad Sci USA 94 (1997) 496–501.

Mafuné F., Kohno J.Y., Takeda Y., Kondow T. Formation of stable platinum nanoparticles by laser ablation in water. J. Phys. Chem. B. 107 (2003) 4218–4242.

Maheswari, R.U., Prabha, A.L., Nandagopalan, V., Anburaja, V. Green synthesis of silver nanoparticles by using rhizome extract of Dioscorea oppositifolia L. and their anti microbial activity against human pathogens. IOSR J Pharm Biol Sci 1 (2012) 38–42.

Malik, S.B., Saggu, J.I., Gul, A., Abbasi, B.A., Iqbal, J., Waris, S., Chalgham, W.. Synthesis and characterization of silver and graphene nanocomposites and their antimicrobial and photocatalytic potentials. Molecules, 27 (16) (2022) 5184.

Mann, S., Frankel, R.B., Blakemore, R.P. Structure, morphology and crystal growth of bacterial magnetite. Nature. 310 (5976) (1984) 405–407.

Mao, C., Flynn, C.E., Hayhurst, A., Sweeney, R., Qi, J., Georgiou, G., Iverson, B., Belcher, A.M. Viral assembly of oriented quantum dot nanowires. Proc. Natl. Acad. Sci. 100 (12) (2003) 6946–6951.

Matsunaga, T., Okamura, Y., Fukuda, Y., Wahyudi, A.T., Murase, Y., Takeyama, H. Complete genome sequence of the facultative anaerobic magnetotactic bacterium Magnetospirillum sp. strain AMB-1. DNA Res. 12 (3) (2005) 157–166.

McCarthy, T.D., Karellas, P., Henderson, S.A., Giannis, M., O'Keefe, D.F. Dendrimers as drugs: discovery and preclinical and clinical development of dendrimer-based mirobiocides for HIV and STI prevention. Mol. Pharm. 2 (2005) 312–318.

McFarland A.D., Van Duyne R.P. Single silver nanoparticles as real-time optical sensors with zeptomole sensitivity. Nano Lett. 3 (2003) 1057–1062.

Mishra, S.K., Tripathy, R., Wahab, S.-H., Jeong, I., Hwang, Y.-B., Yang, Y.-S., Kim, H.-S., Shin, S.-I. Yun, Microbial synthesis of gold nanoparticles using the fungus *Penicillium brevicompactum* and their cytotoxic effects against mouse mayo blast cancer C_2 C_{12} cells. Appl. Microbiol. Biotechnol. 92 (3) (2011) 617–630.

Mohamed, A.A., Fouda, A., Abdel-Rahman, M.A., Hassan, S.E.D., El-Gamal, M.S., Salem, S.S., Shaheen, T.I. Fungal strain impacts the shape, bioactivity and multifunctional properties of green synthesized zinc oxide nanoparticles. Biocatal. Agric. Biotechnol. 19 (2019) 101103.

Mohammadian, A., Habibi, R. M., Shoja, S. A. *Fusarium oxysporum* mediates photogeneration of silver nanoparticles. Scientia Iranica. 14 (4) (2007) 323–326.

Mohanpuria, P., Rana, N.K., Yadav, S.K. Biosynthesis of nanoparticles: technological concepts and future applications. J. Nanopart. Res. 10 (2008) 507–517.

Mohanraj V.J., Chen Y. Nanoparticles – A review. Trop. J. Pharm. Res. 5 (1) (2006) 561–573.

Monowar, T., Rahman, M., Bhore, S.J., Raju, G., Sathasivam, K.V. Silver nanoparticles synthesized by using the endophytic bacterium *Pantoea ananatis* are promising antimicrobial agents against multidrug resistant bacteria. Molecules. 23 (2018) 3220.

Montazer M., Shamei A., Alimohammadi F. Synthesis of nanosilver on polyamide fabric using silver/ammonia complex. Mater. Sci. Eng. C Mater. Biol. Appl. 38 (2014) 170–176.

Mouxing, F., Qingbiao, L., Daohua, S., Yinghua, L., Ning, H., Xu, D., Huixuan, W., Huang, J. Rapid preparation process of silver nanoparticles by bioreduction and their characterizations. Chin. J. Chem. Eng. 14 (1) (2006) 114–117.

Mukherjee P., Roy M., Mandal B., Dey G., Mukherjee P., Ghatak J., Tyagi A., Kale S. Green synthesis of highly stabilized nanocrystalline silver particles by a non-pathogenic and agriculturally important fungus T. asperellum. Nanotechnology. 19 (7) (2008) 075103.

Nair, Pradeep, T. Coalescence of nanoclusters and formation of submicron crystallites assisted by Lactobacillus strains. Cryst. Growth Des. 2 (4) (2002) 293–298.

Nanda, A., Majeed, S. Enhanced antibacterial efficacy of biosynthesized AgNPs from Penicillium glabrum (MTCC1985) pooled with different drugs. Int. J. Pharm. Tech. Res. 6 (2014) 217–223.

Narayanan K.B., Sakthivel N. Biological synthesis of metal nanoparticles by microbes. Adv. Colloid Interface Sci. 156 (2010) 1–13.

Omlor, A.J., Nguyen, J., Bals, R., Dinh, Q.T. Nanotechnology in respiratory medicine. Resp. Res. 16 (2015) 64.

Ovais, M., Khalil, A.T., Islam, N.U., Ahmad, I., Ayaz, M., Saravanan, M., Shinwari, Z.K., Mukherjee, S., Role of plant phytochemicals and microbial enzymes in biosynthesis of metallic nanoparticles. Appl. Microbiol. Biotechnol. 102 (16) (2018) 6799–6814.

Parikh, R.Y., Singh, S., Prasad, B., Patole, M.S., Sastry, M., Shouche, Y.S. Extracellular synthesis of crystalline silver nanoparticles and molecular evidence of silver resistance from Morganella sp.: towards understanding biochemical synthesis mechanism. Chem. Bio. Chem. 9 (9) (2008) 1415–1422.

Patravale, V., Dandekar, P., Jain, R. Nanoparticulate drug delivery: perspectives on the transition from laboratory to market. Elsevier, Oxford (2012).

Peng, X.H., Qian, X., Mao, H., Wang, A.Y., Chen, Z.G., Nie, S., Shin, D.M. Targeted magnetic iron oxide nanoparticles for tumor imaging and therapy. Int. J. Nanomed. 3 (2008) 311–321.

Philipse P., Maas, D., Maas D. Magnetic colloids from magnetotactic bacteria: chain formation and colloidal stability. Langmuir 18 (25) (2002) 9977–9984.

Pugazhenthiran, N., Anandan, S., Kathiravan, G., Prakash, N.K.U., Crawford, S., Ashokkumar, M., J. Nanopart. Res. 11 (2009) 1811. Jafar Ali.et al./Journal of Nanoanalysis No. 02, Issue 01 (2015) 32–38

Qu, Y., You, S., Zhang, X., Pei, X., Shen, W., Li, Z., Li, S., Zhang, Z. Biosynthesis of gold nanoparticles using cell-free extracts of *Magnusiomyces ingens* LH-F1 for nitrophenols reduction. Bioprocess Biosyst. Eng. 41 (2018) 359–367.

Rahman, S., Rahman, L., Khalil, A.T., Ali, N., Zia, D., Ali, M., Shinwari, Z.K. Endophyte-mediated synthesis of silver nanoparticles and their biological applications. Appl. Microbiol. Biotechnol. 103 (6) (2019) 2551–2569.

Rai, M., Deshmukh, S., Ingle, A., Gade, A. Silver nanoparticles: the powerful nanoweapon against multidrug-resistant bacteria. J. Appl. Microbial. 112 (5) (2012) 841–852.

Rao, P., Nallappan, D., Madhavi, K., Rahman, S., Jun Wei, L., Gan, S. Phytochemicals and biogenic metallic nanoparticles as anticancer agents. Oxid. Med. Cell. Longev. 2016 (1) (2016) 3685671.

Reyes, L.R., Gómez, I, Garza, M.T. Biosynthesis of cadmium sulfide nanoparticles by the fungi Fusarium sp. Int. J. Green. Nanotechnol. Biomed. 1 (2009) B90–B95.

Riddin, T.L., Gericke, M., Whiteley, C.G., Analysis of the inter-and extracellular formation of platinum nanoparticles by *Fusarium oxysporum* f. sp. *Lycopersici* using response surface methodology. Nanotechnology 17 (14) (2006) 3482.

Riddin, T.L., Gericke, M., Whiteley, C.G. Analysis of the inter-and extracellular formation of platinum nanoparticles by *Fusarium oxysporum* f. sp. *Lycopersici* using response surface methodology. Nanotechnology. 17 (2006) 3482.

Rupp, R., Rosenthal, S.L., Stanberry, L.R. VivaGel (Spl7013 Gel): a candidate dendrimer-microbicide for the prevention of HIV and HSV infection. Int. J. Nanomed. 2 (2007) 561–566.

Sanna, V., Pala, N., Sechi, M., Targeted therapy using nanotechnology: focus on cancer. Int. J. Nanomedicine. 9 (2014) 467.

Sastry, M., Ahmad, A., Islam Khan, M.I., Kumar, R. Biosynthesis of metal nanoparticles using fungi and actinomycete. Curr. Sci. 85 (2003) 162–170.

Schlinkert, P., Casals, E., Boyles, M., Tischler, U., Hornig, E., Tran, N., Zhao, J., Himly, M., Riediker, M., Oostingh, G.J. et al. The oxidative potential of differently charged silver and gold nanoparticles on three human lung epithelial cell types. J. Nanobiotechnol. 13 (1) (2015) 1–18.

Sebastian, R. Nanomedicine—The future of cancer treatment: A Review. J. Cancer Prev. Curr. Res. 8 (2017) 00265.

Seil, J.T., Webster, T.J., Antimicrobial applications of nanotechnology: Methods and literature. Int. J. Nanomedicine. 7. (2012) 2767.

Senapati, S., Ahmad, A., Khan, M.I., Sastry, M., Kumar, R. Extracellular biosynthesis of bimetallic Au–Ag alloy nanoparticles. Small. 1 (5) (2005) 517–520.

Shahverdi, R., Minaeian, S., Shahverdi, H.R., Jamalifar, H., Nohi, A.-A. Rapid synthesis of silver nanoparticles using culture supernatants of Enterobacteria: a novel biological approach. Process. Biochem. 42 (5) (2007) 919–923.

Sharma, V.K., Yngard, R.A., Lin, Y. Silver nanoparticles: Green synthesis and their antimicrobial activities. Adv. Colloid. Interface. Sci. 145 (2009) 83–96.

Sheikhloo, Z., Salouti, M., Intracellular biosynthesis of gold nanoparticles by the fungus *Penicillium chrysogenum*. Int. J. Nanosci. Nanotechnol. 7 (2) (2011) 102–105.

Shen, W., Qu, Y., Pei, X., Li, S., You, S., Wang, J., Zhang, Z., Zhou, J. Catalytic reduction of 4-nitrophenol using gold nanoparticles biosynthesized by cell-free extracts of *Aspergillus* sp. WL-Au. J. Hazard. Mater. 321 (2017) 299–306.

Siddiqi, N.J., Abdelhalim, M.A.K., El-Ansary, A.K., Alhomida, A.S., Ong, W. Identification of potential biomarkers of gold nanoparticle toxicity in rat brains. J. Neuroinflammation 9 (1) (2012) 1–7.

Singaravelu, G., Arockiamary, J., Kumar, V.G., Govindaraju, K. A novel extracellular synthesis of monodisperse gold nanoparticles using marine alga, Sargassum wightii Greville. Colloids. Surf. B. 57 (1) (2007) 97–101.

Smith, P.R., Holmes, J.D., Richardson, D.J., Russell, D.A., Sodeau, J.R. Photophysical and photochemical characterisation of bacterial semiconductor cadmium sulfide particles. J. Chem. Soc. Faraday Trans. 94 (1998) 1235–1241.

Soliman, H., Elsayed, A., Dyaa, A. Antimicrobial activity of silver nanoparticles biosynthesised by *Rhodotorula* sp. strain ATL72. Egypt. J. Basic Appl. Sci. 5 (2018) 228–233.

Sondi, I., Salopek-Sondi, B. Silver nanoparticles as antimicrobial agent: A case study on E. coli as a model for Gram-negative bacteria. J. Colloid Interface Sci. 275 (2004) 177–182.

Song J.Y., Kim B.S. Rapid biological synthesis of silver nanoparticles using plant leaf extracts. Bioprocess Biosyst. Eng. 32 (2009) 79–84.

Srinath, B.S., Namratha, K., Byrappa, K. Eco-friendly synthesis of gold nanoparticles by *Bacillus subtilis* and their environmental applications. Adv. Sci. Lett. 24 (2018) 5942–5946.

Sriramulu, M., Sumathi, S. Biosynthesis of palladium nanoparticles using *Saccharomyces cerevisiae* extract and its photocatalytic degradation behaviour. Adv. Nat. Sci. Nanosci. Nanotechnol. 9 (2018) 025018.

Subramaniyan, S.A., Sheet, S., Vinothkannan, M., Yoo, D.J., Lee, Y.S., Belal, S.A., Shim, K.S. One-pot facile synthesis of Pt nanoparticles using cultural filtrate of microgravity simulated grown P. chrysogenum and their activity on bacteria and cancer cells. J. Nanosci. Nanotechnol. 18 (2018) 3110–3125.

Suresh, A.K., Pelletier, D.A., Wang, W., Morrell-Falvey, J.L., Gu, B., Doktycz, M.J. Cytotoxicity induced by engineered silver nanocrystallites is dependent on surface coatings and cell types. Langmuir. 28 (2012) 2727–2735.

Suresh K., Prabagaran S.R., Sengupta S., Shivaji S. *Bacillus indicus* sp. nov., an arsenic-resistant bacterium isolated from an aquifer in West Bengal, India. Int J Syst Evol Microbiol 54 (2004) 1369–1375.

Suzuki, Y., Kelly, S.D., Kemner, K.M., Banfield, J. F. Nanometre-size products of uranium bioreduction. Nature. 419 (2002) 134.

Sweeney, R.Y., Mao, C., Gao, X., Burt, J.L., Belcher, A.M., Georgiou, G. Iverson, B.L. Bacterial biosynthesis of cadmium sulfide nanocrystals. Chem Biol. 11 (11) (2004) 1553–1559.

Tang, L., Yang, X., Yin, Q., Cai, K., Wang, H., Chaudhury, I., Yao, C., Zhou, Q., Kwon, M., Hartman, J.A., Investigating the optimal size of anticancer nanomedicine. Proc. Natl. Acad. Sci. 111 (43) (2014) 15344–15349.

Thakkar K.N., Mhatre S.S., Parikh R.Y. Biological synthesis of metallic nanoparticles. Nanomedicine 6 (2010) 257–262.

Thapa, R.K., Kim, J.H., Jeong, J.-H., Shin, B.S., Choi, H.-G., Yong, C.S., Kim, J.O. Silver nanoparticle-embedded graphene oxide-methotrexate for targeted cancer treatment. Colloids Surf. B. 153 2017 95–103.

Thomas, R., Janardhanan, A., Varghese, R.T., Soniya, E.V., Mathew, J., Radhakrishnan, E.K. Antibacterial properties of silver nanoparticles synthesized by marine Ochrobactrum sp. Braz. J. Microbiol. 45 (2014) 1221–1227.

Tripathi, R.M., Shrivastav, B.R., Shrivastav, A. Antibacterial and catalytic activity of biogenic gold nanoparticles synthesised by *Trichoderma harzianum*. IET Nanobiotechnol. 12 (2018) 509–513.

Tuo, Y., Liu, G., Dong, B., Yu, H., Zhou, J., Wang, J. et al. Microbial synthesis of bimetallic PdPt nanoparticles for catalytic reduction of 4- nitrophenol. Environ. Sci. Pollut. Res. Int. 24 (2016) 5249–5258. doi:10.1007/s11356-016-8276-7

Tuo, Y., Liu, G., Dong, B., Yu, H., Zhou, J., Wang, J., Jin, R. Microbial synthesis of bimetallic PdPt nanoparticles for catalytic reduction of 4-nitrophenol. Environ. Sci. Pollut. Res. 2017, 24, 5249–5258.

Uddin, S., Safdar, L.B., Iqbal, J., Yaseen, T., Laila, S., Anwar, S., Quraishi, U.M. Green synthesis of nickel oxide nanoparticles using leaf extract of Berberis balochistanica: Characterization, and diverse biological applications. Microsc. Res. Tech. 84 (9) (2021) 2004–2016.

Vala, S. Shah, R. Patel, Biogenesis of silver nanoparticles by marine-derived fungus *Aspergillus flavus* from Bhavnagar Coast, Gulf of Khambhat, India. J. Mar. Biol. Oceanogr. 3 (1) (2014) 1–3.

Vasanth, K., Ilango, K., MohanKumar, R., Agrawal, A., Dubey, G.P. Anticancer activity of Moringa oleifera mediated silver nanoparticles on human cervical carcinoma cells by apoptosis induction. Colloids Surf. B Biointerfaces. 117 (2014) 354–359.

Venugopal, K., Rather, H.A., Rajagopal, K., Shanthi, M.P., Sheriff, K., Illiyas, M., Rather, R.A., Manikandan, E., Uvarajan, S., Bhaskar, M., et al. Synthesis of silver nanoparticles (AgNPs) for anticancer activities (MCF 7 breast and A549 lung cell lines) of the crude extract of *Syzygium aromaticum*. J. Photochem. Photobiol. B 2017, 167, 282–289.

Virmani, I., Sasi, C., Priyadarshini, E., Kumar, R., Sharma, S.K. Singh, G.P. Pachwarya, R. B. Paulraj, R. Barabadi, H. Saravanan, M., Meena R. Comparative anticancer potential of biologically and chemically synthesized gold nanoparticles. J. Clust. Sci. 31 (2019) 867–876.

Wagner, E., Plank, C., Zatloukal, K., cotton, M., Birnstiel, M.L. Influenza virus hemagglutinin HA-2 N-terminal fusogenic peptides augment gene transfer by transferrin-polylysine-DNA complexes: toward a synthetic virus-like genetransfer vehicle. Proc. Natl. Acad. Sci. 89 (1992) 7934–7938.

Wang, T., Yang, L., Zhang, B., Liu, J. Extracellular biosynthesis and transformation of selenium nanoparticles and application in H2O2 biosensor. Colloids Surf. B. 80 (2010) 94–102.

Wang, W., Zhang, B., He, Z. Bioelectrochemical deposition of palladium nanoparticles as catalysts by *Shewanella oneidensis* MR-1 towards enhanced hydrogen production in microbial electrolysis cells. Electrochim. Acta. 318 (2019) 794–800.

Waszczuk, P., Barnard, T., Rice, M.C., Masel, R.I., Wieckowsky, A. A nanoparticle catalyst with superior activity for electrooxidation of formic acid. Electrochem. Commun. 4 (2002) 599–603.

Weissig, V., Pettinger, T.K., Murdock, N. Nanopharmaceuticals (part 1): products on the market. Int. J. Nanomed. 9 (2014) 4357.

Wiley, B., Herricks, T., Sun, Y., Xia, Y. Polyol synthesis of silver nanoparticles – Use of chloride and oxygen to promote the formation of single-crystal, truncated cubes and tetrahedrons. Nano Lett. 4 (2004) 1733–1739.

Wypij, M., Golinska, P., Dahm, H., Rai, M. Actinobacterial-mediated synthesis of silver nanoparticles and their activity against pathogenic bacteria. IET Nanobiotechnol. 11 (2016) 336–342.

Xu, H., Xiao, Y., Xu, M., Cui, H., Tan, L., Feng, N., et al. Microbial synthesis of Pd–Pt alloy nanoparticles using *Shewanella oneidensis* MR-1 with enhanced catalytic activity for nitrophenol and azo dyes reduction. Nanotechnology. 30 (6) (2019). 065607. doi:10.1088/1361-6528/aaf2a6

Yang, Z., Li, Z., Lu, X., He, F., Zhu, X., Ma, Y., He, R., Gao, F., Ni, W., Yi, Y. Controllable biosynthesis and properties of gold nanoplates using yeast extract. Nano Micro Lett. 9 (2017) 1–13.

Yong, P., Rowson, A.N., Farr, J.P.G., Harris, I.R., Mcaskie, L.E. Bioaccumulation of palladium by *Desulfovibrio desulfuricans*. J. Chem. Technol. Biotechnol. 55 (2002) 593–601.

Zhang, L., Gu, F.X., Chan, J.M., Wang, A.Z., Langer, R.S., Farokhzad, O.C.: Nanoparticles in medicine: the emerging nanomedicine landscape. Nat. Biotechnol. 24 (2008) 1211–1217.

Zhang, H., Li, Q., Lu, Y., Sun, D., Lin, X., Deng, X., He, N., Zheng, S. Biosorption and bioreduction of diamine silver complex by Corynebacterium. J. Chem. Technol. Biotechnol. 80 (3) (2005) 285–290.

Zhang, R., Shao, M., Han, X., Wang, C., Li, Y., Hu, B., Pang, D., Xie, Z. ATP synthesis in the energy metabolism pathway: a new perspective for manipulating CdSe quantum dots biosynthesized in *Saccharomyces cerevisiae*. Int. J. Nanomedicine 12 (2017) 3865.

Zoroddu, M., Medici, S., Ledda, A., Nurchi, V.M., Lachowicz, J.I., Peana, M. Toxicity of nanoparticles. Curr. Med. Chem. 21 (2014) 3837–3853.

6 Novel Polymer-Based Composite Materials Using Biogenic Nanoparticles

Raina Aman Qazi, Nargis Aman, Hajera Gul, Bushra Begum, Nargis Jamila, Naheed Bibi, and Nabi Ullah

6.1 INTRODUCTION

Composite materials have made significant advancements due to the continuous pursuit of improved properties and their diverse applications in various fields such as medicine, diagnosis, and energy production. Composites consist of one or more different materials or polymers that exhibit substantial differences in physical or chemical properties. They integrate multiple materials into a single device (Oréfice, Hench, and Brennan 2001). This integration allows composites to achieve tailored properties by improving or introducing specific characteristics that cannot be attained by individual homogeneous components. Polymer nanocomposite materials form when a polymer matrix combines with nanoscale reinforcement materials (Morgan, Chu, and Harris 2005). Composites can have their properties customized through factors, such as the concentration, type, shape, and size of their constituents. The research interest in polymer-nanoparticle stems from the synergistic characteristics exhibited by fillers and matrices. Polymers serve as excellent matrices for metal and metal oxide nanoparticles (Shifrina, Matveeva, and Bronstein 2019). Nanoparticles as reinforcements for polymer matrices are sought after due to their unique nanoscale properties (Umoren et al. 2022). For example, the small dimensions and significant surface area-to-volume ratio of nanoparticles result in remarkable enhancements in the thermal, mechanical, optical, electrical, and electrochemical characteristics of the resulting composites (Abdallah et al. 2020, Kadam et al. 2019, Asnag, Oraby, and Abdelghany 2019). In recent research, metallic nanoparticles have been reported to suffer from various disadvantages, including reduced mechanical strength and chemical stability (Stanković et al. 2016, Stevanović et al. 2014). Additionally, these nanoparticles are susceptible to aggregation. As a consequence, their surface area-to-volume ratio decreases, eventually leading to reduced activity. Moreover, the scope of these nanostructures is further limited by factors such as cytotoxicity,

DOI: 10.1201/9781003378563-6

genotoxicity, and the production of reactive oxygen species (ROS) (Stevanović et al. 2012). According to Stevanović and coworkers (Stevanović et al. 2014) the exposed/naked metal surface generates toxic effects. However, an effective solution to these issues involves integrating biogenic metallic nanoparticles into the polymeric matrix (Du et al. 2018, Gong, Luo, and Pan 2019, Lin et al. 2017). Numerous studies have demonstrated that encapsulating metallic nanoparticles within the polymeric matrix results in significant antibacterial efficacy (Du et al. 2018, Huh and Kwon 2011, Leng et al. 2018, Stanković et al. 2016). Moreover, metal-polymer composites have exhibited improved adsorption capacities, leading to effective dye removal (Du et al. 2018, Leng et al. 2018). There is a current trend of increasing interest in biogenic nanoparticles sourced from natural origins such as plants, fungi, bacteria, and other bio-organisms. These nanocomposites have been widely applied across diverse fields encompassing medicine (Rónavári et al. 2021), and environment (Huang et al. 2016), including water and wastewater treatment (Talib, Challab, and Hussein Alhameedawi 2021) to food packaging (Vera et al. 2016).

The integration of biogenic nanoparticles with polymers creates synergistic effects in composite materials, combining inherent nanoparticle properties with the processability and flexibility of polymers for unique attributes (Abutalib 2019, Dong et al. 2008). Figure 6.1 shows the properties and applications of polymer-based composite materials using biogenic nanoparticles. Biogenic nanoparticles greatly improve mechanical properties like strength, toughness, and wear resistance and enhance their ability to withstand mechanical stresses (Kadam et al. 2019).

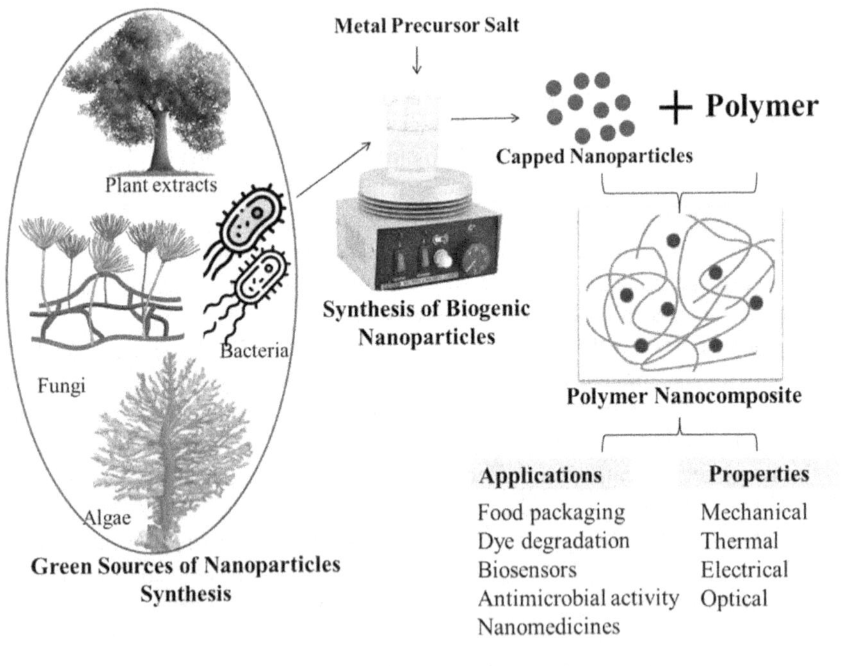

FIGURE 6.1 Properties and applications of polymer-based composite materials using biogenic nanoparticles.

Additionally, when incorporated in polymer matrices, these nanoparticles impart inherent thermal stability to the composites, making them resistant to heat and thermal cycling, ideal for high-temperature applications (Abdallah et al. 2020, Kadam et al. 2019). Certain biogenic nanoparticles display electrical conductivity or interact with electromagnetic fields, offering promising opportunities for developing conductive polymer composites and applications in electronics and sensing devices (Asnag, Oraby, and Abdelghany 2019, Pandey, Goswami, and Nanda 2013).

Biogenic nanoparticles possess functional groups on their surfaces, allowing for tailored surface properties in composite materials and expanding their applications in coatings and sensors (Ahmad et al. 2017, Virkutyte and Varma 2011). Additionally, their use in polymer composites aligns with sustainable and eco-friendly approaches, derived from renewable natural sources, reducing reliance on synthetic materials (Kausar 2021). Moreover, these nanoparticles are inherently biocompatible making them suitable for biomedical applications, enabling use in medical devices, tissue engineering, and drug delivery (Raza et al. 2021).

6.2 IMPACT OF CHEMICALLY SYNTHESIZED POLYMER MATRICES ON NANOPARTICLE PROPERTIES

The chemical, physical, and biological properties of nanoparticles are influenced by factors such as the experimental conditions for the reaction, synthesis pathway, and the type of stabilizing and/or reducing agent used (Ahmadi and Ahmad 2020, Behera and Ram 2014). Chemically synthesized polymers are often used as capping agents because they can provide steric hindrance and electrostatic repulsion, preventing nanoparticles from aggregating (Fayaz et al. 2010, Tao, Habas, and Yang 2008). Nanoparticles have been combined into various polymers matrices, including poly(methyl methacrylate) (Philip et al. 2021, Siddiqui et al. 2018), polyvinyl alcohol (PVA) (Aktürk et al. 2019, Wei et al. 2021), polylactic acid (Alippilakkotte, Kumar, and Sreejith 2017, Zhang et al. 2020), polyurethanes (Jafari et al. 2018, Jatoi 2020), polysiloxanes (Ditaranto et al. 2015), polytetrafluoroethylene (Sportelli, Picca, and Cioffi 2016) and poly(lactic-co-glycolic acid) (PLGA) (Zare et al. 2019). In a specific study (Stevanović et al. 2015), biogenic selenium nanoparticles (SeNPs) were combined and immobilized in spherical PLGA particles. SeNPs-PLGA nanocomposites were utilized in bioactive glass-based scaffolds. Wilson et al. (Wilson et al. 2017) investigated the use of poly(2-ethyl-2-oxazoline) and poly(2-oxazoline) coated particles. By employing the strategies of poly(2-oxazoline) or its derivative grafting to inorganic particles, researchers successfully prepared core-shell hybrid nanoparticles. For instance, silica nanoparticles were grafted covalently with a poly(2-ethyl-2-oxazoline) shell (Wilson et al. 2017).

6.3 IMPACT OF BIOGENIC POLYMER MATRICES ON NANOPARTICLE PROPERTIES

Natural polymer-based materials have shown potential as organic supports for inorganic nanostructures (Jamróz, Kulawik, and Kopel 2019). The issue of non-degradable synthetic polymer waste is escalating, posing an increasingly serious environmental problem. The development of alternative natural solutions is of significant interest in addressing this issue. The growing emphasis on biodegradable materials, particularly biopolymers derived from natural sources, has become increasingly important in various industries due to the benefits they offer for both the environment and sustainable resource management. These biopolymers offer a promising avenue for creating environmentally friendly materials and reducing the impact of non-degradable waste (John and Thomas 2008). Biopolymers have demonstrated their efficacy as favorable host matrices for a variety of nanostructures. Biopolymer-based nanocomposites are a compelling and promising field of study (Jamróz, Kulawik, and Kopel 2019). Biocomposites incorporating polymers from renewable resources and biogenic inorganic fillers are eco-friendly and efficient green materials (Khan et al. 2021, Trache et al. 2016). Moreover, biopolymers demonstrate high efficiency in stabilizing nanoparticles. Currently, chitosan (CS), a cationic polysaccharide that ranks second in terms of abundance in nature, is the extensively investigated biopolymer for nanoparticle incorporation (Frank et al. 2020).

6.4 SYNTHESIS STRATEGIES FOR POLYMER COMPOSITES REINFORCED WITH BIOGENIC NANOPARTICLES

There are various methods for the preparation of polymer-based composite materials using biogenic nanoparticles (Figure 6.2). Some of the most commonly used important methods are discussed below.

6.4.1 Solution Mixing

Solution mixing employs a solvent system in which the prepolymer or polymer is soluble and the nanofillers (for stacked nanofillers) remain expandable (Nathawat et al. 2023). Polymer composite preparation through solution mixing requires intense agitation to disperse nanofiller in a polymer solution, controlled solvent evaporation, and composite film casting. Shear mixing, magnetic stirring, reflux, or sonication can all serve as effective methods to agitate and mix the components. For example, Lalitha et al. (Ediyilyam et al. 2022) synthesize biogenic silver nanoparticles (AgNps) utilizing the medicinal plant *Mussaenda frondosa* and then using a solution mixing technique for the synthesis of CS-AgNPs composite scaffolds. The composite materials have a wide range of uses, particularly in biomedical fields like bone regeneration, tissue engineering, and wound treatment. The low viscosity of the solution, which enables particle dispersion, is an advantage of solution mixing (Rane et al. 2018). This method has been employed in numerous studies to provide a high level of contact and dispersion between the nanoparticles and polymer matrices.

6.4.2 Melt Mixing

The process of "melt mixing", also known as "melt blending", involves mechanically shearing nanoparticles into the molten polymer matrix (Brandenburg et al. 2017, Das, Bhardwaj, and Pandey 2021). It is the easiest and most common method for producing different polymer hybrids with inorganic nanoparticles (Kango et al. 2015).

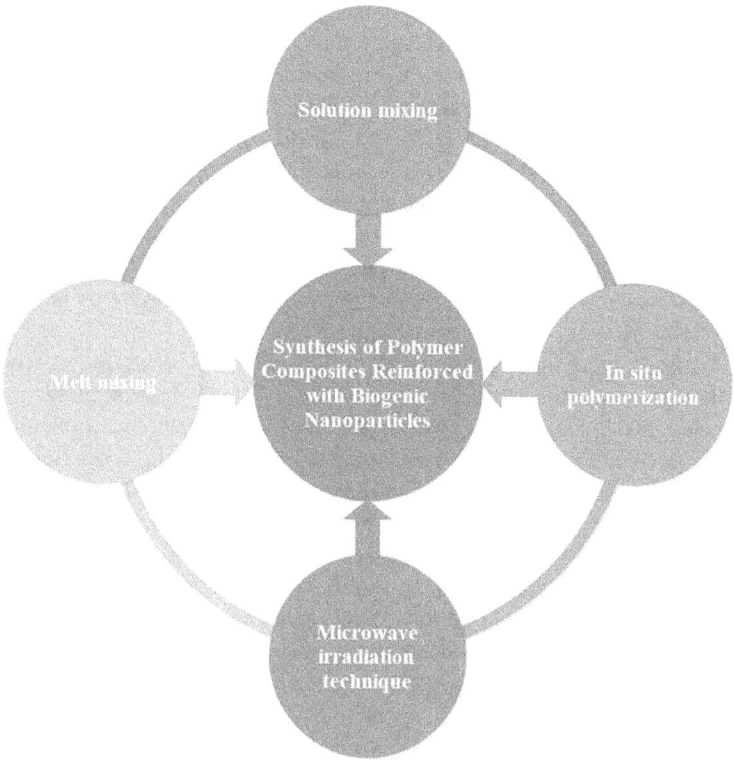

FIGURE 6.2 Synthesis of polymer composites reinforced with biogenic nanoparticles.

An inert gas, such as neon, nitrogen, or argon, is used during the melt blending process (Rane et al. 2018). Alternately, the polymer and intercalant may be dry mixed to achieve the required nanocomposites. Subsequently, the resulting mixture is heated in a mixer and sheared to the required degree. Due to the absence of organic solvents, melt blending is eco-friendly. Melt mixing aligns with contemporary industrial processes, contributing to the rising popularity of this technique for potential industrial applications. Chen et al. doped AgNPs into CS using melt blending route (Chen et al. 2020). According to their results, AgNPs-CS/polypropylene nanocomposite has the potential to be used in a variety of disciplines and is a promising material for biomedical and industrial applications. It also has outstanding mechanical and antibacterial properties. The method is especially useful for synthesis which involves some insoluble polymeric materials and cannot be processed using the solution mixing and other method.

6.4.3 *In Situ* Polymerization

In situ polymerization is the process of producing polymer chains while nanomaterials are present in the same environment (Nguyen and Yang 2018). *In situ* polymerization initiates through various methods, such as applying heat, radiation, initiator diffusion, or introducing an organic initiator or catalyst. In this method, the nanoparticles are better dispersed and more uniformly distributed within the polymer matrix. As a result, *in situ* polymerization has been extensively employed over the past decades for the preparation of nanocomposite materials with excellent dispersion and distribution of nanoparticles. The direct integration of well-dispersed nanoparticles in bulk polymer composites contributes to the improved performance and properties of these nanocomposite materials (Nguyen and Yang 2018). *In situ* polymerization can be carried out in either a solvent-free system (bulk phase) or a solvent-based system, which includes different types such as purely solvent-, emulsion-, or suspension-based systems. The solvent-based systems may involve aqueous phases as well.

This versatile method of polymerization is widely used for the synthesis of polymer-based composite materials, particularly when incorporating biogenic nanoparticles. The *in situ* polymerization process allows for the direct integration of biogenic nanoparticles into the polymer matrix, resulting in the formation of composite materials with enhanced properties and potential applications in various fields.

In a study, biogenic AgNPs were synthesized using the leaf extract of the fireball lily. The leaf extract of the fireball lily served as a natural and eco-friendly reducing and stabilizing agent for the formation of silver nanoparticles (Mane et al. 2020). Then the polypyrrole nanotubes (PPy-NTs) were synthesized through *in situ* chemical oxidative polymerization and decorated with biogenically synthesized AgNPs. The resulting nanocomposite material showed promise for the photodegradation of methyl violet dye, indicating its potential as a photocatalyst for environmental applications. This study has opened new avenues for the development of other polymer-metal nanocomposites with enhanced photocatalytic activity. By combining the unique properties of polymers and biogenic AgNPs, researchers can explore various nanocomposite materials using *in situ* polymerization method for their applications in the fields of environmental remediation, catalysis, and more.

6.4.4 Microwave Irradiation Technique

The microwave-assisted method offers significant advantages compared to conventional heating methods. Some of the key benefits include the following (Fawaz and Mittal 2014, Kempe, Becer, and Schubert 2011):

i. *Rapid Volumetric Heating*: Microwaves penetrate and heat materials volumetrically. The entire reaction volume is heated simultaneously, resulting in faster and more uniform heating than conventional methods. This efficiency leads to reduced reaction times and enhanced overall performance.
ii. *High Reaction Rate*: Due to efficient volumetric heating, chemical reactions can occur at accelerated rates in a microwave-assisted process. This can lead to quicker reaction kinetics and higher yields, making it particularly beneficial for time-sensitive or high-throughput applications.
iii. *Enhanced Reaction Selectivity*: Microwave heating can lead to better selectivity in certain reactions, as it can specifically activate certain chemical bonds or functional groups. This can result in cleaner and more precise reactions with fewer unwanted byproducts.

iv. *Energy-Saving Behavior*: Microwave-assisted methods can be more energy-efficient compared to conventional heating methods, especially in certain reactions that require high temperatures or prolonged reaction times. The rapid heating reduces the overall energy consumption and makes the process more environmentally friendly.
v. *Improved Scalability*: Microwave-assisted reactions are easily scalable and can be adapted for both laboratory-scale and industrial-scale processes. This versatility makes them suitable for various applications in research and manufacturing.

The integration of biogenic nanoparticles into polymer matrices using microwave irradiation opens up new possibilities for various useful applications. The resulting polymer composites can exhibit unique properties and functionalities derived from the combination of the polymer matrix and the biogenic nanoparticles. These composite materials have great potential in a wide range of applications, such as in the fields of medicine, electronics, catalysis, and environmental remediation. Raza et al. (Raza et al. 2021) utilize fungal biomass for synthesis of biogenic AgNPs. The pre-synthesized AgNPs were employed for fabrication of a bipolymer-based nanocomposite system, through microwave irradiation technique. The resulting nanocomposite material exhibited promising applications in nanomedicines or nanocarrier systems. Furthermore, a microwave-assisted synthesis method was employed to create a biocompatible CS/PVA/AgNPs gel specifically intended for wound dressing applications (Hiep et al. 2016) owing to its antibacterial activity. Overall, these findings highlight the potential of microwave-assisted techniques in producing advanced nanocomposite materials with enhanced antibacterial properties, which could be used in various medical applications, including wound dressing and nanomedicines.

6.5 PROPERTIES AND APPLICATIONS OF BIOGENIC NANOPARTICLES/POLYMER COMPOSITES

6.5.1 Mechanical Properties

The mechanical characteristics pertain to the load-bearing attributes exhibited by polymer nanocomposites are influenced by their interphase properties. The addition of metal nanoparticles to a polymer matrix improves mechanical properties like stiffness, tensile modulus, and tensile strength (TS) through reinforcement. The effectiveness of load transfer across the metal-polymer interface relies on the interfacial interactions between the polymer matrix and nanoparticles (Pramanik and Das 2019). Previous reports have highlighted that the mechanical properties are influenced by factors e.g. shape, and size of the nanomaterial, the nature (synthetic/biogenic) of a polymer matrix, and the interactions among its components. The elongation (% E) and TS values represent the maximum elongation and stress that polymeric composite film can withstand while maintaining its structural integrity (Liang et al. 2017). Kadam and co-authors (Kadam et al. 2019) successfully synthesize biogenic AgNPs utilizing *Nigella sativa* extract and performed a comprehensive analysis of the mechanical properties of AgNPs/CS nanocomposites. The incorporation of AgNPs led to improvements in the TS and % E of the films by

8.4%–22.5% and 4.4%–9.2%, respectively, with variations in the concentration of AgNPs. The results indicated a robust molecular-level interaction between AgNPs and CS chains, facilitated by the formation of Hydrogen bonds and ester linkages. This interaction resulted in the production of films with enhanced strength, rigidity, and significant stretching capability (Ahmed, Mulla, and Arfat 2016). The increasing concentration of nanoparticle in the casting solution cause an enhancement in the TS (Yang et al. 2007). Nevertheless, an excessive concentration of nanoparticles may result in aggregation and hinder their uniform dispersion on the polymeric support, ultimately compromising the mechanical strength (Goswami et al. 2018). Similarly, incorporating biogenic ZnO-NPs into the polymeric PLGA matrix further enhanced both the chemical stability and mechanical properties of the material. The polymeric matrix served as a protective environment for the ZnO-NPs, shielding them from deactivation and preventing undesired agglomeration (Parmar et al. 2019). The addition of biogenic Pd-NPs in the modified polysulfone membrane leads to an improvement in TS with increasing nanoparticle concentration in the casting solution (Goswami et al. 2018). However, an excessive concentration of nanoparticles can cause aggregation, hindering their uniform dispersion on the polymeric support, and ultimately weakening the mechanical strength. Plasticizers, such as glycerol, are frequently employed to modify the mechanical properties of a composite film (Kumari et al. 2016, Laohakunjit and Noomhorm 2004). Plasticizers work by reducing intermolecular interactions between neighboring polymeric chains, thereby enhancing film flexibility and facilitating better dispersion of nanoparticles within the matrix (Lavorgna et al. 2010). Mohseni MS et al. enhanced the mechanical properties of thermoplastic maize starch films by incorporating AgNPs, StNPs (starch nanoparticles), and plasticizers in Ag-starch nanocomposites (Mohseni et al. 2020).

6.5.2 Thermal Properties

The response of a material to being heated or cooled (or, in certain situations, kept isothermally) is measured by a set of measuring techniques known as thermal analysis (Prime et al. 2009). The objective is to establish a relationship between temperature and particular material characteristics. Understanding the thermal stability of polymeric composite materials is crucial because it is essential for a variety of applications. CS-based nanocomposite films with biogenic AgNPs (*N. sativa* extract-mediated) displayed higher melting temperatures than pure CS films due to the formation of stronger chemical bonds between the AgNPs and the CS polymer matrix. The study also found that the films showed improved thermal stability, which can be attributed to their higher degree of crystallinity, requiring more energy to disrupt the stronger crystalline arrangement (Kadam et al. 2019). The presence of AgNPs in flaxseed protein-alginate films had a significant effect on their thermal degradation behavior, which can be attributed to the incorporation of more heat-stable metallic silver (Dash et al. 2021). Goswami et al. (Goswami et al. 2018) studied the thermal behavior of biogenic synthesized Pd-NPs incorporated antifouling polymeric membrane. They found that adding biogenic Pd-NPs does not significantly improve the thermal stability of polymeric membranes. Another study examined the thermal stability of nanocomposites consisting of biogenic AgNPs (mediated by CS) loaded into a CS matrix (Chen et al. 2020). The study found that biogenic Ag nanoparticle-loaded CS-based membranes degrade significantly more quickly than pure CS does. The following factors could be used to explain this phenomenon: (1) Ag nanoparticles have a higher thermal conductivity than CS, which allows heat to be transferred more effectively in the polymer matrix; and (2) the insertion of Ag nanoparticles could cause intra- and intermolecular hydrogen bonds of CS to break down. Abdallah OM et al. (Abdallah et al. 2020) synthesized AgNPs using *Enterobacter cloacae* and prepared nanocomposites (PVA-Ag and CS-Ag). Thermogravimetric analysis (TGA) showed improved thermal stability and delayed thermal decomposition in both nanocomposites compared to their pure polymers. However, CS-Ag exhibited higher thermal stability than PVA-Ag, attributed to its higher AgNPs concentration (Ali, Rajendran, and Joshi 2011). The higher concentration of AgNPs hindered the mobility of PVA and CS chains in the polymer matrix, leading to a delayed degradation process. Consequently, the nanocomposite degraded at higher temperatures compared to the pure polymer (Loo et al. 2014). Atomic absorption spectroscopy analysis and standard curve measurements showed that in the PVA-Ag nanocomposite, the concentration of AgNPs per 6 mm disc was found to be 14 ± 2 μg, while in the CS-Ag nanocomposite, it was 21 ± 3 μg. This indicates that the two nanocomposites had different amounts of AgNPs present in them. The variation in the concentration of AgNPs can be attributed to two factors: the type of polymer used and the ability of the nanoparticles to interact and integrate into the polymer matrix. Different polymers may have varying affinities for nanoparticles, leading to different levels of nanoparticle incorporation. Additionally, the ability of nanoparticles to interact with the polymer matrix can affect their dispersion and concentration within the material. Overall, these findings suggest that the choice of polymer and its compatibility with the nanoparticles are crucial factors in determining the concentration of nanoparticles in the nanocomposite material. These variations in nanoparticle concentration can have implications for tailoring the properties and performance of nanocomposite materials for specific applications.

6.5.3 Electrical and Optical Properties

The inclusion of nanoparticles resulted in a significant enhancement of the electrical conductivity of the composite. This improvement is attributed to the high carrier density and electron mobility of the nanoparticles. Asnag GM et al. (Asnag, Oraby, and Abdelghany 2019) incorporated *Ricinus communis* extract-synthesized AuNPs at varying doping levels into modified carboxymethyl cellulose (CMC) for polymeric film preparation using the casting method. They demonstrated that the DC conduction of the films improved as the AuNPs content increased, ranging from 9.86×10^{-12} to 4.18×10^{-9} Scm^{-1} at 353 K. Moreover, with an increase in AuNPs content, the optical energy gap (Eg) decreased. The reduction in Eg values is due to defects generated upon adding AuNPs and the interaction between CMC and AuNPs, forming charge transfer complexes. In a study rice powder was used to synthesize graphene quantum dots, which were then doped with polyacrylonitrile and polyaniline to create polymer-based nanocomposites. The electrical conductivity of the higher-loading graphene quantum dot composites showed a gradual increase from 1.090×10^7 S m^{-1} to 2.362×10^{-6} S m^{-1} at room temperature. The fabricated composites exhibited promising supercapacitor applications. During 5000 cycles of testing, the specific capacitance of the composites remained at approximately 86.9%, indicating good cycling stability and retention of their energy storage capacity over a prolonged period (Arthisree and Madhuri 2020). CS/polyacrylamide nanocomposites with biogenic SeNPs were prepared using a casting technique. The addition of SeNPs improved the AC conductivity and complex permittivity of the nanocomposite samples. The pure blend initially had Egs of 4.98 eV (indirect) and 5.18 eV (direct). When SeNPs were added, these gaps decreased to 2.55 eV (indirect) and 2.78 eV (direct) in the nanocomposite sample with the highest SeNPs concentration. This decrease happened because SeNPs affected the polymer's crystallinity, leading to changes in its structure. The CS/PAM blend has wide band gaps, allowing for the engineering of energy levels. The embedded SeNPs created new energy levels between the valence and conduction bands, enabling optical transitions at lower energy values. This improvement could be attributed to the high density of the polymeric films, which reduced interatomic spacing and increased packing density and molecular weight distributions. This enhancement in properties makes the material promising for energy storage device applications (Gami et al. 2022).

6.5.4 Food Packaging

Food packaging is vital for preserving food quality. With increased awareness of disease transmission through food, there's a rising demand for antibacterial packaging that prevents contamination without compromising stability (Park and Zhao 2004). Proper food packaging should prevent contamination, control moisture, and act as a barrier against the diffusion of water vapor, carbon dioxide, oxygen, and volatile compounds. A preferred food packaging material should display remarkable thermal, strong mechanical and antimicrobial properties, chemical stability, and can maintain its shape. Additionally, it is advantageous if the material is both recyclable and biodegradable, promoting environmental sustainability (Moreno et al. 2019). Ojha N et al. (Ojha and Das 2020) synthesized biogenic silica nanoparticles (bSNPs) from groundnut shells and used them to fabricate biodegradable poly(3-hydroxybutyrate-co-3-hydroxyvalerate/silicon dioxide) (PHBV/SiO_2) nanocomposites through a simple solution casting method. The incorporation of bSNPs led to significant improvements in the mechanical, heat resistance, thermal, and barrier properties of the PHBV matrix, owing to their uniform dispersion and strong interfacial adhesion facilitated by hydrogen bonding interactions. These PHBV/SiO_2 nanocomposites exhibited biocompatibility, biodegradability, and antimicrobial properties, effectively reducing the growth of *Escherichia coli* and *Staphylococcus aureus*. Consequently, nanocomposites are promising in extending the shelf-life of food and enhancing its overall quality. The bactericidal effect was attributed to the severe damage to the bacterial outer membrane and disruption of cellular integrity upon contact with bSNPs. The water-repellent and positively charged nature of bSNPs played a crucial role in the bacterial eradication process (Chitra and Annadurai 2013). AgNP/polymer nanocomposites have demonstrated antimicrobial activity against common food contaminants such as *E. coli, Aspergillus niger,*

Candida albicans, and S. aureus (Kumar and Münstedt 2005, Kumar, Howdle, and Münstedt 2005, Sánchez-Valdes et al. 2009). Tripathi RM et al. (Tripathi et al. 2018) demonstrated similar outcomes of a biodegradable PVA/biogenic-AgNPs film as an antibacterial packaging material. Marrez et al. (Marrez, Abdelhamid, and Darwesh 2019) designed an eco-friendly cellulose acetate and biosynthesized AgNPs for antibacterial food packaging. A strong antibacterial functionality of the fabricated non-toxic cellulose acetate-AgNPs nanocomposite was found against *S. aureus, Bacillus cereus, Salmonella typhi, E. coli, and Klebsiella pneumoniae.* Natural biopolymers e.g. CS, gelatin (GL), agar, starch, and zein are commonly utilized as eco-friendly alternatives to synthetic plastic food packaging (, Kumar, Mitra, and Halder 2017, Kumar et al. 2019, Moreno et al. 2019, Sharma, Malik, and Jain 2018). According to scientific research (Tripathi et al. 2018), CS exhibits advantageous mechanical properties; however, it does suffer from a propensity to absorb moisture and lacks adequate barrier characteristics. To address these limitations, scientists have explored methods to augment its mechanical strength and barrier performance. To improve its performance, researchers blend CS with other biopolymers and reinforce it with nanomaterials, leading to promising CS-based nanocomposite films for biodegradable food packaging (Kumar, Mukherjee, and Dutta 2020, Kumar et al. 2019, Moreno et al. 2019, Priyadarshi and Rhim 2020). Biogenic ZnO NPs are favored for food and biomedical applications (Bandeira et al. 2020, Chakraborty et al. 2020). Kumar S et al. (Kumar et al. 2020) developed biodegradable films containing CS, GL, and biogenic ZnO-NPs. These nanocomposite films demonstrated significant antimicrobial activity against *E. coli* as compared to *S. aureus* (Figure 6.3). These amine groups bind to negatively charged microbial cell membranes, resulting in membrane damage, which, in turn, hampers nutrient intake and causes leakage of cytoplasmic fluid (Al-Naamani, Dutta, and Dobretsov 2018). The antimicrobial activity of the CS and GL matrix was further enhanced by incorporating ZnO-NPs (Munnawar et al. 2017). The primary antimicrobial mechanism of ZnO-NPs is the release of Zn^{2+} ions, which have a strong affinity for the negatively charged bacterial cell wall, leading to membrane damage (Król et al. 2017). Another important mechanism by which ZnO-NPs

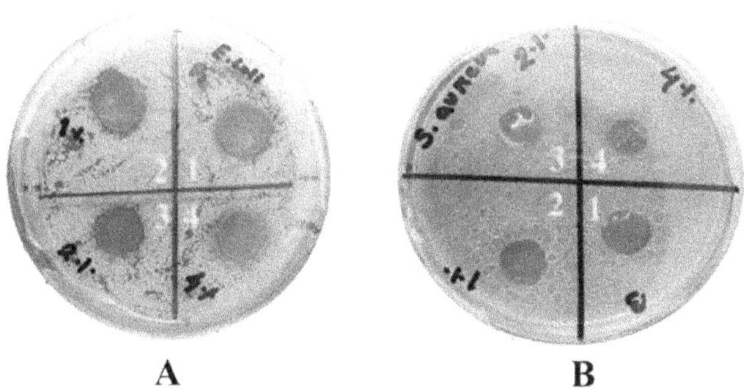

FIGURE 6.3 Bactericidal activity of (1) CS/GL, (2) CS/GL/ZnO NPs (1%), (3) CS/GL/ZnO-NPs (2%), and, (4) CS/GL/ZNO-NPs (4%), (a) against *Escherichia coli*, (b) against *Staphylococcus aureus*, after 24 hours of incubation (Kumar et al. 2020).

exert their antimicrobial properties is through the generation of ROS. These ROS cause oxidative stress within the microbial cells and interact with various cellular components, such as proteins, DNA, enzymes, and lipids. This interaction hampers cell growth and may ultimately lead to cell death (Qi et al. 2017).

Moreover, Ediyilyam et al. (Ediyilyam et al. 2021) synthesized biodegradable food packaging films using CS-GL-AgNPs composites. The incorporation of AgNPs (obtained using *M. frondosa* leaves extract) with the polymer matrix reduced swelling extent, vapor transmission rate, and moisture retention capability, while magnificently increasing the antimicrobial activity of CS-GL-AgNPs composite. Pandian et al. (Pandian et al. 2023) prepared a nanocomposite film by using *Azadirachta indica* leaf extract-assisted AgNPs, starch, protein, and microcrystalline cellulose (MCC) derived from sugarcane bagasse (AgNPs/MCC/starch/whey protein). The prepared composite material exhibited excellent antibacterial activity with improved life of food up to 15 days. Agar/copper oxide (CuO) nanocomposite-based food packaging films offer a strong UV barrier property along with bactericidal activity (Roy and Rhim 2019). Considering the above discussion, such biodegradable and safe food packaging substances based on biosynthesized metal nanoparticles and polymer composites are not only potentially antimicrobial packing materials but also better alternatives for toxic conventional food packaging. These can be considerably used for protecting food for a long time.

6.5.5 Dye Degradation

Wastewater pollution due to various organic and inorganic toxic materials needs to be cured owing to their potentially high risk to human and aquatic systems. The integrated biogenic nanoparticles in a polymeric matrix are the focus of research interest in the field of wastewater remediation (Krishnaswamy et al. 2020, Ltaief, Jabli, and Abdessalem 2021). Embedding of biogenic nanoparticles into polymer improves the structural, mechanical, optical, and surface properties, consequently leading to better adsorption and degradation capabilities (Mane et al. 2020, Parmar et al. 2019). Mane et al. (2020) have extensively investigated the photocatalytic properties of biogenic AgNPs (obtained using *Scadoxus multiflorus* leaf extract) decorated PPy-NTs for the methyl orange dye. The composite was found to exhibit better photocatalytic degradation efficiency of 88.78% as compared to that of bare AgNPs (39.69%). Likewise, Ltaief et al. (Ltaief, Jabli, and Abdessalem 2021) used *Pistacia vera* hull extract synthesized CuO and CS for the fabrication of an advanced and eco-friendly composite. Their fabricated CuO-CS composite was simple to synthesize, easy to recover, and efficient for the degradation of naphthol blue black with 86% of efficiency.

In this sense, Krishnaswamy et al. (Krishnaswamy et al. 2020) thoroughly studied the synthesis and photocatalytic activity of ZnO and sodium iodide (NaI) integrated PPy (ZnO/NaI-PPy) nanocomposite. ZnO was obtained from cucumber fruit pulp extract. They reported not only an improved conductivity (1.85×10^{-3} S/cm), small particle size (100 nm), and low optical bandgap (2.16 eV) for ZnO/NaI-PPy nanohybrid but a higher photocatalytic efficiency of 87% for Rhodamine 6G dye as well. Additionally, CuO nanotubes, biosynthesized from *Robinia pseudoacacia* leaf extract, embedded in aspartic acid cross-linked poly (vinyl alcohol) (PVA/CuO/Asp) composite were designed by Yamanappagouda et al. (Amaregouda, Gasti, and Kamanna 2022). PVA/CuO/Asp presented 77% degradation efficiency for the target methylene blue dye in small time duration of 50 min. Similarly, an incessant attempt was made by Parmar and coworkers (Parmar et al. 2019) for the photocatalytic degradation of malachite green dye and antibacterial activity against gram (+) ve (*S. aureus*) and gram (−) ve (*E. coli*) strain. They prepared a composite based on biogenic ZnO-NPs (obtained from *A. indica* leaves), and poly (DL-lactide-co-glycolide) with remarkable optical, structural, morphological, photocatalytic, and antibacterial properties. AgNPs were synthesize using curcumin powder. The synthesized AgNPs were decorated on δ-FeOOH/PUF. The fabricated nanocomposite led to the degradation of congo red and methyl orange dyes with catalytic efficiency up to 90% (Mahmoud et al. 2021). Henceforth, the fabrication and use of biogenic metal nanoparticles incorporated polymer nanocomposites with enhanced features, low-cost, and environment-friendly nature is a potent solution for combating wastewater contamination and bacterial infractions.

6.5.6 Biosensors

Integrating nanoparticles and polymers in biosensor development significantly enhances performance and sensitivity (Moulahoum et al. 2022, Noah and Ndangili 2022). A study demonstrates the successful fabrication of an optical sensor based on *Cyamopsis tetragonoloba* gum-mediated synthesis of AgNPs and guar gum which detected 1 ppm of aqueous ammonia in 2–3 seconds. This nanocomposite was applied for ammonia biosensing (Pandey, Goswami, and Nanda 2012). Biosensors are devices that use a sensitive biological element to analyze the concentration of a specific target component. A typical biosensor consists of a substrate that is chemically modified to immobilize biorecognition elements (such as antibodies, enzymes, nucleic acids, hormones, cells, microorganisms, or tissues). Biologically derived materials and biomimetic components can also serve as biorecognition elements (Andryukov et al. 2020, Naresh and Lee 2021). Proper immobilization and stabilization of the biological component on the surface of the transducer is a key challenge in biosensor development, impacting sensitivity and selectivity (Mhd Haniffa et al. 2016). Bio-coatings offer a promising solution to address these challenges. Bio-coatings encompass a diverse range of neutral and synthetic materials used for immobilizing and protecting the biological component in biosensors. They enhance the performance and prevent non-specific interactions with the sample matrix (Pradhan, Brooks, and Yadavalli 2020). In this context, researchers have investigated the use of green-synthesized metallic nanoparticles and biopolymers as environmentally friendly alternatives for bio-coatings. These materials play a crucial role in biosensing applications within the health and food industries. Bio-coatings in healthcare modify biosensor surfaces for specific binding to disease markers or pathogens. Additionally, bio-coating can be used to enhance the stability and biocompatibility of implantable biosensors for long-term monitoring of health parameters (Onuki et al. 2008). Bio-coatings in the food industry modify biosensor surfaces to detect contaminants like toxins, chemicals, and bacteria and indicate food spoilage through changes in volatile organic compounds and pH levels (Figure 6.4). They enhance biosensor stability and shelf-life, enabling more reliable and accurate food quality control (Neethirajan et al. 2018).

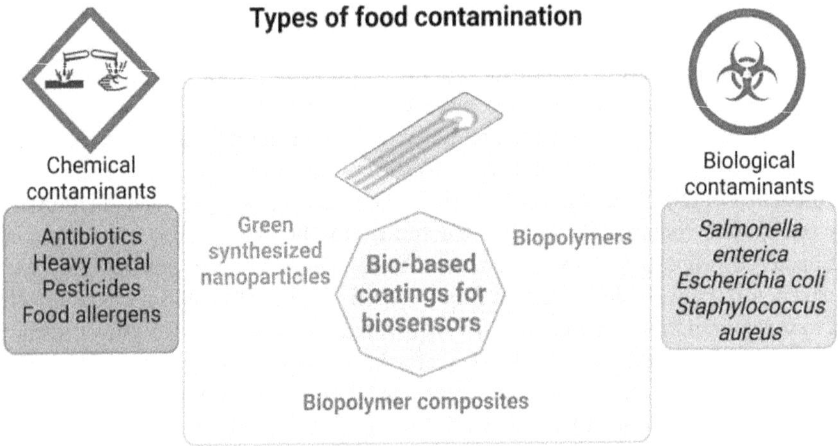

FIGURE 6.4 Harnessing bio-based coatings in biosensors for food contamination monitoring (Popescu and Ungureanu 2023).

A review study by Popescu, M et al. (Popescu and Ungureanu 2023) provides in-depth information on the application of bio-coatings based on biogenic nanoparticles and biopolymers for the food and healthcare industries.

Additionally, a polymeric nanocomposite composed of myoglobin, AuNPs, polydopamine, and graphene, inspired by mussel adhesive proteins, was successfully synthesized. The nanocomposite demonstrated excellent electrocatalytic activity attributed to its high electron transfer capacity, rendering it highly suitable for various biosensing applications (Liu et al. 2016). da Silva et al. (da Silva et al. 2019) fabricated a novel sensor to detect tyramine, a biogenic amine, using a biogenic AuNPs-polymer-modified gold electrode. The sensor had a broad linear range from 0.8 to 80 μM and a low detection limit of 0.04 μM.

6.6 ANTIMICROBIAL AND BIOMEDICAL APPLICATIONS

Ye et al. (Ye, Cheng, and Yu 2019) reported an effective composite consisting of GL and biosynthesized AgNPs. GL was employed both as a reducing agent and stabilizer to facilitate the *in situ* formation of AgNPs. Tannic acid was used as a cross-linker between AgNPs and CS. The resultant freeze-dried composite GL/CS/Ag (Figure 6.5) exhibited a densely interconnected pore structure, better water absorption, moisture retention, and mechanical properties. CS possesses antibacterial properties and, in this study, these AgNPs were shown to improve the antibacterial potential of CS. The antibacterial activity was shown to be proportional to the increasing concentration of AgNPs against *S. aureus* and *E. coli*. This composite further enhanced the antibacterial activity of AgNPs against *S. aureus* and showed good wound healing when tested on animals. The antibacterial activities of the composite against *E. coli* were greater than those against *S. aureus*. This may be explained by a higher content of charged peptidoglycans in the cell wall of

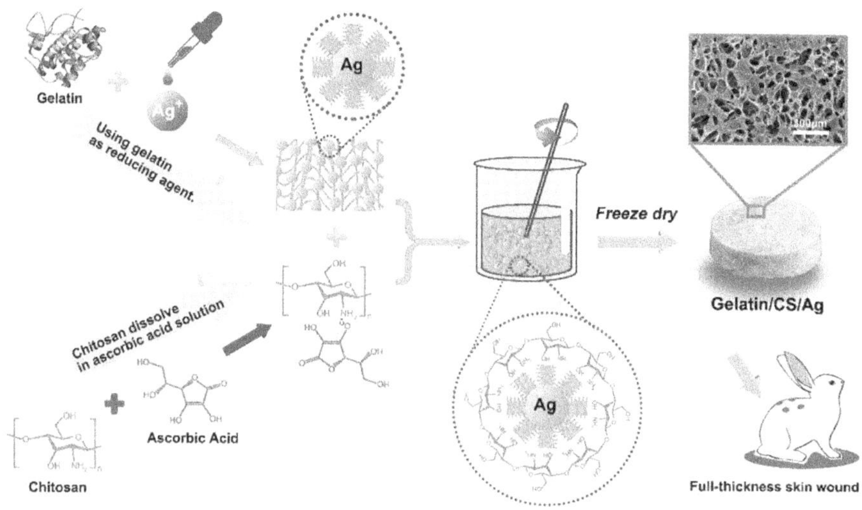

FIGURE 6.5 Preparation process and the principle of composite gelatin/CS/Ag (Ye, Cheng, and Yu 2019).

S. aureus than in *E. coli* (Ye, Cheng, and Yu 2019). The fabrication of polycaprolactone nanocomposites integrated with AgNPs (synthesized using *Bacillus* sp.) offers the potential to serve as a proficient material for wound dressing, as it has shown notable capability in effectively inhibiting the growth of *Staphylococcus haemolyticus* and *Staphylococcus epidermidis* bacteria (Thomas et al. 2015). A nanocomposite synthesized from Kappa Carrageenan and AgNPs (synthesized using *Citrullus colocynthis* seed extract) exhibited antimicrobial activity and cytotoxicity below 50% (Azizi et al. 2017). The nanocomposite films composed of agar/ZnO-NPs and CS/GL/AgNPs demonstrated remarkable antimicrobial efficacy (Kumar et al. 2020). Nanocomposites of guar gum/GL polymer/biogenic AgNPs exhibited antibacterial activity against both gram-negative bacteria *Pseudomonas aeruginosa* and *E. coli* and gram-positive bacteria *S. aureus* (Khan, Kumar, and Kumar 2020).

For an efficacious wound healing Pathak et al., synthesized a nanocomposite based on biogenic AgNPs (synthesized by using *Praecitrullus fistulosus* extract) via the solvent casting method. These nanocomposites were found efficient for wound healing via reducing bacterial infection at the wound site. The composite films showed higher water absorption, suitable porosity, and improved folding endurance as compared to the control GL film. They possessed better thermal stability, improved roughness, efficient antibacterial activity against *E. coli*. All these properties were making it an ideal dressing material for skin applications (Sheokand et al. 2022). Nanocomposites based on CS-functionalized AgNPs demonstrated antimicrobial and anticoagulant activity, along with low toxicity, indicating their potential in nanomedicine development (Asghar et al. 2020). AgNPs, synthesized from white tea extract and added to the montmorillonite polymer, displayed remarkable antioxidant efficacy, suggesting their potential as effective radical scavengers. In vitro cytotoxic assays were conducted on MOLT-4 cells and compared with anticancer agents like cisplatin and doxorubicin. The nanocomposite's IC_{50} values were notably lower than those of the anticancer drugs (Haghparasti and Shahri 2018). Honorato et al. (Artunduaga Bonilla et al. 2021) reported that biogenic AgNPs, incorporated into a CS matrix, create AgNPs@CS. This novel composite showed promise as a potential treatment against *Candida* spp.

6.7 CONCLUSION

Composite materials, especially polymer nanocomposites with biogenic metallic nanoparticles, have shown significant advancements in diverse applications. The integration of biogenic nanoparticles with polymers results in unique attributes, combining inherent nanoparticle properties with the processability and flexibility of polymers. In this chapter, we have discussed various synthesis routes that can be used for the synthesis of polymer composites with biogenic nanoparticles. Each of these methods has its own advantages and limitations. The method of synthesis can be adopted based on the starting material and the properties desired in the end product. Biogenic nanoparticles, when incorporated into polymer matrices, demonstrate enhanced mechanical properties. However, excessive nanoparticle concentrations may lead to aggregation, compromising mechanical strength. Plasticizers can further modify mechanical properties, enhancing film flexibility and nanoparticle dispersion within the matrix. The incorporation of nanoparticles into the composite significantly enhances its electrical conductivity, primarily due to the increased carrier density and improved electron mobility imparted by the nanoparticles. This improvement in optical properties is complemented by the observed reduction in Eg values, which can be attributed to the generation of defects induced by the nanoparticles and the formation of charge transfer complexes within the composite material. However, novel approaches could optimize nanoparticle dispersion and interactions for improved thermal performance.

Polymer-based composite materials incorporating biogenic nanoparticles have emerged as promising solutions in diverse applications. In food packaging, these composites offer enhanced mechanical properties and effective antimicrobial effects, ensuring food safety and shelf-life extension. In wastewater treatment, they exhibit remarkable adsorption capacities and efficient pollutant removal, contributing to environmental remediation. In energy storage, the composites show promise as electrode materials, enhancing energy storage capabilities for supercapacitors. In biosensors, they offer biocompatibility and sensitivity, enabling rapid and accurate detection of various biomolecules and analytes. In the field of nanomedicines use of biogenic nanocomposites has been an emerging and preferable option for research nowadays. Their less toxicity and effective therapeutic application make them an appealing field to work in.

REFERENCES

Abdallah, Omnia M, Khaled Z EL-Baghdady, Mostafa MH Khalil, Mervat I El Borhamy, and Gamal A Meligi. 2020. "Antibacterial, antibiofilm and cytotoxic activities of biogenic polyvinyl alcohol-silver and chitosan-silver nanocomposites." *Journal of Polymer Research* no. 27:1–9.

Abutalib, Maymona M. 2019. "Effect of zinc oxide nanorods on the structural, thermal, dielectric and electrical properties of polyvinyl alcohol/carboxymethyl cellulose composites." *Physica B: Condensed Matter* no. 557:108–116.

Ahmadi, Younes, and Sharif Ahmad. 2020. "Recent progress in the synthesis and property enhancement of waterborne polyurethane nanocomposites: promising and versatile macromolecules for advanced applications." *Polymer Reviews* no. 60 (2):226–266.

Ahmad, Aftab, Yun Wei, Fatima Syed, Kamran Tahir, Aziz Ur Rehman, Arifullah Khan, Sadeeq Ullah, and Qipeng Yuan. 2017. "The effects of bacteria-nanoparticles interface on the antibacterial activity of green synthesized silver nanoparticles." *Microbial Pathogenesis* no. 102:133–142.

Ahmed, Jasim, Mehrajfatema Zafar Mulla, and Yasir Ali Arfat. 2016. "Thermo-mechanical, structural characterization and antibacterial performance of solvent casted polylactide/cinnamon oil composite films." *Food Control* no. 69:196–204.

Aktürk, Ayşen, Melek Erol Taygun, Funda Karbancıoğlu Güler, Gultekin Goller, and Sadriye Küçükbayrak. 2019. "Fabrication of antibacterial polyvinylalcohol nanocomposite mats with soluble starch coated silver nanoparticles." *Colloids and Surfaces A: Physicochemical and Engineering Aspects* no. 562:255–262.

Alippilakkotte, Shebi, Sanjeev Kumar, and Lisa Sreejith. 2017. "Fabrication of PLA/Ag nanofibers by green synthesis method using *Momordica charantia* fruit extract for wound dressing applications." *Colloids and Surfaces A: Physicochemical and Engineering Aspects* no. 529:771–782.

Ali, S Wazed, Subbiyan Rajendran, and Mangala Joshi. 2011. "Synthesis and characterization of chitosan and silver loaded chitosan nanoparticles for bioactive polyester." *Carbohydrate Polymers* no. 83 (2):438–446.

Al-Naamani, Laila, Joydeep Dutta, and Sergey Dobretsov. 2018. "Nanocomposite zinc oxide-chitosan coatings on polyethylene films for extending storage life of okra (*Abelmoschus esculentus*)." *Nanomaterials* no. 8 (7):479.

Amaregouda, Yamanappagouda, Tilak Gasti, and Kantharaju Kamanna. 2022. Optoelectronic, microstructural, ecofriendly and photo-catalytic evaluation of Aspartic acid cross-linked Poly (vinyl alcohol)/Copper Oxide nanotubes composite films. Paper read at IOP Conference Series: Materials Science and Engineering.

Andryukov, BG, IN Lyapun, EV Matosova, and LM Somova. 2020. "Biosensor technologies in medicine: from detection of biochemical markers to research into molecular targets." *Современные технологии в медицине* no. 12 (6 (eng)):70–83.

Arthisree, Devendran Lakshmi, and Wuppulluri Madhuri. 2020. "Optically active polymer nanocomposite composed of polyaniline, polyacrylonitrile and green-synthesized graphene quantum dot for supercapacitor application." *International Journal of Hydrogen Energy* no. 45 (16):9317–9327.

Artunduaga Bonilla, Jhon Jhamilton, Leandro Honorato, Douglas Felipe Cordeiro de Oliveira, Rodrigo Araújo Gonçalves, Allan Guimarães, Kildare Miranda, and Leonardo Nimrichter. 2021. "Silver chitosan nanocomposites as a potential treatment for superficial candidiasis." *Medical Mycology* no. 59 (10):993–1005. doi: 10.1093/mmy/myab028

Asghar, Muhammad Arif, Rabia Ismail Yousuf, Muhammad Harris Shoaib, Muhammad Asif Asghar, Sabah Ansar, Mehrukh Zehravi, and Ahad Abdul Rehman. 2020. "RETRACTED ARTICLE: synergistic nanocomposites of different antibiotics coupled with green synthesized chitosan-based silver nanoparticles: characterization, antibacterial, in vivo toxicological and biodistribution studies." *International Journal of Nanomedicine* no. 15:7841–7859.

Asnag, Ghalib M, Ahmad Hamza Oraby, and Amr Mohamed Abdelghany. 2019. "Green synthesis of gold nanoparticles and its effect on the optical, thermal and electrical properties of carboxymethyl cellulose." *Composites Part B: Engineering* no. 172:436–446.

Azizi, Susan, Rosfarizan Mohamad, Raha Abdul Rahim, Reza Mohammadinejad, and Arbakariya Bin Ariff. 2017. "Hydrogel beads bio-nanocomposite based on Kappa-Carrageenan and green synthesized silver nanoparticles for biomedical applications." *International Journal of Biological Macromolecules* no. 104:423–431.

Bandeira, Marina, Marcelo Giovanela, Mariana Roesch-Ely, Declan M Devine, and Janaina da Silva Crespo. 2020. "Green synthesis of zinc oxide nanoparticles: a review of the synthesis methodology and mechanism of formation." *Sustainable Chemistry and Pharmacy* no. 15:100223.

Behera, Manoranjan, and Shanker Ram. 2014. "Inquiring the mechanism of formation, encapsulation, and stabilization of gold nanoparticles by poly (vinyl pyrrolidone) molecules in 1-butanol." *Applied Nanoscience* no. 4:247–254.

Brandenburg, Ricardo Fischer, Carlos Mauricio Lepienski, Daniela Becker, and Luiz Antonio Ferreira Coelho. 2017. "Influence of mixing methods on the properties of high density polyethylene nanocomposites with different carbon nanoparticles." *Matéria (Rio de Janeiro)* no. 22: e-11888.

Chakraborty, Sohini, J Joselin Farida, Remya Simon, S Kasthuri, and N L Mary. 2020. "*Averrhoa carambola* fruit extract assisted green synthesis of ZnO nanoparticles for the photodegradation of congo red dye." *Surfaces and Interfaces* no. 19:100488.

Chen, Jixi, Liangbiao Fan, Chaolong Yang, Shengcai Wang, Man Zhang, Jing Xu, and Shuiyuan Luo. 2020. "Facile synthesis of Ag nanoparticles-loaded chitosan antibacterial nanocomposite and its application in polypropylene." *International Journal of Biological Macromolecules* no. 161:1286–1295.

Chitra, Kethirabalan, and Gurusamy Annadurai. 2013. "Fluorescent silica nanoparticles in the detection and control of the growth of pathogen." *Journal of Nanotechnology* no. 2013:509628.

da Silva, Wanderson, Mariana Emilia Ghica, Rachel F Ajayi, Emmanuel I Iwuoha, and Christopher MA Brett. 2019. "Impedimetric sensor for tyramine based on gold nanoparticle doped-poly (8-anilino-1-naphthalene sulphonic acid) modified gold electrodes." *Talanta* no. 195:604–612.

Das, Samvidha, Aman Bhardwaj, and Lalit M Pandey. 2021. "Functionalized biogenic nanoparticles for use in emerging biomedical applications: a review." *Current Nanomaterials* no. 6 (2):119–139.

Dash, Kshirod Kumar, Ashwani Kumar, Sonam Kumari, and Mudasir Ahmad Malik. 2021. "Silver nanoparticle incorporated flaxseed protein-alginate composite films: effect on physicochemical, mechanical, and thermal properties." *Journal of Polymers and the Environment* no. 29:3649–3659.

Ditaranto, Nicoletta, Inez Dorothé van der Werf, Rosaria Anna Picca, Maria Chiara Sportelli, Lorena Carla Giannossa, Elisabetta Bonerba, Giuseppina Tantillo, and Luigia Sabbatini. 2015. "Characterization and behaviour of ZnO-based nanocomposites designed for the control of biodeterioration of patrimonial stoneworks." *New Journal of Chemistry* no. 39 (9):6836–6843.

Dong, Qun, Huilan Su, Wei Cao, Jie Han, Di Zhang, and Qixin Guo. 2008. "Biogenic synthesis of hierarchical hybrid nanocomposites and patterning of silver nanoparticles." *Materials Chemistry and Physics* no. 110 (1):160–165.

Du, Yang, Zhike Huang, Shuyi Wu, Kairong Xiong, Xingcai Zhang, Bingna Zheng, Reddeppa Nadimicherla, Ruowen Fu, and Dingcai Wu. 2018. "Preparation of versatile yolk-shell nanoparticles with a precious metal yolk and a microporous polymer shell for high-performance catalysts and antibacterial agents." *Polymer* no. 137:195–200.

Ediyilyam, Sreelekha, Bini George, Sarojini Sharath Shankar, Thomas Thuruthiyil Dennis, Stanisław Wacławek, Miroslav Černík, and Vinod VT Padil. 2021. "Chitosan/gelatin/silver nanoparticles composites films for biodegradable food packaging applications." *Polymers* no. 13 (11):1680.

Ediyilyam, Sreelekha, Mahesh M Lalitha, Bini George, Sarojini Sharath Shankar, Stanisław Wacławek, Miroslav Černík, and Vinod Vellora Thekkae Padil. 2022. "Synthesis, characterization and physicochemical properties of biogenic silver nanoparticle-encapsulated chitosan bionanocomposites." *Polymers* no. 14 (3):463.

Fawaz, Joel, and Vikas Mittal. 2014. "Synthesis of polymer nanocomposites: review of various techniques." *Synthesis Techniques for Polymer Nanocomposites* no. 24:1–30.

Fayaz, Amanulla Mohammed, Kulandaivelu Balaji, Morukattu Girilal, Ruchi Yadav, Pudupalayam Thangavelu Kalaichelvan, and Ramasamy Venketesan. 2010. "Biogenic synthesis of silver nanoparticles and their synergistic effect with antibiotics: a study against gram-positive and gram-negative bacteria." *Nanomedicine: Nanotechnology, Biology and Medicine* no. 6 (1):103–109.

Frank, Luiza Abrahão, Giovana Ravizzoni Onzi, Andrey Silva Morawski, Adriana Raffin Pohlmann, Silva Stanisçuaski Guterres, and Renata Vidor Contri. 2020. "Chitosan as a coating material for nanoparticles intended for biomedical applications." *Reactive and Functional Polymers* no. 147:104459.

Gami, Fatma, Norah Algethami, Hanan Mahmoud E Ragab, and Ahmed E Tarabiah. 2022. "Structural, optical and electrical studies of chitosan/polyacrylamide blend filled with synthesized selenium nanoparticles." *Journal of Molecular Structure* no. 1257:132631.

Gong, Cui-Ping, Ying Luo, and Ying-Ying Pan. 2019. "Novel synthesized zinc oxide nanoparticles loaded alginate-chitosan biofilm to enhanced wound site activity and anti-septic abilities for the management of complicated abdominal wound dehiscence." *Journal of Photochemistry and Photobiology B: Biology* no. 192:124–130.

Goswami, Rajiv, Monti Gogoi, Hirok J Borah, Pravin G Ingole, and Swapnali Hazarika. 2018. "Biogenic synthesized Pd-nanoparticle incorporated antifouling polymeric membrane for removal of crystal violet dye." *Journal of Environmental Chemical Engineering* no. 6 (5):6139–6146.

Haghparasti, Zeinab, and Mahnaz Mahdavi Shahri. 2018. "Green synthesis of water-soluble nontoxic inorganic polymer nanocomposites containing silver nanoparticles using white tea extract and assessment of their in vitro antioxidant and cytotoxicity activities." *Materials Science and Engineering: C* no. 87:139–148.

Hiep, Nguyen Thi, Huynh Chan Khon, Vo Van Thanh Niem, Vo Van Toi, Tran Ngoc Quyen, Nguyen Dai Hai, and Mai Ngoc Tuan Anh. 2016. "Microwave-assisted synthesis of chitosan/polyvinyl alcohol silver nanoparticles gel for wound dressing applications." *International Journal of Polymer Science* no. 2:1584046.

Huang, Zhong-Hua, Yan-Na Yin, Gu-li-mi-la Aikebaier, and Yan Zhang. 2016. "Preparation of a novel positively charged nanofiltration composite membrane incorporated with silver nanoparticles for pharmaceuticals and personal care product rejection and anti-bacterial properties." *Water Science and Technology* no. 73 (8):1910–1919.

Huh, Ae Jung, and Young Jik Kwon. 2011. ""Nanoantibiotics": a new paradigm for treating infectious diseases using nanomaterials in the antibiotics resistant era." *Journal of Controlled Release* no. 156 (2):128–145.

Jafari, Arman, Shadi Hassanajili, Mohammad Bagher Karimi, Amir Emami, Farnaz Ghaffari, and Negar Azarpira. 2018. "Effect of organic/inorganic nanoparticles on performance of polyurethane nanocomposites for potential wound dressing applications." *Journal of the Mechanical Behavior of Biomedical Materials* no. 88:395–405.

Jamróz, Ewelina, Piotr Kulawik, and Pavel Kopel. 2019. "The effect of nanofillers on the functional properties of biopolymer-based films: a review." *Polymers* no. 11 (4):675.

Jatoi, Abdul Wahab. 2020. "Polyurethane nanofibers incorporated with ZnAg composite nanoparticles for antibacterial wound dressing applications." *Composites Communications* no. 19:103–107.

John, Maya Jacob, and Sabu Thomas. 2008. "Biofibres and biocomposites." *Carbohydrate Polymers* no. 71 (3):343–364.

Kadam, Deepak, Bilal Momin, Shanooba Palamthodi, and Smith S Lele. 2019. "Physicochemical and functional properties of chitosan-based nano-composite films incorporated with biogenic silver nanoparticles." *Carbohydrate Polymers* no. 211:124–132.

Kango, Sarita, Susheel Kalia, Pankaj Thakur, Bandna Kumari, and Deepak Pathania. 2015. "Semiconductor–polymer hybrid materials." In *Organic-Inorganic Hybrid Nanomaterials*, 283–311. Springer Cham. https://doi.org/10.1007/978-3-319-13593-9

Kausar, Ayesha. 2021. "Progress in green nanocomposites for high-performance applications." *Materials Research Innovations* no. 25 (1):53–65.

Kempe, Kristian, C Remzi Becer, and Ulrich S Schubert. 2011. "Microwave-assisted polymerizations: recent status and future perspectives." *Macromolecules* no. 44 (15): 5825–5842.

Khan, Muhammad Umar Aslam, Sajjad Haider, Adnan Haider, Saiful Izwan Abd Razak, Mohammed Rafiq Abdul Kadir, Saqlain A Shah, Aneela Javed, Imran Shakir, and Ateyah A Al-Zahrani. 2021. "Development of porous, antibacterial and biocompatible GO/n-HAp/bacterial cellulose/β-glucan biocomposite scaffold for bone tissue engineering." *Arabian Journal of Chemistry* no. 14 (2):102924.

Khan, Nida, Deepak Kumar, and Pramendra Kumar. 2020. "Silver nanoparticles embedded guar gum/gelatin nanocomposite: green synthesis, characterization and antibacterial activity." *Colloid and Interface Science Communications* no. 35:100242.

Krishnaswamy, Srimathi, Puspamitra Panigrahi, Senthil Kumaar, and Ganapathi Subramanian Nagarajan. 2020. "Effect of conducting polymer on photoluminescence quenching of green synthesized ZnO thin film and its photocatalytic properties." *Nano-Structures & Nano-Objects* no. 22:100446.

Król-Górniak, Anna, Pawel Pomastowski, Katarzyna Rafińska, Viorica Railean-Plugaru, and Bogusław Buszewski. 2017. "Zinc oxide nanoparticles: synthesis, antiseptic activity and toxicity mechanism." *Advances in Colloid and Interface Science* no. 249: 37–52.

Kumar, Radhesh, Steve Howdle, and Helmut Münstedt. 2005. "Polyamide/silver antimicrobials: effect of filler types on the silver ion release." *Journal of Biomedical Materials Research Part B: Applied Biomaterials:* no. 75 (2):311–319.

Kumar, Santosh, Atanu Mitra, and Dipankar Halder. 2017. "Centella asiatica leaf mediated synthesis of silver nanocolloid and its application as filler in gelatin based antimicrobial nanocomposite film." *Lebensmittel-Wissenschaft & Technologie* no. 75:293–300.

Kumar, Santosh, Abhinab Mudai, Barnali Roy, Indra Bhusan Basumatary, Avik Mukherjee, and Joydeep Dutta. 2020. "Biodegradable hybrid nanocomposite of chitosan/gelatin and green synthesized zinc oxide nanoparticles for food packaging." *Foods* no. 9 (9):1143.

Kumar, Santosh, Avik Mukherjee, and Joydeep Dutta. 2020. "Chitosan based nanocomposite films and coatings: emerging antimicrobial food packaging alternatives." *Trends in Food Science & Technology* no. 97:196–209.

Kumar, Radhesh, and Helmut Münstedt. 2005. "Silver ion release from antimicrobial polyamide/silver composites." *Biomaterials* no. 26 (14):2081–2088.

Kumar, Santosh, Fei Ye, Sergey Dobretsov, and Joydeep Dutta. 2019. "Chitosan nanocomposite coatings for food, paints, and water treatment applications." *Applied Sciences* no. 9 (12):2409.

Kumari, Sangeeta, Raj Pal Singh, Nayaku N Chavan, and Pratheep K Annamalai. 2016. "Chitosan-based bionanocomposites for biomedical application." *Bioinspired, Biomimetic and Nanobiomaterials* no. 7 (4):219–227.

Laohakunjit, Natta, and Athapol Noomhorm. 2004. "Effect of plasticizers on mechanical and barrier properties of rice starch film." *Starch-Stärke* no. 56 (8):348–356.

Lavorgna, Marino, Filomena Piscitelli, Pasqualina Mangiacapra, and Giovanna G Buonocore. 2010. "Study of the combined effect of both clay and glycerol plasticizer on the properties of chitosan films." *Carbohydrate Polymers* no. 82 (2):291–298.

Leng, Kunyi, Weicong Mai, Xingcai Zhang, Ruliang Liu, Xidong Lin, Junlong Huang, He Lou, Yiming Xie, Ruowen Fu, and Dingcai Wu. 2018. "Construction of functional nanonetwork-structured carbon nitride with Au nanoparticle yolks for highly efficient photocatalytic applications." *Chemical Communications* no. 54 (52):7159–7162.

Liang, Jin, Hua Yan, Jiuya Zhang, Wenzhong Dai, Xueling Gao, Yibin Zhou, Xiaochun Wan, and Pradeep Puligundla. 2017. "Preparation and characterization of antioxidant edible chitosan films incorporated with epigallocatechin gallate nanocapsules." *Carbohydrate Polymers* no. 171:300–306.

Lin, Xidong, Yeru Liang, Zhitao Lu, He Lou, Xingcai Zhang, Shaohong Liu, Bingna Zheng, Ruliang Liu, Ruowen Fu, and Dingcai Wu. 2017. "Mechanochemistry: a green, activation-free and top-down strategy to high-surface-area carbon materials." *American Chemical Society Sustainable Chemistry & Engineering* no. 5 (10):8535–8540.

Liu, Pan, Feng-Qiao Bai, Dan-Wei Lin, Hua-Ping Peng, Yan Hu, Yan-Jie Zheng, Wei Chen, Ai-Lin Liu, and Xin-Hua Lin. 2016. "One-pot green synthesis of mussel-inspired myoglobin–gold nanoparticles–polydopamine–graphene polymeric bionanocomposite for biosensor application." *Journal of Electroanalytical Chemistry* no. 764:104–109.

Loo, Ching-Yee, Paul M Young, Wing-Hin Lee, Rosalia Cavaliere, Cynthia B Whitchurch, and Ramin Rohanizadeh. 2014. "Non-cytotoxic silver nanoparticle-polyvinyl alcohol hydrogels with anti-biofilm activity: designed as coatings for endotracheal tube materials." *Biofouling* no. 30 (7):773–788.

Ltaief, Syrine, Mahjoub Jabli, and Saber Ben Abdessalem. 2021. "Immobilization of copper oxide nanoparticles onto chitosan biopolymer: application to the oxidative degradation of Naphthol blue black." *Carbohydrate Polymers* no. 261:117908.

Mahmoud, Mohamed E, Mohamed F Amira, Magda E Abouelanwar, and Mohamed Abdel Salam. 2021. "Green synthesis and surface decoration of silver nanoparticles onto δ-FeOOH-Polymeric nanocomposite as efficient nanocatalyst for dyes degradation." *Journal of Environmental Chemical Engineering* no. 9 (1):104697.

Mane, Suyog S, Sunil M Patil, Krishna K Pawar, Manjunath D Salgaonkar, Pallavi Jagdale, Trupti Kamble, and Mahesh Agharkar. 2020. "Biogenic synthesized silver nanoparticles decorated polypyrrole nanotubes as promising photocatalyst for methyl violet dye degradation." *Materials Today: Proceedings* no. 28:2311–2317.

Marrez, Diaa A, Ahmed E Abdelhamid, and Osama M Darwesh. 2019. "Eco-friendly cellulose acetate green synthesized silver nano-composite as antibacterial packaging system for food safety." *Food Packaging and Shelf Life* no. 20:100302.

Mhd Haniffa, Mhd Abd Cader, Yern Chee Ching, Luqman Chuah Abdullah, Sin Chew Poh, and Cheng Hock Chuah. 2016. "Review of bionanocomposite coating films and their applications." *Polymers* no. 8 (7):246.

Mohseni, Maedeh Sadat, Mohammad A Khalilzadeh, Moein Mohseni, Fariba Zamani Hargalani, Muhammad Ibrahim Getso, Vahid Raissi, and Omid Raiesi. 2020. "Green synthesis of Ag nanoparticles from pomegranate seeds extract and synthesis of Ag-Starch nanocomposite and characterization of mechanical properties of the films." *Biocatalysis and Agricultural Biotechnology* no. 25:101569.

Moreno, María Alejandra, María Eugenia Orqueda, Laura G Gómez-Mascaraque, María Inés Isla, and Amparo López-Rubio. 2019. "Crosslinked electrospun zein-based food packaging coatings containing bioactive chilto fruit extracts." *Food Hydrocolloids* no. 95:496–505.

Morgan, Alexander B, Lih-Long Chu, and Joseph D Harris. 2005. "A flammability performance comparison between synthetic and natural clays in polystyrene nanocomposites." *Fire and Materials: An International Journal* no. 29 (4):213–229.

Moulahoum, Hichem, Faezeh Ghorbanizamani, Emine Guler Celik, and Suna Timur. 2022. "Nano-scaled materials and polymer integration in biosensing tools." *Biosensors* no. 12 (5):301.

Munnawar, Iqra, Sadia S Iqbal, Muhammad N Anwar, Mehwish Batool, Sheraz Tariq, Nosheen Faitma, Asim L Khan, Asad U Khan, Umair Nazar, and Tahir Jamil. 2017. "Synergistic effect of Chitosan-Zinc Oxide Hybrid Nanoparticles on antibiofouling and water disinfection of mixed matrix polyethersulfone nanocomposite membranes." *Carbohydrate Polymers* no. 175:661–670.

Naresh, Varnakavi, and Nohyun Lee. 2021. "A review on biosensors and recent development of nanostructured materials-enabled biosensors." *Sensors* no. 21 (4):1109.

Nathawat, Rashi, Satyapal S Rathore, Poonam R Kharangarh, Reena Devi, and Anita Kumari. 2023. "Synthesis and application of carbon-based nanocomposite." In *Carbon Nanomaterials and Their Nanocomposite-Based Chemiresistive Gas Sensors*, 169–203. Elsevier.

Neethirajan, Suresh, Vasanth Ragavan, Xuan Weng, and Rohit Chand. 2018. "Biosensors for sustainable food engineering: challenges and perspectives." *Biosensors* no. 8 (1):23.

Nguyen, Thien-Phap, and Sheng-Hsiung Yang. 2018. "Hybrid materials based on polymer nanocomposites for environmental applications." In *Polymer-Based Nanocomposites for Energy and Environmental Applications*, 507–551. Elsevier.

Noah, Naumih M, and Peter M Ndangili. 2022. "Green synthesis of nanomaterials from sustainable materials for biosensors and drug delivery." *Sensors International* no. 3:100166.

Ojha, Nupur, and Nilanjana Das. 2020. "Fabrication and characterization of biodegradable PHBV/SiO2 nanocomposite for thermo-mechanical and antibacterial applications in food packaging." *Institution of Engineering and Technology Nanobiotechnology* no. 14 (9):785–795.

Onuki, Yoshinori, Upkar Bhardwaj, Fotios Papadimitrakopoulos, and Diane J Burgess. 2008. "A review of the biocompatibility of implantable devices: current challenges to overcome foreign body response." *Journal of Diabetes Science and Technology* no. 2 (6):1003–1015.

Oréfice, Rodrigo Lambert, Larry L Hench, and Anthony B Brennan. 2001. "Effect of particle morphology on the mechanical and thermo-mechanical behavior of polymer composites." *Journal of the Brazilian Society of Mechanical Sciences* no. 23:1–8.

Pandey, Sadanand, Gopal K Goswami, and Karuna K Nanda. 2012. "Green synthesis of biopolymer–silver nanoparticle nanocomposite: an optical sensor for ammonia detection." *International Journal of Biological Macromolecules* no. 51 (4):583–589.

Pandey, Sadanand, Gopal K Goswami, and Karuna K Nanda. 2013. "Green synthesis of polysaccharide/gold nanoparticle nanocomposite: an efficient ammonia sensor." *Carbohydrate Polymers* no. 94 (1):229–234.

Pandian, Harish, Kandasamy Senthilkumar, M Naveenkumar, and S Samraj. 2023. "*Azadirachta indica* leaf extract mediated silver nanoparticles impregnated nano composite film (AgNP/MCC/starch/whey protein) for food packaging applications." *Environmental Research* no. 216:114641.

Park, Su-il, and Yanyun Zhao. 2004. "Incorporation of a high concentration of mineral or vitamin into chitosan-based films." *Journal of Agricultural and Food Chemistry* no. 52 (7):1933–1939.

Parmar, Ankush, Gurpreet Kaur, Shikha Kapil, Vipasha Sharma, and Shweta Sharma. 2019. "Biogenic PLGA-Zinc oxide nanocomposite as versatile tool for enhanced photocatalytic and antibacterial activity." *Applied Nanoscience* no. 9:2001–2016.

Philip, Princy, Tomlal Jose, Manoj Parameswaran, and Sajini Thankaraj. 2021. "Structurally modified poly (methyl methacrylate) electrospun nanofibers as better host matrix for noble metal nanoparticles." *Journal of Applied Polymer Science* no. 138 (15):50210.

Popescu, Melania, and Camelia Ungureanu. 2023. "Biosensors in Food and Healthcare Industries: bio-coatings based on biogenic nanoparticles and biopolymers." *Coatings* no. 13 (3):486.

Pradhan, Sayantan, Anne Katherine Brooks, and Vamsi K Yadavalli. 2020. "Nature-derived materials for the fabrication of functional biodevices." *Materials Today Bio* no. 7: 100065.

Pramanik, Sujata, and Pankaj Das. 2019. "Metal-based nanomaterials and their polymer nanocomposites." In *Nanomaterials and Polymer Nanocomposites*, 91–121. Elsevier.

Prime, R Bruce, Harvey E Bair, Sergey Vyazovkin, Patrick K Gallagher, and Alan Riga. 2009. "Thermogravimetric analysis (TGA)." *Thermal Analysis of Polymers: Fundamentals and Applications* no. 3:241–317.

Priyadarshi, Ruchir, and Jong-Whan Rhim. 2020. "Chitosan-based biodegradable functional films for food packaging applications." *Innovative Food Science & Emerging Technologies* no. 62:102346.

Qi, Kezhen, Bei Cheng, Jiaguo Yu, and Wingkei Ho. 2017. "Review on the improvement of the photocatalytic and antibacterial activities of ZnO." *Journal of Alloys and Compounds* no. 727:792–820.

Rane, Ajay Vasudeo, Krishnan Kanny, Vayyaprontavida Kaliyathan Abitha, and Sabu Thomas. 2018. "Methods for synthesis of nanoparticles and fabrication of nanocomposites." In *Synthesis of Inorganic Nanomaterials*, 121–139. Elsevier.

Raza, Sadaf, Asma Ansari, Nadir Naveed Siddiqui, Fariha Ibrahim, Muhammad Ishaque Abro, and Afsheen Aman. 2021. "Biosynthesis of silver nanoparticles for the fabrication of non cytotoxic and antibacterial metallic polymer based nanocomposite system." *Scientific Reports* no. 11 (1):10500.

Rónavári, Andrea, Nóra Igaz, Dora I Adamecz, Bettina Szerencsés, Csaba Molnar, Zoltán Kónya, Ilona Pfeiffer, and Monika Kiricsi. 2021. "Green silver and gold nanoparticles: biological synthesis approaches and potentials for biomedical applications." *Molecules* no. 26 (4):844.

Roy, Swarup, and Jong-Whan Rhim. 2019. "Melanin-mediated synthesis of copper oxide nanoparticles and preparation of functional agar/CuO NP nanocomposite films." *Journal of Nanomaterials* no. 2019:1–10.

Sánchez-Valdés, Saúl, Hortensia Ortega-Ortiz, Luis Francisco Ramos-de Valle, Francisco J Medellín-Rodríguez, and R Guedea-Miranda. 2009. "Mechanical and antimicrobial properties of multilayer films with a polyethylene/silver nanocomposite layer." *Journal of Applied Polymer Science* no. 111 (2):953–962.

Sharma, Bhasha, Parul Malik, and Purnima Jain. 2018. "Biopolymer reinforced nanocomposites: a comprehensive review." *Materials Today Communications* no. 16:353–363.

Sheokand, Bharti, Seema R Pathak, Chandra Mohan Srivastava, Anand Kumar, Indra Bahadur, and M Vats. 2022. "Green silver nanoparticles functionalised gelatin nanocomposite film for wound healing: construction and characterization." *Journal of Molecular Liquids* no. 368:120561. doi: 10.1016/j.molliq.2022.120561.

Shifrina, Zinaida B, Valentina G Matveeva, and Lyudmila M Bronstein. 2019. "Role of polymer structures in catalysis by transition metal and metal oxide nanoparticle composites." *Chemical Reviews* no. 120 (2):1350–1396.

Siddiqui, Nayma, Arpan Bhardwaj, Renu Hada, Vishwajeet Singh Yadav, and Deepti Goyal. 2018. "Synthesis, characterization and antimicrobial study of poly (methyl methacrylate)/Ag nanocomposites." *Vacuum* no. 153:6–11.

Sportelli, Maria Chiara, Rosaria Anna Picca, and Nicola Cioffi. 2016. "Recent advances in the synthesis and characterization of nano-antimicrobials." *TrAC Trends in Analytical Chemistry* no. 84:131–138.

Stanković, Ana, Meltem Sezen, Marina Milenković, Sonja Kaišarević, Nebojša Andrić, and Magdalena Stevanović. 2016. "PLGA/nano-ZnO composite particles for use in biomedical applications: preparation, characterization, and antimicrobial activity." *Journal of Nanomaterials* no. 2016:9425289.

Stevanović, Magdalena, Ines Bračko, Marina Milenković, Nenad Filipović, Jana Nunić, Metka Filipič, and Dragan P Uskoković. 2014. "Multifunctional PLGA particles containing poly (l-glutamic acid)-capped silver nanoparticles and ascorbic acid with simultaneous antioxidative and prolonged antimicrobial activity." *Acta Biomaterialia* no. 10 (1):151–162.

Stevanović, Magdalena, Nenad Filipović, Jelena Djurdjević, Miodrag Lukić, Marina Milenković, and Aldo Boccaccini. 2015. "45S5Bioglass®-based scaffolds coated with selenium nanoparticles or with poly (lactide-co-glycolide)/selenium particles: processing, evaluation and antibacterial activity." *Colloids and Surfaces B: Biointerfaces* no. 132:208–215.

Stevanović, Magdalena M, Srečo D Škapin, Ines Bračko, Marina Milenković, Jana Petković, Metka Filipič, and Dragan P Uskoković. 2012. "Poly (lactide-co-glycolide)/silver nanoparticles: synthesis, characterization, antimicrobial activity, cytotoxicity assessment and ROS-inducing potential." *Polymer* no. 53 (14):2818–2828.

Talib, Salam Hussein, Mohammed Khlaif Challab, and Firas Ali Hussein Alhameedawi. 2021. Using Nano Composites to Purify Water from Phenol Pollutants. Paper read at Journal of Physics: Conference Series.

Tao, Andrea R, Susan Habas, and Peidong Yang. 2008. "Shape control of colloidal metal nanocrystals." *Small* no. 4 (3):310–325.

Thomas, Roshmi, Kandammuriyil Radhakrishnan Soumya, Jyothis Mathew, and Edayileveetil Krishnankutty Radhakrishnan. 2015. "Electrospun polycaprolactone membrane incorporated with biosynthesized silver nanoparticles as effective wound dressing material." *Applied Biochemistry and Biotechnology* no. 176:2213–2224.

Trache, Djalal, M Hazwan Hussin, Caryn Tan Hui Chuin, Sumiyyah Sabar, MR Nurul Fazita, Owolabi F A Taiwo, T M Hassan, and MK Mohamad Haafiz. 2016. "Microcrystalline cellulose: isolation, characterization and bio-composites application—A review." *International Journal of Biological Macromolecules* no. 93:789–804.

Tripathi, Ravi Mani, Ramesh Namdeo Pudake, Brij Raj Shrivastav, and Archana Shrivastav. 2018. "Antibacterial activity of poly (vinyl alcohol)—Biogenic silver nanocomposite film for food packaging material." *Advances in Natural Sciences: Nanoscience and Nanotechnology* no. 9 (2):025020.

Umoren, Peace Saviour, Doga Kavaz, Alexis Nzila, Saravanan Sankaran, and Saviour A Umoren. 2022. "Biogenic synthesis and characterization of chitosan-cuo nanocomposite and evaluation of antibacterial activity against gram-positive and-negative bacteria." *Polymers* no. 14 (9):1832.

Vera, Paula, Yolanda Echegoyen, Elena Canellas, Cristina Nerín, María Palomo, Yolanda Madrid, and Carmen Cámara. 2016. "Nano selenium as antioxidant agent in a multilayer food packaging material." *Analytical and Bioanalytical Chemistry* no. 408: 6659–6670.

Virkutyte, Jurate, and Rajender S Varma. 2011. "Green synthesis of metal nanoparticles: biodegradable polymers and enzymes in stabilization and surface functionalization." *Chemical Science* no. 2 (5):837–846.

Wei, Xiaohui, Junjie Cai, Song Lin, Fan Li, and Feng Tian. 2021. "Controlled release of monodisperse silver nanoparticles via in situ cross-linked polyvinyl alcohol as benign and antibacterial electrospun nanofibers." *Colloids and Surfaces B: Biointerfaces* no. 197:111370.

Wilson, Paul, Pu Chun Ke, Thomas P Davis, and Kristian Kempe. 2017. "Poly (2-oxazoline)-based micro-and nanoparticles: a review." *European Polymer Journal* no. 88:486–515.

Yang, Yanan, Huixuan Zhang, Peng Wang, Qingzhu Zheng, and Jun Li. 2007. "The influence of nano-sized TiO_2 fillers on the morphologies and properties of PSF UF membrane." *Journal of Membrane Science* no. 288 (1–2):231–238.

Ye, Hongli, Junwen Cheng, and Kun Yu. 2019. "In situ reduction of silver nanoparticles by gelatin to obtain porous silver nanoparticle/chitosan composites with enhanced antimicrobial and wound-healing activity." *International Journal of Biological Macromolecules* no. 121:633–642.

Zare, Ehsan Nazarzadeh, Pooyan Makvandi, Assunta Borzacchiello, Franklin R Tay, Behnaz Ashtari, and Vinod VT Padil. 2019. "Antimicrobial gum bio-based nanocomposites and their industrial and biomedical applications." *Chemical Communications* no. 55 (99):14871–14885.

Zhang, Hao Yang, Heng Bo Jiang, Ji-Eun Kim, ShuXin Zhang, Kwang-Mahn Kim, and Jae-Sung Kwon. 2020. "Bioresorbable magnesium-reinforced PLA membrane for guided bone/tissue regeneration." *Journal of the Mechanical Behavior of Biomedical Materials* no. 112:104061.

7 Nanobiotechnological Interventions for Sustainable Food Supplies and Agricultural Production

Sara Najeeb, Sumaira Miskeen, Muhammad Jahangir, and Imran Khan

7.1 INTRODUCTION

The European Commission recognizes nanotechnology as a "key enabling" technology, and its influence extends to nearly every aspect of our daily lives (1). Nanotechnology is described as an interdisciplinary, innovative scientific field surrounding the design, development, and application of devices and materials at the molecular and nanoscale levels. Basic concept of nanotechnology is to work particle by particle to create structures through novel molecular organization and then utilize these characteristics to develop unique nanodevices (2).

The combination of biotechnology and nanotechnology, in addition to physical and chemical processing, nanoscience, and systems analysis, create innovative products which have been previously modified at the nanometer scale and may have improved properties than conventional materials, including food, agriculture, pharmacy, health, electronics, energy, and mechatronics (3).

Nanobiotechnology stands out as one of the most important technologies of the 21st century. As a relatively novel field, terms like nanobiotechnology, bionanotechnology, and nanobiology are often used interchangeably (4). Essentially, it involves the use of enhanced biomolecules like cells, nucleic acids, cellular components, and enzymes to craft functional nanostructured and microstructured materials from both natural and synthetic sources (5).

7.2 NANOBIOTECHNOLOGY INTERVENTIONS

The pioneering technology can tackle global issues including food supply, sustainable farming, and climate variability. Plant-based nanoparticles that have been specifically created can produce salinity-resistant and drought plants with increased

DOI: 10.1201/9781003378563-7

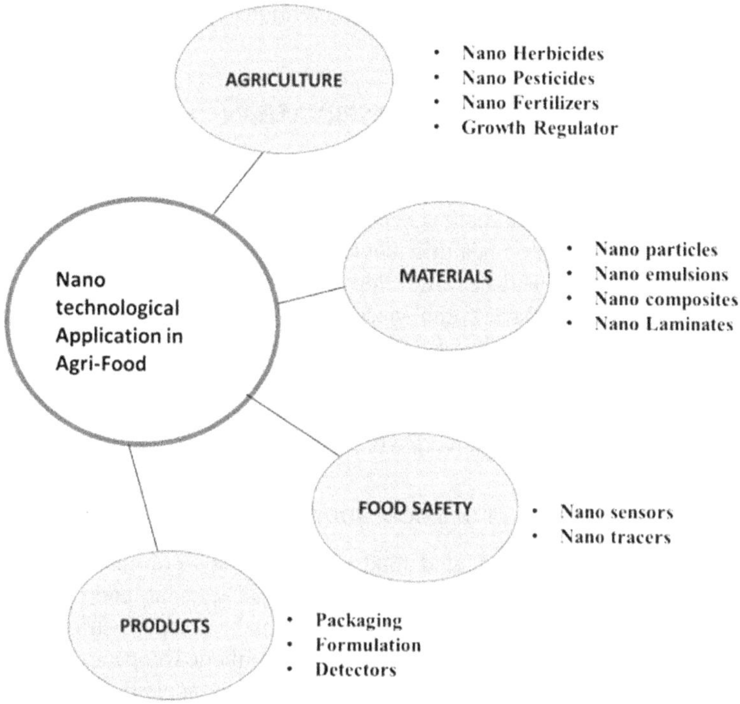

FIGURE 7.1 A graphical illustration of applicability of nanotechnology in agri-food sector.

enzymatic efficacy to combat stress (6). Methods for efficient delivery of drug and pharmacogenomics, devices, Nano diagnostics, bio-compatible nanodevices, nanobiosensors, and nanobioconstructs have all benefited from nanobiotechnology (7). Figure 7.1 shows a representation of the practical use of nanotechnology in the field of agri-food, as highlighted in reference (8).

7.3 NANOTECHNOLOGY INTERVENTIONS IN SUSTAINABLE FOOD PRODUCTION

If humanity wants to feed a growing and wealthier world population, it will need to boost the production and efficiency of the food system. The potential exists for a considerable transformation in food preparation and consumption through the implementation of nanobiotechnological interventions in food production. It could also be utilized to make food that is more attractive in both flavor and texture. Nanobiotechnology is applied to enhance food packaging and storage. Innovative packing techniques may be utilized to prevent spoilage and waste while also making food more easily shipped and stored. Furthermore, food made with nanobiotechnology is much more resistant to disease and other environmental conditions. It holds the potential to substantially decrease the amount of food lost to spoilage and contamination. Nanobiotechnology enhances food security and safety as well (9,10).

The applications of nanobiotechnology interventions in sustainable food production are explained below.

7.4 FOOD PACKAGING AND PRESERVATION

7.4.1 Modifying the Structure of Food

The presence of fat imparts a distinct sensation that is unique to high-fat products. Because fat-rich diets aren't a healthier choice for everyone, for this purpose nano-structured lipids have indeed been developed to emulate the creamy consistency of foodstuffs for unique flavors and viscoelastic and mechanical characteristics qualities (11). These nanostructures find application in various food products like nanostructured mayonnaise, frozen desserts, frozen foods, yogurt, cheese, and milk spreads. Despite being lower in fat, they manage to replicate the taste of traditional high-fat products, providing customers with a healthier alternative (12).

7.4.2 Nano Encapsulation of Food Additives

The utilization of nano-encapsulated food additives helps prevent adverse alterations in food materials due to reduced usage of sugar, salt, fat, and preservatives. Additionally, it introduces the potential for the gradual release of various bioactive components (13). Encapsulation ingredients such as synthetic lycopene, benzoic acid, omega-3 fatty acids, isoflavones, and enzymes are widely applied in food products owing to their excellent sensory acceptance, uptake, absorption, and bioavailability (14). Nanocarriers can help protect against the use of certain difficult-to-detect food additives. Commonly produced nano-encapsulated food additives encompass omega-3 and omega-6 fatty acids, lycopene, benzoic acid, citric acid, ß-carotene, ascorbic acid, and fat-soluble vitamins A and E (13).

7.4.3 Food Packaging Nanocoating

The usage of nanocoating in packaged foods can be extremely beneficial because of their reduced thickness can increase membrane qualities by trading resistivity, elasticity, and food appearance.

Furthermore, they have additional dynamic functions including antimicrobial activity. The mechanical properties, gas permeability, swelling, and wetting characteristics of nanolaminate coatings are influenced by the choice of solubilizing materials, the sequence, the overall number of layers, and processing conditions (such as pH, temperature, and ionic strength). These nanocoating are derived from consumable carbohydrates, proteins, and lipids. Research indicates that nanocoating based on polysaccharides and proteins serve as robust barriers against oxygen and carbon dioxide, yet exhibit weaker resistance to moisture (15).

7.5 NANOBIOSENSORS

Nanobiosensors refer to sensors capable of monitoring physical parameters and converting them into signals, which are subsequently identified and analyzed. Such biosensors are known to be diagnostics instruments used to evaluate food quality as well as food security (16). Such nanobiosensors find application in food supply chains for the detection of extremely low levels of chemical contaminants, viruses, or bacteria.

It has been asserted that within an hour, a detector could discern different quantities of an infectious strain of *Escherichia coli* O157:H7 during milk assessment (17). Such sensors will be useful in identifying environmental pollution and controlling it in the food industry. In the food packaging industry, two types of nanosensors commonly employed are time-temperature integrators and gas detectors (18).

7.6 NANOCOMPOSITES

Nanocomposite packaged foods are a viable alternative to conventional packaging (19). Nanocomposites enhance the mechanical and thermal characteristics of packaging materials while also augmenting their barrier properties (20). The application of nanocomposites in packaged foods enables them to resist the thermal stress of food during production, transportation, and preservation of food (21). Nanocomposite food packaging offers additional advantages such as sensory stability, including taste, color, and texture. Furthermore, it contributes to greater product stability throughout the food chain, leading to reduced spoilage (22). In addition to serving as materials for food packaging, nanocomposites can also function as delivery systems, facilitating the migration of functional additives such as minerals, probiotics, and vitamins into food (23).

7.7 INTELLIGENT FOOD PACKAGING

According to Kerry, O'Grady (24) intelligent packaging communicates information about the state of food and its immediate environment to the processor, retailer, and customer (24). Purpose of intelligent packaging is to keep an eye on the food that has been packaged or the environment in which it is stored. Data regarding the product's quality or status can be detected, sensed, recorded, traced, or communicated by an intelligent packaging system throughout the food supply chain. In addition to the item itself, information about the item's composition, origin, and expiration date will be included in the package (25).

7.8 SELF-CLEANING NANOPARTICLES AND EDIBLE COATINGS

Due to the altered respiration mechanism, vegetables and fruits coated in nanosilver remain active throughout the transportation and storage processes. To enhance bioavailability and absorption, heavy metal nanoparticles are currently being investigated as a component of edible coating materials. The emerging research trend focuses on self-cleaning nanomaterials, driven by the novel self-cleaning functions exhibited by nanocomposites under specific storage conditions. Flores-López, Cerqueira (26) found that hybrid edible films with a thickness of less than 100 nm can enhance barrier properties against humidity, carbon dioxide, water, and oxygen. Additionally, these thinner films contribute to extending food shelf life and improving sensory features (27).

7.9 NANOBIOTECHNOLOGICAL INTERVENTIONS IN AGRICULTURAL PRODUCTION

Agriculture plays a crucial and sustainable role as it generates and provides raw materials, primarily for the food and animal industries. Scarcity of environmental assets (productive land, irrigation, soils), worldwide population growth, rising climate change and food security, demands for agricultural expansion to be financially, ecologically, and technically sustainable. The demand for novel technology interventions has been intended to allow revolutionizing the agricultural sectors for molecular disease control, early diagnosis, improving plant nutrient absorption, promoting sustainable agricultural products, and efficiency with reduced cost and less waste. Nanobiotechnology holds the potential to enhance agricultural productivity through various applications, including the development of nano-formulations for agrochemicals, such as fertilizers and pesticides, to optimize crop production. Additionally, the use of nanosensors and nanobiosensors in plant protection allows for the detection of diseases and agrochemical residues. Nanotechnologies can also be applied in gene engineering of plants, the creation of diagnostic tools for crop diseases, and the implementation of precision agricultural techniques, all aiming to increase crop yields while minimizing water and soil damage (28). The plants assimilate nanoparticles through both shoot and root entry points, subsequently penetrating the roots (refer to Figure 7.2).

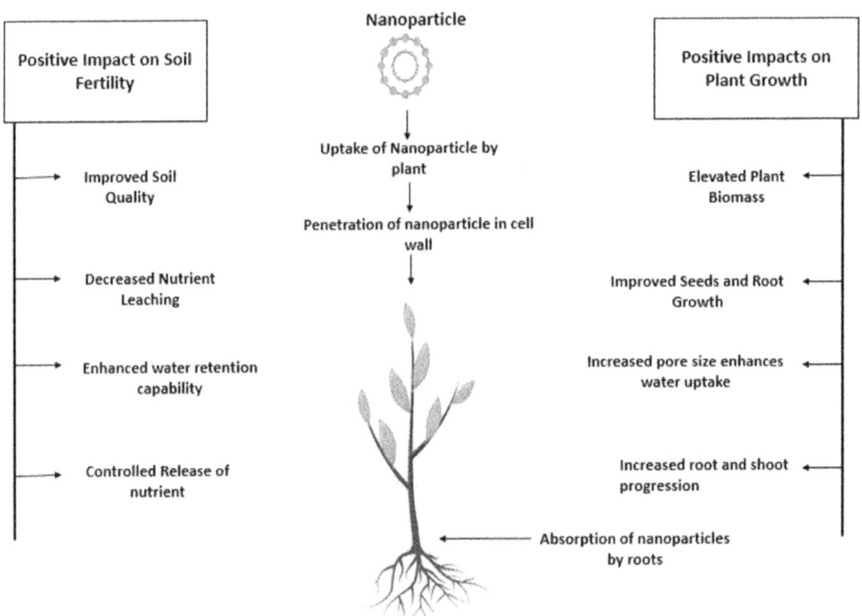

FIGURE 7.2 Mechanisms of nanoparticle absorption and favorable effects on both plant growth and soil fertility.

7.10 ADVANCEMENTS IN AGRICULTURAL DEVELOPMENT: EXPLORING NANO-PESTICIDES AND NANOFERTILIZERS

In agriculture, insecticides were routinely used to increase agricultural efficiency and productivity. Urea is by far the most generally used agricultural fertilizer as well as a producer of nitrates, ammonia, and urease, which are pervasive pollutants in groundwater that cause harmful algal blooms and have ecological impacts (29).

Nano-pesticides offer a method for efficiently managing pesticide distribution and achieving improved results with lower chemical doses. The initial category of nano-pesticides is envisioned as potent agents produced from nanomaterials, including metallic nanoparticles such as silver, copper, or metal oxides like zinc oxide, copper oxide, manganese dioxide, silicon dioxide, and titanium dioxide (30).

7.10.1 Nanofertilizer

The utilization of nanofertilizer technology represents a cutting-edge approach. Micro- or macronutrients can be encapsulated within nanomaterials, coated with a thin protective layer, or incorporated as emulsifiers or nanostructures in nanofertilizers (31). The use of nanofertilizers enables nutrients to be released slowly and in a regulated manner into the soil, reducing eutrophication and damage to water supplies (32). Nanofertilizers improve crop quality and yield by maximizing nutrient resource usage while minimizing production costs, so contributing to crop production. Nanofertilizers, in comparison to traditional fertilizers, exhibit a decreased nutrient release rate, heightened stability via agglomeration or absorption, and modified chemical speciation influenced by pH-dependent phenomena, redox potential, and soil access to ligands.

In comparison to soybeans (*Glycine max* L.) treated with conventional fertilizer, those treated with phosphatic nanofertilizer have shown a notable 32% increase in growth rate and a 20% increase in seed production (33).

7.11 CROP IMPROVEMENT

7.11.1 Gene Therapy in Plants

The production of better crops can be accelerated through gene editing. In 1996, the first genetically modified (GM) crops became available for sale. Since then, 28 countries have scaled up production of these crops, covering a tenfold larger area, utilizing 3-nm mesoporous silica nanoparticles to deliver DNA into cells. There's ongoing exploration into the feasibility of DNA sequencing using a fluidic system. However, a challenge arises from the fact that the introduction of additional genes through this process may interfere with or deactivate the expression of nearby important genes (34).

7.11.2 Seed Technology

Seeds represent the most critical input for the productivity of any crop. Typically, farmers receive seeds for sowing after rigorous testing for germination in well-equipped laboratories. However, the challenge arises during rainy conditions, where the lack of moisture makes it challenging to replicate the testing conditions in the field.

Based on the data, carbon nanotubes (CNTs) facilitate the transfer of water from the substrate into the seeds and serve as additional pores, enhancing water penetration through the seed coat. These processes contribute to the facilitation of seed germination, particularly beneficial in rain-fed farming systems. The utilization of CNTs improves moisture penetration, ultimately leading to an increase in germination rates (35).

7.12 PRECISION FARMING

7.12.1 Soil Remediation

Sequoia Pacific Research in Utah employs a nanotech-based soil binder named Soil Set to effectively mitigate soil erosion. Soil Set, a nanotech-based soil binder, offers a new answer to the problems of soil erosion. Soil Set is a soil binder using nanotechnology that employs bio-polymer molecules to greatly enhance soil structure stability, prevent erosion, and minimize flow. Soil Set provides several advantages over standard soil binding methods. It is harmless and inexpensive, holds soil particles more firmly, gives extra soil health advantages, aids in the reduction of harmful chemicals, and is simple to apply (36).

7.12.2 Bar Coding

Microbe recognition, including pathogens like Ebola and SARS, can be monitored through bacterial systems using DNA nano barcoding, functioning as a "nanosensor." This innovative approach allows for the assessment of chemical levels, soil nutrient levels, and water levels. Bio molecular tracing enhances efficiency, providing high-quality agricultural goods (37).

Additionally, DNA barcoding offers a rapid, straightforward, and cost-effective method for detecting the presence of various disease targets and microorganisms in food and the environment (38). Encoded nano barcodes, which are machine-readable, durable, and smaller than a micron in size, are now employed to track specific genes inserted into seeds before they're sold. This helps in monitoring the seeds' uniqueness in the market. It's worth noting that pathogens have a tendency to destroy stored seeds, and seeds themselves can sometimes carry diseases (39).

7.13 CONCLUSION

Nanobiotechnology holds immense potential for revolutionizing the food industry and sustainable agriculture. It offers a wide array of innovative solutions to tackle pressing global challenges such as food supply, sustainable farming practices, and climate variability. The applications of nanobiotechnology in areas like food packaging, preservation, processing, agricultural production, pest management, and seed technology bring numerous benefits. These include enhanced food safety and security, improved nutritional value, reduced food waste, and increased crop yields.

Furthermore, nanobiotechnology can play a vital role in addressing the critical issue of food security, which is paramount in meeting the needs of a rapidly growing global population. However, it is crucial to emphasize responsible research and development practices to ensure the safe and ethical deployment of nanobiotechnology. Collaboration among scientists, policymakers, industry stakeholders, and researchers is essential in order to fully unlock the transformative potential of this technology. It is also imperative to ensure that the benefits of nanobiotechnology are distributed equitably and accessible to all, fostering a more inclusive and sustainable future for our food systems.

REFERENCES

1. Gwarda-Gruszczyńska E, Starzyńska D, Gortatowicz K. Key enabling technologies in the European Union and in Poland. Optimum Economic Studies. 2020;2(100):47–62.
2. Chen H. Mapping nanotechnology innovations and knowledge: Global and longitudinal patent and literature analysis. 2008.
3. Chen A, Chatterjee S. Nanomaterials based electrochemical sensors for biomedical applications. Chemical Society Reviews. 2013;42(12):5425–38.
4. Abdussalam-Mohammed W. Review of therapeutic applications of nanotechnology in medicine field and its side effects. Journal of Chemical Reviews. 2019;1(3):243–51.
5. Sharma AR, Lee Y-H, Bat-Ulzii A, Bhattacharya M, Chakraborty C, Lee S-S. Recent advances of metal-based nanoparticles in nucleic acid delivery for therapeutic applications. Journal of Nanobiotechnology. 2022;20(1):1–21.
6. Tripathi A, Prakash S. Nanobiotechnology: Emerging trends, prospects, and challenges. Agricultural Nanobiotechnology. 2022;1:1–21.
7. Seil JT, Webster TJ. Antimicrobial applications of nanotechnology: Methods and literature. International Journal of Nanomedicine. 2012;7:2767.
8. Pramanik P, Krishnan P, Maity A, Mridha N, Mukherjee A, Rai V. Application of nanotechnology in agriculture. Environmental Nanotechnology 2020;4:317–48.
9. Nazir M, Arif S, Khan RS, Nazir W, Khalid N, Maqsood S. Opportunities and challenges for functional and medicinal beverages: Current and future trends. Trends in Food Science & Technology. 2019;88:513–26.
10. Imathiu S. Benefits and food safety concerns associated with consumption of edible insects. NFS Journal. 2020;18:1–11.
11. Posey KA, Clegg DJ, Printz RL, Byun J, Morton GJ, Vivekanandan-Giri A, et al. Hypothalamic proinflammatory lipid accumulation, inflammation, and insulin resistance in rats fed a high-fat diet. American Journal of Physiology-Endocrinology and Metabolism. 2009;296(5):E1003–12.
12. Singh L, Varshney JG, Agarwal T. Polycyclic aromatic hydrocarbons' formation and occurrence in processed food. Food Chemistry. 2016;199:768–81.
13. Watkins R, Castle L, Chaudhry Q. Nanotechnologies in food: RSC; 2010.
14. Bastarrachea L, Dhawan S, Sablani SS. Engineering properties of polymeric-based antimicrobial films for food packaging: A review. Food Engineering Reviews. 2011; 3(2):79–93.
15. Vasile C. Polymeric nanocomposites and nanocoatings for food packaging: A review. Materials. 2018;11(10):1834.
16. Naresh V, Lee N. A review on biosensors and recent development of nanostructured materials-enabled biosensors. Sensors. 2021;21(4):1109.
17. Omanović-Mikličanina E, Maksimović M. Nanosensors applications in agriculture and food industry. Bulletin of the Chemists and Technologists of Bosnia and Herzegovina. 2016;47:59–70.

18. Pradhan N, Singh S, Ojha N, Shrivastava A, Barla A, Rai V, et al. Facets of nanotechnology as seen in food processing, packaging, and preservation industry. BioMed Research International. 2015;1(2015):365672.
19. Huang J-Y, Li X, Zhou W. Safety assessment of nanocomposite for food packaging application. Trends in Food Science & Technology. 2015;45(2):187–99.
20. Avella M, De Vlieger JJ, Errico ME, Fischer S, Vacca P, Volpe MG. Biodegradable starch/clay nanocomposite films for food packaging applications. Food Chemistry. 2005; 93(3):467–74.
21. Ray SS, Okamoto M. Polymer/layered silicate nanocomposites: A review from preparation to processing. Progress in Polymer Science. 2003;28(11):1539–641.
22. Sorrentino A, Gorrasi G, Vittoria V. Potential perspectives of bio-nanocomposites for food packaging applications. Trends in Food Science & Technology. 2007;18(2):84–95.
23. Rodriguez F, Sepulveda HM, Bruna J, Guarda A, Galotto MJ. Development of cellulose eco-nanocomposites with antimicrobial properties oriented for food packaging. Packaging Technology and Science. 2013;26(3):149–60.
24. Kerry J, O'Grady M, Hogan S. Past, current and potential utilisation of active and intelligent packaging systems for meat and muscle-based products: A review. Meat Science. 2006;74(1):113–30.
25. Yam KL, Takhistov PT, Miltz J. Intelligent packaging: concepts and applications. Journal of Food Science. 2005;70(1):R1–10.
26. Flores-López ML, Cerqueira MA, de Rodríguez DJ, Vicente AA. Perspectives on utilization of edible coatings and nano-laminate coatings for extension of postharvest storage of fruits and vegetables. Food Engineering Reviews. 2016;8:292–305.
27. López M, Cerqueira MÂPR, de Rodríguez DJ, Vicente AA. Perspectives on utilization of edible coatings and nano-laminate coatings for extension of postharvest storage of fruits and vegetables. 2016.
28. Adegbeye M, Reddy PRK, Obaisi A, Elghandour M, Oyebamiji K, Salem A, et al. Sustainable agriculture options for production, greenhouse gasses and pollution alleviation, and nutrient recycling in emerging and transitional nations–An overview. Journal of Cleaner Production. 2020;242:118319.
29. Gržinić G, Piotrowicz-Cieślak A, Klimkowicz-Pawlas A, Górny RL, Ławniczek-Wałczyk A, Piechowicz L, et al. Intensive poultry farming: A review of the impact on the environment and human health. Science of the Total Environment. 2022;858:160014.
30. Vasseghian Y, Arunkumar P, Joo S-W, Gnanasekaran L, Kamyab H, Rajendran S, et al. Metal-organic framework-enabled pesticides are an emerging tool for sustainable cleaner production and environmental hazard reduction. Journal of Cleaner Production. 2022;373:133966.
31. Haris M, Hussain T, Mohamed HI, Khan A, Ansari MS, Tauseef A, et al. Nanotechnology–A new frontier of nano-farming in agricultural and food production and its development. Science of the Total Environment. 2022;857:159639.
32. Manjunatha S, Biradar D, Aladakatti YR. Nanotechnology and its applications in agriculture: A review. Journal of Farm Sciences. 2016;29(1):1–13.
33. Li Q, Ma C, White JC, Xing B. Effects of phosphorus ensembled nanomaterials on nutrient uptake and distribution in *Glycine max* L. under simulated precipitation. Agronomy. 2021;11(6):1086.
34. Krajewska K, Gołkowska AM, Nowak M, Kozakiewicz-Latała M, Pudło W, Żak A, et al. Molecular level characterisation of the surface of carbohydrate-functionalised mesoporous silica nanoparticles (MSN) as a potential targeted drug delivery system via high resolution magic angle spinning (HR-MAS) NMR spectroscopy. International Journal of Molecular Sciences. 2022;23(11):5906.
35. Khodakovskaya M, Dervishi E, Mahmood M, Xu Y, Li Z, Watanabe F, et al. Carbon nanotubes are able to penetrate plant seed coat and dramatically affect seed germination and plant growth. American Chemical Society Nano. 2009;3(10):3221–7.

36. Hatti V, Raghavendra M, Singh Y, Sharma S, Halli HM, Goud B. Application of nano technology in crop production to enhance resource use efficiency with special reference to soil and water conservation: A review. Journal of Soil and Water Conservation. 2020;19(4):375–81.
37. Vijayalakshmi C, Chellaram C, Kumar SL, editors. Trendy usage of nanotechnology in agriculture—A review. 2014 International Conference on Science Engineering and Management Research (ICSEMR); 2014: IEEE.
38. Jinbo U, Kato T, Ito M. Current progress in DNA barcoding and future implications for entomology. Entomological Science. 2011;14(2):107–24.
39. Agrawal S, Rathore P. Nanotechnology pros and cons to agriculture: a review. International Journal of Current Microbiology and Applied Science. 2014;3(3):43–55.

8 Nanomaterials in the Treatment of Bacterial Infections

Shumaila Ijaz, Javed Iqbal, Banzeer Ahsan Abbasi, Tabassum Yaseen, Rashid Iqbal, Ghulam Murtaza, Tariq Mahmood, Zaman Ashraf, Hussan Ara Begum, and Ikram Ullah

8.1 INTRODUCTION

Bacterial infections are a major global health concern due to their broad frequency, propensity for serious disease, and rising antibiotic resistance. They have a significant global impact on morbidity and death, impacting people of all ages and socioeconomic statuses. Bacterial infections may present themselves in a variety of ways, including bloodstream infections, gastrointestinal infections, urinary tract infections, and respiratory infections. In particular, vulnerable groups, such as young children, elderly adults, and those with compromised immune systems, are susceptible to these infections, which may result in severe consequences such as pneumonia, sepsis, meningitis, and endocarditis (Govindaraj Vaithinathan and Vanitha, 2018, Shrivastava et al., 2018). A serious concern to world health is the establishment and spread of bacterial antibiotic resistance. Antibiotic overuse and abuse in healthcare and agriculture have accelerated the emergence of resistant bacterial strains. Because of this resistance, current medicines are less effective, making it harder or even impossible to treat illnesses that were formerly manageable. It also restricts the kind of treatments available, which causes illnesses to last longer, costs more money to cure, and raises fatality rates (Centers for Disease Control and Prevention, 2021).

Bacterial infections need to be addressed as a global health issue in several ways. It entails encouraging proper antibiotic use, increasing infection prevention and control procedures, bolstering healthcare systems, and supporting the development of fresh antibiotics and nontraditional treatments. In order to properly address this expanding hazard, cooperation between healthcare professionals, researchers, governments, and communities is also essential (Ventola, 2015). Due to the rising incidence of germs that are resistant to antibiotics and the ineffectiveness of current treatment choices, it is critical that new methods be developed to fight bacterial illnesses. To successfully address this global health issue, new approaches and treatments must be developed. For many years, the cornerstone of bacterial illness therapy has been

 DOI: 10.1201/9781003378563-8

conventional antibiotics. However, many antibiotics are no longer effective due to the establishment and spread of antibiotic resistance. As a result, there are now far fewer treatment choices available, putting patients at risk for infections that cannot be treated and adding to the strain on healthcare systems. Researchers are looking at several cutting-edge strategies to tackle bacterial infections in an effort to solve this problem (Bebell and Muiru, 2014, Iqbal et al., 2018a).

Scientists are trying to find and create new antibiotics that can efficiently eradicate germs that are resistant to antibiotics. This entails finding fresh targets for bacteria, looking into other places to find antibiotics, and using creative approaches to drug creation (Terreni et al., 2021). Combination medicines may improve treatment effectiveness and circumvent resistance mechanisms by mixing several antibiotics or antibiotics with non-antibiotic substances. Drug combinations that work well together may help treat patients better and reduce resistance (Huh and Kwon, 2011, Iqbal et al., 2018b). As a possible complement or alternative to antibiotics, monoclonal antibodies that target certain bacterial infections or their virulence characteristics are being studied. These antibodies strengthen the immune system's capacity to fight infections by inhibiting bacterial adhesion, neutralizing bacterial toxins, and neutralizing germs (Motley et al., 2019). Viruses known as bacteriophages may infect and eradicate bacteria. Using certain bacteriophages to target and eliminate bacterial illnesses is known as phage treatment. This method, especially in confined settings, offers promise for treating illnesses resistant to antibiotics (Shan et al., 2023). Improving the body's immune response to infections by focusing on host components or pathways rather than the germs themselves. Treatment results may be enhanced by altering bacterial virulence pathways or the human immune system (Yang et al., 2023, Ali et al., 2021).

In terms of treating bacterial infections, nanomaterials, which are distinguished by their distinctive features at the nanoscale, have drawn much interest. These substances have unique physicochemical characteristics, such as a high surface area-to-volume ratio, programmable surface properties, and improved reactivity, which make them intriguing candidates for cutting-edge treatment strategies against bacterial infections. By directly interacting with bacteria or by carefully controlling the release of antimicrobial chemicals, nanomaterials may demonstrate strong antibacterial characteristics. In order to successfully kill or stop the development of bacteria, they may damage bacterial cell membranes, impede enzyme function, cause oxidative stress, and interfere with bacterial adhesion (Joshi et al., 2014, Hameed et al., 2020). Nanomaterials may operate as medication carriers or delivery systems to increase the effectiveness of traditional antibiotics. They may enhance drug solubility, shield antibiotics from deterioration, and enable targeted distribution to the location of infection, enhancing treatment effectiveness and minimizing adverse effects (Jia et al., 2023, Cai et al., 2022). Because of their growing antibiotic resistance, bacterial biofilms provide a major problem in the treatment of infections. Biofilms may be penetrated and disturbed by nanomaterials, which can also destroy already existing biofilms or stop their production. They may reduce biofilm matrices' adherence, block biofilm adhesion, and boost the effectiveness of antimicrobial drugs in combating bacteria found in biofilms (Mi et al., 2018). The development of sensitive and focused diagnostic techniques for bacterial illnesses has shown potential for nanomaterials. To identify and visualize bacterial infections, they may be functionalized using targeting ligands and imaging probes. This helps with early diagnosis and treatment response monitoring (Khalil et al., 2021, Khan et al., 2022). Metal nanoparticles (NPs) (such as silver and gold), metal oxides (such as zinc oxide and titanium dioxide), carbon-based nanomaterials (such as graphene and carbon nanotubes), and polymeric NPs are just a few of the kinds of nanomaterials that have been investigated for the treatment of infections. The stability, biocompatibility, and targeted distribution of these NPs may all be improved by surface modification and engineering (Ahmad et al., 2022, Huang et al., 2023). Nanomaterials have a great deal of promise for treating infections, but their long-term impacts and safety must be carefully considered. The effective introduction of nanomaterial-based medicines into clinical practice depends on factors such as biocompatibility, biodegradability, possible toxicity, and regulatory issues (Duran et al., 2016, Sachan et al., 2023). Nanomaterials have emerged as promising tools for infection treatment due to their unique properties and advantages. The nanoscale size of these materials offers distinct characteristics that can be leveraged to combat infections caused by bacteria, viruses, and other pathogens (Figure 8.1).

Bacterial infections are a significant global health concern, causing a wide range of diseases with varying clinical presentations. The mechanisms of bacterial pathogenesis involve adherence, colonization, invasion, evasion of host defenses, toxin production, and immune evasion. Understanding these mechanisms is essential for

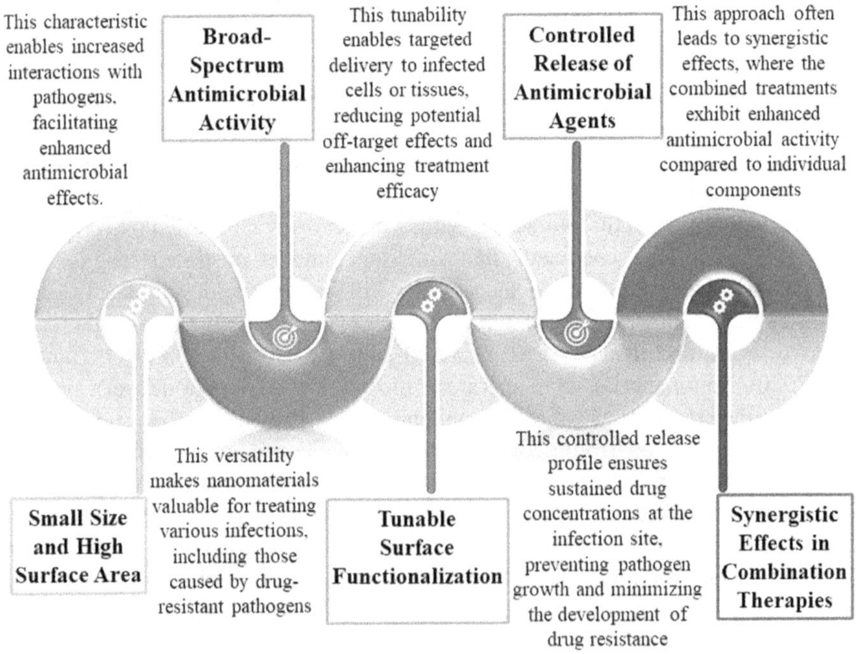

FIGURE 8.1 Properties and advantages of nanomaterials for bacterial infection treatment.

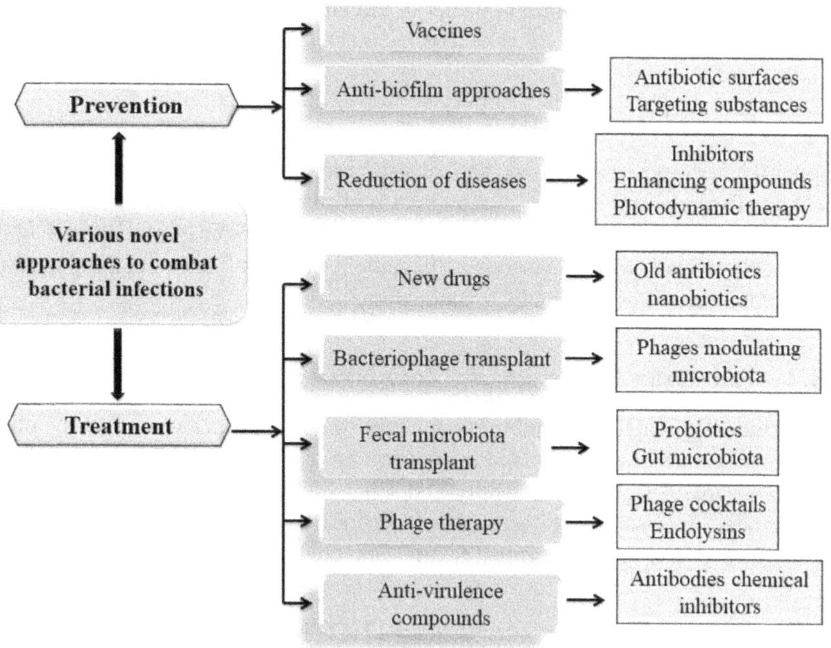

FIGURE 8.2 Various novel approaches to combat bacterial infections.

developing effective treatment and prevention strategies to combat bacterial infections (Figure 8.2) and improve public health.

8.2 TYPES OF NANOMATERIALS WITH INHERENT ANTIBACTERIAL PROPERTIES

Nanomaterials are engineered structures with nanoscale dimensions, which endow them with unique properties and applications. Among these properties, some nanomaterials possess inherent antibacterial properties, making them promising candidates for combating bacterial infections. This chapter provides an overview of various types of nanomaterials with intrinsic antibacterial properties, supported by relevant research references.

8.3 NANOMATERIALS WITH INTRINSIC ANTIBACTERIAL PROPERTIES

8.3.1 Silver Nanoparticles (AgNPs)

Silver NPs (AgNPs) are one of the most extensively studied nanomaterials with inherent antibacterial activity. The small size and high surface area of AgNPs facilitate interactions with bacterial cells, leading to membrane damage and disruption of essential cellular processes (Morones et al., 2005, Abbasi et al., 2019). AgNPs have demonstrated broad-spectrum antibacterial effects against both Gram-positive and Gram-negative bacteria.

8.3.2 Copper Nanoparticles (CuNPs)

Copper (CuNPs) possess intrinsic antibacterial properties due to the release of copper ions. The released copper ions interact with bacterial cells, causing oxidative stress, DNA damage, and disruption of cell membranes (Feng et al., 2000). CuNPs have shown efficacy against various pathogenic bacteria and have been explored for antimicrobial coatings and medical applications.

8.3.3 Zinc Oxide Nanoparticles (ZnO NPs)

Zinc oxide NPs (ZnONPs) exhibit inherent antibacterial properties primarily attributed to the release of zinc ions and the generation of reactive oxygen species (ROS) upon exposure to light (Padmavathy and Vijayaraghavan, 2008). ZnO NPs have demonstrated antibacterial effects against various bacterial strains and have been investigated for use in antimicrobial textiles and coatings.

8.3.4 Titanium Dioxide Nanoparticles (TiO_2 NPs)

Titanium dioxide NPs possess photocatalytic antibacterial properties. Under ultraviolet (UV) light irradiation, TiO_2 NPs generate ROS, which exhibits strong antibacterial activity (Foster et al., 2011). TiO_2 NPs have been investigated for various applications, including antimicrobial coatings, water treatment, and air purification.

8.3.5 Chitosan Nanoparticles

Chitosan NPs are derived from chitin, a natural polysaccharide found in the exoskeleton of crustaceans. Chitosan NPs possess inherent antibacterial properties due to their positive charge, which allows them to interact with negatively charged bacterial cell membranes and disrupt their integrity (Kumar et al., 2010). Chitosan NPs have shown antimicrobial activity against a range of bacterial pathogens. Nanomaterials with inherent antibacterial properties offer promising solutions for combating bacterial infections. AgNPs, CuNPs, ZnONPs, titanium dioxide NPs, and chitosan NPs are among the types of nanomaterials that have demonstrated intrinsic antibacterial activity. The unique properties of these nanomaterials make them attractive candidates for various antimicrobial applications, including medical devices, antimicrobial coatings, water purification, and air filtration.

8.4 MECHANISMS OF ACTION OF ANTIBACTERIAL NANOMATERIALS

Antibacterial nanomaterials have emerged as promising agents for combating bacterial infections due to their unique properties and mechanisms of action. These nanomaterials can effectively target bacterial cells and disrupt their vital processes, leading to antibacterial effects (Figure 8.3).

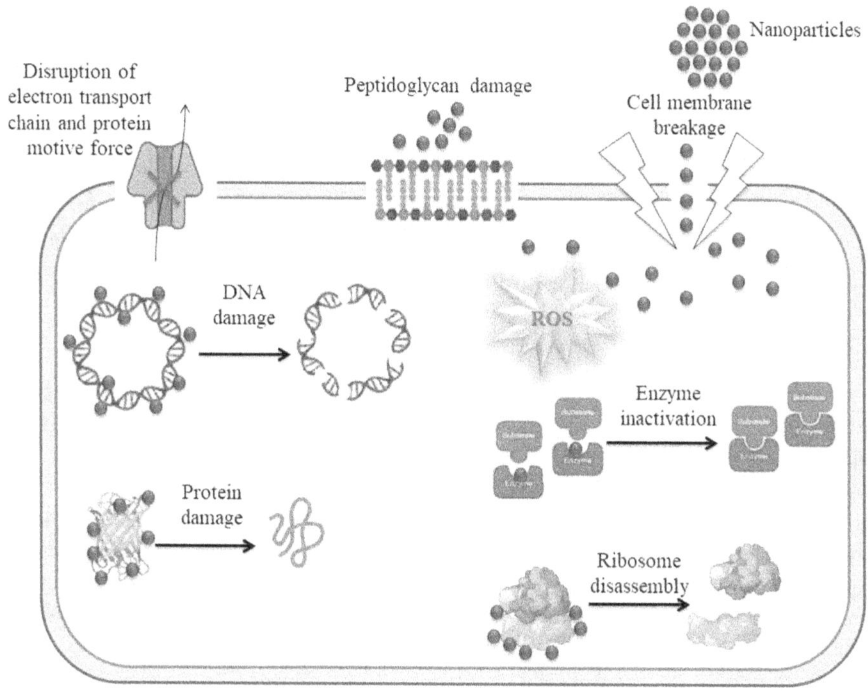

FIGURE 8.3 Schematic representation of general mechanisms for the antibacterial mode of action of nanoparticles.

8.4.1 Reactive Oxygen Species (ROS) Generation

Many antibacterial nanomaterials, such as AgNPs, CuNPs, and ZnONPs, can generate ROS upon contact with bacterial cells. ROS, such as superoxide radicals, hydrogen peroxide (H_2O_2), and hydroxyl radicals, induce oxidative stress within bacterial cells, causing damage to proteins, lipids, and DNA (Feng et al., 2000). This oxidative stress disrupts essential cellular processes, leading to bacterial cell death.

8.4.2 Membrane Disruption

Antibacterial nanomaterials, including AgNPs and chitosan NPs, can interact with bacterial cell membranes. Positively charged nanomaterials, such as chitosan NPs, can bind to negatively charged bacterial cell membranes, leading to membrane disruption and leakage of cellular contents (Rabea et al., 2003). AgNPs can also interact with the bacterial membrane, resulting in destabilization and increased permeability.

8.4.3 Metal Ion Release

Certain antibacterial nanomaterials, such as CuNPs and ZnONPs, release metal ions in their surroundings. These metal ions, particularly copper and zinc ions, can penetrate bacterial cells and interfere with essential enzymatic processes (Morones et al., 2005). The disruption of cellular processes leads to bacterial cell death and inhibits bacterial growth.

8.4.4 Nanomaterial-Induced Stress

Some antibacterial nanomaterials exert direct stress on bacterial cells due to their unique physical and chemical properties. For instance, titanium dioxide NPs can induce stress on bacterial cells under UV light irradiation, leading to the generation of ROS and bacterial cell damage (Foster et al., 2011).

8.4.5 Disruption of Cellular Functions

Antibacterial nanomaterials can interfere with various cellular functions of bacteria. Some nanomaterials can disrupt bacterial DNA replication, inhibit enzyme activity, and interfere with cellular signaling pathways, leading to bacterial cell death (Pulit-Prociak et al., 2011). Antibacterial nanomaterials exhibit multiple mechanisms of action to combat bacterial infections. ROS generation, membrane disruption, metal ion release, nanomaterial-induced stress, and interference with cellular functions contribute to their antibacterial effects. Understanding these mechanisms of action is crucial for developing effective antibacterial nanomaterials and harnessing their potential in various biomedical and environmental applications.

8.5 NANOMATERIALS FOR ENHANCED ANTIBACTERIAL DELIVERY

The emergence of antibiotic-resistant bacteria has become a significant global health concern. Conventional antibiotic therapies often face challenges in achieving sufficient drug concentrations at the infection site while minimizing systemic side effects. Nanocarriers for targeted antibiotic delivery have garnered considerable attention as promising strategies to overcome these limitations. These nanoscale drug carriers can improve the therapeutic efficacy of antibiotics by selectively delivering them to the site of infection, enhancing drug stability, and reducing toxicity. This chapter provides an introduction to nanocarriers for targeted antibiotic delivery, supported by relevant research references.

8.5.1 Nanocarriers in Antibacterial Delivery

Nanocarriers are nanoscale platforms designed to encapsulate and deliver therapeutic agents, including antibiotics, to specific targets in the body. They can be formulated from various materials, such as lipids, polymers, and inorganic NPs. The size and surface properties of nanocarriers can be tailored to improve drug loading, stability, and controlled release (Ventola, 2017).

8.5.2 Targeted Drug Delivery

Targeted antibiotic delivery aims to deliver antimicrobial agents directly to the site of infection while minimizing exposure to healthy tissues. By achieving site-specific

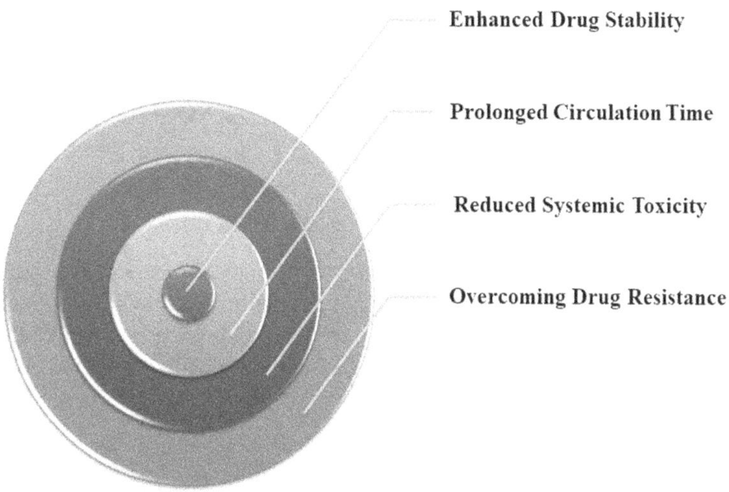

FIGURE 8.4 Benefits of nanocarrier-based antibiotic delivery.

drug release, targeted delivery enhances the local concentration of antibiotics, overcoming bacterial resistance and reducing the risk of systemic side effects (Wang et al., 2019).

8.5.3 Benefits of Nanocarrier-Based Antibiotic Delivery

Nanocarriers protect antibiotics from enzymatic degradation and premature drug release, ensuring stability during circulation and storage. Nanocarriers can extend the circulation time of antibiotics, enabling sustained release and prolonged exposure to the infection site (Wang et al., 2019) (Figure 8.4). Targeted antibiotic delivery minimizes drug accumulation in healthy tissues, reducing systemic toxicity and side effects associated with conventional antibiotic treatments (Jain et al., 2015). Nanocarrier-based delivery strategies can enhance the efficacy of existing antibiotics against drug-resistant bacteria, as targeted delivery increases local drug concentrations (Vishwakarma et al., 2019). Nanocarriers for targeted antibiotic delivery represent a promising approach to address the challenges associated with conventional antibiotic therapies. By improving drug stability, enhancing drug efficacy, and reducing systemic toxicity, these nanoscale drug carriers offer a potential solution to combat antibiotic-resistant infections effectively. Ongoing research in nanomedicine is continuously exploring new formulations and strategies to optimize targeted antibiotic delivery for improved patient outcomes.

8.6 STRATEGIES TO IMPROVE ANTIBIOTIC EFFICACY USING NANOMATERIALS

The rise of antibiotic-resistant bacteria poses a significant challenge to healthcare, necessitating the development of innovative strategies to enhance antibiotic efficacy.

Nanomaterials have emerged as promising tools to improve antibiotic effectiveness by addressing drug delivery challenges, overcoming bacterial resistance, and enhancing targeted drug delivery. This chapter discusses various strategies that utilize nanomaterials to improve antibiotic efficacy, supported by relevant research references.

8.6.1 Encapsulation of Antibiotics in Nanoparticles

NP-based encapsulation of antibiotics improves drug stability and bioavailability, leading to enhanced therapeutic efficacy. By entrapping antibiotics within NPs, such as liposomes, polymer NPs, or metal NPs, the drugs are protected from enzymatic degradation and premature clearance, resulting in sustained release and prolonged circulation time (Lu et al., 2018).

8.6.2 Synergistic Antibacterial Combinations

Nanomaterials can be utilized to deliver multiple antibiotics or combine antibiotics with other therapeutic agents, creating synergistic effects against bacteria. For instance, metal NPs such as AgNPs can be combined with antibiotics to potentiate their antibacterial activity through the generation of ROS and disruption of bacterial cell membranes (Panacek et al., 2018).

8.6.3 Overcoming Drug Resistance

Nanomaterials can help overcome bacterial drug resistance by enhancing the delivery and accumulation of antibiotics at the site of infection. Targeted antibiotic delivery using nanocarriers can improve the drug's local concentration, allowing for effective action against resistant bacterial strains and reducing the risk of resistance development (Alavala et al., 2021).

8.6.4 Combination Therapy with Photothermal or Photodynamic Effects

Nanomaterials possessing photothermal or photodynamic properties can be combined with antibiotics to achieve synergistic effects against bacteria. Photothermal agents, such as gold NPs (AuNPs), can generate heat upon exposure to light, which can enhance the permeability of bacterial cell membranes, improving antibiotic uptake and efficacy (Li et al., 2019).

8.6.5 Targeted Drug Delivery

Nanomaterials can be functionalized with ligands or antibodies that specifically bind to bacterial cell surface receptors, allowing for targeted antibiotic delivery. This approach enables the selective accumulation of antibiotics at the infection site, increasing local drug concentrations and minimizing off-target effects (Jin et al., 2020). Nanomaterials offer versatile and effective strategies to improve antibiotic efficacy, addressing the challenges posed by antibiotic resistance and conventional drug delivery methods. By encapsulating antibiotics, creating synergistic combinations, targeting specific bacterial cells, and combining with photothermal or photodynamic effects, nanomaterials hold great promise in revolutionizing antibiotic therapy. The continued research and development of nanomaterial-based approaches will play a vital role in combating antibiotic-resistant bacteria and improving patient outcomes.

8.7 EXAMPLES OF NANOMATERIAL-BASED SYSTEMS FOR CONTROLLED AND SUSTAINED DRUG RELEASE

Nanomaterial-based systems have shown great potential for controlled and sustained drug release, offering advantages such as improved drug stability, reduced dosing frequency, and enhanced therapeutic efficacy. These systems utilize nanoscale carriers to encapsulate and release drugs in a controlled manner, allowing for targeted delivery and prolonged drug release at the site of action. This chapter provides examples of nanomaterial-based systems for controlled and sustained drug release, supported by relevant research references.

8.7.1 Liposomes

Liposomes are lipid-based nanocarriers composed of phospholipid bilayers. They have been extensively used for controlled drug release due to their ability to entrap both hydrophilic and hydrophobic drugs within their aqueous core and lipid layers. Liposomes offer sustained drug release by modifying their lipid composition, size, and surface charge (Torchilin, 2011).

8.7.2 Polymeric Nanoparticles

Polymeric NPs, such as poly(lactic-*co*-glycolic acid) (PLGA) NPs, are widely used for controlled drug release. PLGA NPs can be engineered to control drug release rates by adjusting the polymer composition and particle size. Drug release from polymeric NPs can be sustained over extended periods, depending on the degradation rate of the polymer (Jain et al., 2019).

8.7.3 Mesoporous Silica Nanoparticles

Mesoporous silica NPs (MSNs) possess a highly ordered porous structure that allows for high drug-loading capacities and controlled drug release. The porous nature of MSNs enables drug molecules to be adsorbed within the nanopores, and the release can be tailored by modifying the pore size and surface functionalization (He et al., 2010).

8.7.4 Dendrimers

Dendrimers are branched, tree-like macromolecules with well-defined sizes and structures. They can be used as carriers for controlled drug release by encapsulating drugs in their interior voids or through covalent conjugation. Dendrimer-based drug delivery systems enable sustained drug release and can be designed to respond to specific stimuli, such as pH or temperature changes (Madaan et al., 2014).

8.7.5 Hydrogels

Hydrogels are three-dimensional networks of hydrophilic polymers that can absorb and retain large amounts of water. They are used as carriers for controlled drug release due to their ability to hold drugs within their structure and release them gradually over time. Hydrogels can be administered as injectables or topical formulations, providing sustained drug release at the site of application (Shah et al., 2016). Nanomaterial-based systems for controlled and sustained drug release offer promising solutions to overcome the limitations of conventional drug delivery. Liposomes, polymeric NPs, MSNs, dendrimers, and hydrogels are examples of nanocarriers that can encapsulate drugs and release them in a controlled manner. These systems provide enhanced therapeutic efficacy, reduced dosing frequency, and targeted drug delivery, making them valuable tools for improving patient outcomes in various medical applications.

8.8 NANOMATERIALS FOR IMAGING AND DIAGNOSIS OF BACTERIAL INFECTIONS

Bacterial infections remain a significant global health challenge, and early and accurate diagnosis is crucial for effective treatment. Nanomaterial-based imaging techniques have emerged as powerful tools for visualizing bacterial infections at the molecular and cellular levels. These techniques offer advantages such as high sensitivity, specificity, and noninvasiveness, enabling real-time monitoring of infections. This chapter provides an introduction to nanomaterial-based imaging techniques for bacterial infections, supported by relevant research references.

8.8.1 Quantum Dots (QDs) Imaging

Quantum dots (QDs) are nanoscale semiconductor crystals that fluoresce upon excitation with light. QDs have unique optical properties, including tunable emission wavelengths and high photostability, making them ideal for bacterial imaging (Kumar et al., 2020). Functionalized QDs can target specific bacterial species or surface biomarkers, allowing for the specific detection and imaging of bacteria in infected tissues.

8.8.2 Gold Nanoparticles (AuNPs) Imaging

AuNPs have unique plasmonic properties, enabling them to scatter and absorb light at specific wavelengths. AuNPs can be functionalized with targeting ligands or antibodies to selectively bind to bacterial cells (Dykman and Khlebtsov, 2017). By using techniques such as dark-field microscopy and surface-enhanced Raman scattering (SERS), AuNPs can visualize bacteria with high sensitivity and provide valuable information about bacterial localization and distribution in infected tissues.

8.8.3 Iron Oxide Nanoparticles (IONPs) Imaging

Iron oxide NPs (IONPs) possess superparamagnetic properties, making them suitable for magnetic resonance imaging (MRI) of bacterial infections (Lee et al., 2015). IONPs can be coated with specific targeting agents to accumulate at the site of bacterial infection, allowing for the sensitive and noninvasive detection of bacteria in deep tissues.

8.8.4 Silica Nanoparticles Imaging

Silica NPs possess excellent biocompatibility and can be easily functionalized with targeting moieties for bacterial imaging (Liu et al., 2017). Silica NPs can encapsulate imaging agents, such as fluorophores or contrast agents, and deliver them to bacterial cells. These NPs offer high stability, low toxicity, and versatile surface modification options for specific bacterial targeting.

8.8.5 Carbon-Based Nanomaterials Imaging

Carbon-based nanomaterials, such as carbon nanotubes and graphene oxide, have unique optical and electronic properties that enable their application in bacterial imaging (Wang et al., 2019). These nanomaterials can be conjugated with targeting ligands to selectively bind to bacterial cells and provide contrast for imaging modalities such as fluorescence imaging and photoacoustic imaging. Nanomaterial-based imaging techniques offer innovative and versatile approaches for visualizing bacterial infections. QDs, AuNPs, IONPs, silica NPs, and carbon-based nanomaterials provide various imaging modalities for sensitive and specific detection of bacterial cells in infected tissues. As research in nanotechnology advances, nanomaterial-based imaging techniques hold great promise for improving the diagnosis and monitoring of bacterial infections, leading to more targeted and effective therapeutic interventions.

8.9 APPLICATIONS OF NANOMATERIALS IN RAPID AND ACCURATE BACTERIAL DETECTION

The timely and accurate detection of bacterial pathogens is critical for effective disease management and infection control. Nanomaterials have revolutionized bacterial detection methods by offering rapid, sensitive, and specific approaches for identifying bacterial species. This chapter highlights the diverse applications of nanomaterials in the rapid and accurate detection of bacterial pathogens, supported by relevant research references.

8.9.1 Nanoparticle-Based Biosensors

NP-based biosensors have emerged as powerful tools for bacterial detection due to their high surface area and sensitivity. These biosensors can be functionalized with specific antibodies, aptamers, or peptides that selectively bind to target bacterial cells (Li et al., 2019). When the target bacteria are captured, the biosensor undergoes a detectable signal change, such as fluorescence, colorimetry, or electrochemical response, enabling rapid and specific bacterial identification.

8.9.2 Nanomaterial-Enhanced PCR Assays

Polymerase chain reaction (PCR) is a widely used technique for bacterial detection but can be limited by low sensitivity. The incorporation of nanomaterials, such as AuNPs or carbon nanotubes, into PCR assays can enhance signal amplification and improve detection sensitivity (Loo et al., 2015). Nanomaterials can act as carriers for DNA probes or enhance fluorescence signals, enabling the rapid and accurate identification of low-abundance bacterial targets.

8.9.3 Nanomaterial-Based Lateral Flow Assays

Lateral flow assays (LFAs) are simple and rapid tests widely used for point-of-care bacterial detection. Nanomaterials, such as AuNPs or QDs, can be incorporated into LFAs to improve the sensitivity and quantification of bacterial targets (Nagarajan et al., 2021). The unique optical properties of nanomaterials allow for direct visualization of LFA results, enabling rapid and accurate identification of bacterial pathogens.

8.9.4 Surface-Enhanced Raman Scattering (SERS)

SERS is a sensitive technique that can identify bacterial species based on their unique Raman spectral signatures. Nanomaterials, such as silver or AuNPs, can act as SERS substrates, enhancing Raman signals and improving the sensitivity of bacterial detection (Li et al., 2017). SERS-based bacterial detection offers label-free, rapid, and highly specific identification of pathogens.

8.9.5 Nanomaterial-Based Microfluidic Devices

Microfluidic devices integrated with nanomaterials have emerged as powerful tools for on-chip bacterial detection. Nanomaterials can be immobilized within microfluidic channels to capture bacterial cells, and the devices can be equipped with optical or electrical sensors for real-time monitoring (Liu et al., 2020). These nanomaterial-based microfluidic devices enable rapid, sensitive, and automated bacterial detection. Nanomaterials offer diverse and innovative applications in the rapid and accurate detection of bacterial pathogens. From NP-based biosensors and nanomaterial-enhanced PCR assays to nanomaterial-based LFAs and SERS, these technologies provide sensitive, specific, and rapid bacterial identification. Nanomaterial-based microfluidic devices further enhance the potential for point-of-care bacterial detection. As nanotechnology continues to advance, the integration of nanomaterials into bacterial detection methods holds great promise for improving disease diagnosis, infection control, and public health outcomes.

8.10 ADVANCEMENTS IN NANOMATERIAL-BASED DIAGNOSTIC PLATFORMS FOR BACTERIAL INFECTIONS

The accurate and timely diagnosis of bacterial infections is crucial for effective treatment and infection control. Nanomaterial-based diagnostic platforms have witnessed remarkable advancements in recent years, offering innovative approaches for sensitive, rapid, and specific detection of bacterial pathogens. Advancements in NP-based biosensors have revolutionized bacterial detection. Researchers have developed biosensors that employ functionalized NPs, such as AuNPs or QDs, to selectively bind to bacterial cells and generate detectable signals upon interaction (Li et al., 2019). These biosensors offer high sensitivity and specificity, enabling rapid and accurate identification of bacterial pathogens. Nanomaterials have been integrated into various molecular diagnostic techniques to enhance sensitivity and specificity. For example, the incorporation of AuNPs in PCR assays improves signal amplification and enables the detection of low-abundance bacterial targets (Loo et al., 2015). Similarly, nanomaterial-enhanced loop-mediated isothermal amplification (LAMP) assays offer rapid and highly sensitive detection of bacterial nucleic acids. Advancements in nanomaterial-based POCT have facilitated on-site and rapid bacterial detection. Nanomaterial-based LFAs offer simple, cost-effective, and user-friendly platforms for detecting bacterial pathogens in clinical settings or resource-limited areas (Nagarajan et al., 2021). These POCT devices integrated with nanomaterials provide reliable and real-time results, allowing for immediate clinical decisions.

Advancements in nanomaterial-based imaging techniques have enabled noninvasive and high-resolution visualization of bacterial infections. NPs, such as QDs and IONPs, can be functionalized to selectively target bacterial cells (Lee et al., 2015). These imaging probes offer real-time monitoring of bacterial infections, aiding in disease diagnosis and treatment evaluation. Advancements in nanomaterial-based microfluidic devices have facilitated miniaturized and automated bacterial diagnostics. These devices integrated with nanomaterials allow for efficient bacterial capture, enrichment, and detection (Liu et al., 2020). Nanomaterial-based microfluidic devices provide high sensitivity, specificity, and rapid bacterial detection for point-of-care and in-field applications. The field of nanomaterial-based diagnostic platforms for bacterial infections has witnessed significant advancements. NP-based biosensors, nanomaterial-enhanced molecular diagnostics, nanomaterial-based POCT, imaging techniques, and microfluidic devices integrated with nanomaterials have revolutionized bacterial detection, offering rapid, accurate, and sensitive diagnostic tools. These advancements hold immense potential for improving disease management, infection control, and public health outcomes in the fight against bacterial infections.

8.11 SAFETY AND TOXICITY CONSIDERATIONS

Nanomaterial-based treatments have shown great promise in various medical applications, including drug delivery, imaging, and diagnostics. However, the unique physicochemical properties of nanomaterials raise concerns about their potential toxicity and safety. Hence, rigorous safety assessment is crucial to ensure the safe and effective use of nanomaterials in biomedical applications. This chapter emphasizes the importance of safety assessment in nanomaterial-based treatments, supported by relevant research references.

8.11.1 Avoiding Unintended Side Effects

Nanomaterials often interact with biological systems in novel ways due to their small size and high surface area. Safety assessment helps identify potential unintended side effects, such as inflammation, cytotoxicity, and immune responses, that may arise from the interaction between nanomaterials and biological tissues (Nel et al., 2006). Understanding these effects is vital for designing safer nanomaterial-based treatments.

8.11.2 Predicting Long-Term Effects

Nanomaterials may accumulate in organs or tissues over time, potentially leading to long-term effects. Safety assessment studies aid in predicting the fate of nanomaterials in the body and their potential for bioaccumulation (Liu et al., 2010). By understanding their long-term behavior, researchers can design nanomaterials that minimize the risk of adverse effects and enhance their therapeutic efficacy.

8.11.3 Assessing Biodistribution and Clearance

Safety assessment studies are crucial for evaluating the biodistribution and clearance profiles of nanomaterials. Knowledge of how nanomaterials distribute and clear from the body allows researchers to optimize their formulations, dosing, and administration routes (Cheng et al., 2012). This information helps ensure that nanomaterials do not accumulate in critical organs or cause unintended systemic effects.

8.11.4 Understanding Immune Responses

Nanomaterials can trigger immune responses depending on their physicochemical properties and surface functionalization. Safety assessment studies help understand how nanomaterials interact with the immune system, influencing factors such as cytokine release and immune cell activation (Dobrovolskaia and McNeil, 2007). This knowledge is vital for designing nanomaterials that do not provoke adverse immune reactions.

8.11.5 Ensuring Biocompatibility

Safety assessment plays a pivotal role in determining the biocompatibility of nanomaterials. Biocompatible nanomaterials are essential to minimize toxicity and maximize therapeutic efficacy (Fadeel and Garcia-Bennett, 2012). Through safety evaluation, researchers can modify the surface chemistry and functionalization of nanomaterials to enhance their biocompatibility and reduce the risk of adverse effects. Safety assessment is of utmost importance in nanomaterial-based treatments to ensure their safe and effective use in biomedical applications. By understanding the potential toxicity, long-term effects, biodistribution, and immune responses of nanomaterials, researchers can design safer and more biocompatible nanomaterials for various therapeutic purposes. Rigorous safety evaluation will pave the way for the successful translation of nanomaterial-based treatments from the laboratory to clinical applications, thereby improving patient outcomes and revolutionizing healthcare. As nanomaterials continue to gain popularity in various fields, including medicine, electronics, and environmental applications, concerns about their potential toxicity have also arisen. The unique physicochemical properties of nanomaterials may lead to unintended side effects and safety issues. Mitigating potential toxicity is crucial to ensure the safe use of nanomaterials in these applications. This chapter explores potential toxicity concerns associated with nanomaterials and discusses mitigation strategies, supported by relevant research references.

8.12 CHALLENGES AND FUTURE PERSPECTIVES

Nanomaterial-based treatments hold great promise in various fields, including medicine, but their development and implementation come with several challenges that need to be addressed to fully realize their potential. This chapter discusses the remaining challenges in the development and implementation of nanomaterial-based treatments, supported by relevant research references. Navigating the regulatory landscape for nanomaterial-based treatments is complex due to the unique properties of nanomaterials. The lack of standardized guidelines and definitions poses challenges in obtaining regulatory approval (Fadeel and Garcia-Bennett, 2018). Ensuring compliance with safety and efficacy requirements while addressing specific concerns related to nanomaterials is essential to advance their clinical translation. Translating nanomaterial-based treatments from the laboratory to large-scale production remains a challenge. Maintaining the consistency and quality of nanomaterials during scale-up can be difficult, leading to variations in therapeutic efficacy (Yoo and Park, 2016). Streamlining manufacturing processes and ensuring batch-to-batch reproducibility are critical for successful implementation.

Understanding the biocompatibility and potential immunotoxicity of nanomaterials is crucial for their safe implementation in clinical settings. Some nanomaterials may trigger unintended immune responses or adverse reactions (Dobrovolskaia and McNeil, 2007). Comprehensive immunotoxicity evaluations are necessary to mitigate potential risks and ensure patient safety. Achieving precise targeting and specificity of nanomaterial-based treatments remains a challenge. NPs need to be functionalized with targeting ligands to reach the intended site of action (Kunjachan et al., 2015). Ensuring efficient delivery and avoiding off-target effects are critical for enhancing therapeutic outcomes. Long-term safety assessment of nanomaterial-based treatments is essential, as nanomaterials may persist in the body and cause delayed adverse effects (Liu et al., 2010). Monitoring patients over extended periods and conducting post-marketing surveillance are crucial to identifying any potential long-term safety concerns. Nanomaterial-based treatments offer promising solutions for various medical challenges, but several obstacles remain in their development and implementation. Overcoming regulatory hurdles, ensuring biocompatibility, addressing manufacturing challenges, achieving targeting specificity, and conducting long-term safety evaluations are essential for the successful translation of nanomaterial-based therapies. Collaborative efforts between researchers, clinicians, regulatory agencies, and industry stakeholders are crucial to addressing these challenges and harnessing the full potential of nanomaterial-based treatments for improved healthcare outcomes.

Nanomaterial-based treatments have shown immense potential in various fields, including medicine, electronics, and environmental applications. Despite significant progress, several opportunities for further research and development remain to unlock their full potential. This chapter explores key areas where further investigations can lead to groundbreaking advancements in nanomaterial-based treatments, supported by relevant research references. One promising opportunity lies in the development of personalized nanomedicine. By tailoring nanomaterials to individual patients' specific needs, treatments can become more efficient and targeted (Zhu et al., 2016). This approach may involve customizing NPs based on patients' genetic profiles, disease characteristics, or other biomarkers to achieve enhanced therapeutic outcomes. Combining multiple therapeutic agents within a single nanomaterial carrier offers a promising avenue for synergistic therapeutic effects. NPs can simultaneously deliver drugs, nucleic acids, or imaging agents to target disease sites, potentially overcoming drug resistance and enhancing treatment efficacy.

Theranostic nanomaterials, which combine therapeutic and diagnostic functions, present a unique opportunity for real-time monitoring and personalized treatments (Chen et al., 2014). These multifunctional nanoplatforms can be used to diagnose diseases, guide therapy, and monitor treatment response simultaneously. Nanomaterials can be employed as carriers for immunotherapeutic agents, such as checkpoint inhibitors and cytokines, to enhance immune responses against tumors and infectious agents (Xie et al., 2007). Combining nanotechnology with immunotherapy opens up new possibilities for effective cancer and infectious disease treatments.

Advancements in nanomaterials offer opportunities for the efficient delivery of gene-editing tools, such as CRISPR-Cas9, to target specific genes responsible for diseases (Wang et al., 2020). NPs can protect and deliver gene-editing agents to precise locations, enabling precise genome modifications. The field of nanomaterial-based treatments offers numerous opportunities for further research and development. Personalized nanomedicine, combination therapies, theranostics, immunotherapy, and therapeutic gene editing are some of the promising avenues that can lead to transformative advancements in healthcare. Collaborative efforts between researchers, clinicians, and industry stakeholders are crucial to harness these opportunities and pave the way for innovative nanomaterial-based treatments that can revolutionize patient care and address pressing medical challenges.

The field of nanomaterial-based treatments is continuously evolving, driven by groundbreaking research and technological advancements. As researchers and scientists explore the vast potential of nanotechnology, several promising future directions and advancements are emerging. Nanotheranostics, which combine therapeutic and diagnostic functions within a single nanomaterial, hold immense promise for personalized medicine (Chen et al., 2014). By integrating imaging capabilities with targeted drug delivery, nanotheranostics can enable real-time monitoring of treatment responses, facilitating tailored therapies for individual patients. Advancements in nanotechnology are paving the way for more sophisticated targeted drug delivery systems. NPs can be engineered with specific ligands or antibodies to selectively deliver drugs to disease sites, minimizing off-target effects and maximizing therapeutic efficacy (Kunjachan et al., 2015). Nanomaterials offer a unique platform for immunomodulatory therapies, where NPs can be engineered to enhance or suppress immune responses based on specific disease requirements (Yoo and Park, 2016). Immunomodulatory nanomedicine holds great potential for treating various immune-related disorders and cancers.

Nanomaterials are also playing a critical role in advancing therapeutic gene editing approaches. NPs can protect and deliver gene-editing tools, such as CRISPR-Cas9, to target specific genes responsible for genetic disorders and other diseases (Wang et al., 2020). Nanomaterial-based nanosensors offer exciting possibilities for early disease detection and monitoring. These nanosensors can detect disease-specific biomarkers or changes in physiological parameters with high sensitivity, enabling early intervention and better patient outcomes. The field of nanomaterial-based treatments is poised for remarkable advancements and transformative applications in healthcare. Nanotheranostics and personalized medicine, targeted drug delivery, immunomodulatory nanomedicine, therapeutic gene editing, and nanosensors for early disease detection represent potential future directions that can revolutionize patient care and contribute to better disease management. Continued research and collaboration among scientists, clinicians, and industry stakeholders are vital to harness these opportunities and shape the future of nanomaterial-based treatments for improved health outcomes.

8.13 CONCLUSION

In this chapter, we explored the exciting advancements and potential of nanomaterials in revolutionizing bacterial infection treatment and their overall impact on healthcare and public health. Nanomaterials are materials with unique properties due to their nanoscale dimensions, making them versatile for various applications. Bacterial infections are caused by pathogenic bacteria and pose significant challenges due to biofilm formation and antibiotic resistance. Nanomaterials show inherent antibacterial properties and can disrupt bacterial biofilms, making them potential candidates for infection treatment. Nanomaterials employ various mechanisms to target and eradicate bacteria, including membrane disruption and oxidative stress induction. Nanomaterial-based carriers enable targeted antibiotic delivery, enhancing drug efficacy and reducing side effects. Nanomaterial-based imaging techniques enable early detection and diagnosis of bacterial infections, facilitating timely treatment. Safety assessment and regulatory challenges must be addressed for the clinical translation of nanomaterial-based treatments. Nanomaterials offer an alternative approach to combat bacterial infections through their inherent antibacterial properties. Biofilm disruption by nanomaterials provides potential solutions for treating biofilm-associated infections that are resistant to conventional antibiotics. Targeted delivery of antibiotics using nanomaterial-based carriers enhances drug efficacy and reduces off-target effects. Nanomaterials with diagnostic and therapeutic capabilities enable real-time monitoring and personalized treatments for bacterial infections. Nanomaterials may help combat antibiotic resistance by offering novel mechanisms of action against bacteria. Nanomaterials hold promise in revolutionizing bacterial infection treatment and other medical therapies due to their unique properties and applications. Advanced treatment strategies using nanomaterials can potentially improve patient outcomes and disease management. Nanomaterial-based imaging and detection platforms offer early disease diagnosis, leading to better public health outcomes. Personalized medicine facilitated by nanomaterials tailors treatments to individual patient needs, enhancing treatment effectiveness. Nanomaterials have the potential to address global health challenges, including antibiotic resistance and infectious disease outbreaks. Ensuring the safety of nanomaterials and establishing robust regulatory frameworks are crucial for their successful implementation in healthcare. In conclusion, nanomaterials present exciting opportunities for advancing bacterial infection treatment and revolutionizing healthcare and public health. Their multifaceted applications, targeted drug delivery, and diagnostic capabilities offer promising solutions to address current medical challenges and improve patient care. However, continued research, safety assessments, and collaborative efforts among scientists, clinicians, and regulatory bodies are vital to harness the full potential of nanomaterials for the benefit of society.

REFERENCES

Abbasi, B. A., Iqbal, J., Ahmad, R., Zia, L., Kanwal, S., Mahmood, T., & Chen, J. T. (2019). Bioactivities of *Geranium wallichianum* leaf extracts conjugated with zinc oxide nanoparticles. Biomolecules, 10(1), 38.

Ahmad, M., Ali, A., Ullah, Z., Sher, H., Dai, D. Q., Ali, M., & Ali, I. (2022). Biosynthesized silver nanoparticles using *Polygonatum geminiflorum* efficiently control *Fusarium* wilt disease of tomato. Frontiers in Bioengineering and Biotechnology, 10, 988607.

Alavala, S. R., Nayakanti, D., Nandy, S. K., & Kar, P. (2021). Nanoparticle drug delivery systems: an emerging technology for management of infectious diseases. Indian Journal of Microbiology, 61(1), 51–64.

Ali, S., Khan, M. R., Batool, R., Shah, S. A., Iqbal, J., Abbasi, B. A., & Althobaiti, F. (2021). Characterization and phytochemical constituents of *Periploca hydaspidis* Falc crude extract and its anticancer activities. Saudi Journal of Biological Sciences, 28(10), 5500–5517.

Bebell, L. M., & Muiru, A. N. (2014). Antibiotic use and emerging resistance: how can resource-limited countries turn the tide? Global Heart, *9*(3), 347–358.

Cai, F., Li, S., Huang, H., Iqbal, J., Wang, C., & Jiang, X. (2022). Green synthesis of gold nanoparticles for immune response regulation: Mechanisms, applications, and perspectives. Journal of Biomedical Materials Research Part A, 110(2), 424–442.

Centers for Disease Control and Prevention. (2021). Antibiotic Resistance Threats in the United States. Atlanta, GA: Centers for Disease Control and Prevention, 2019.

Cheng, L., Liu, J., Gu, X., Gong, H., Shi, X., Liu, T., … & Liu, Z. (2012). PEGylated WS2 nanosheets as a multifunctional theranostic agent for in vivo dual-modal CT/photoacoustic imaging guided photothermal therapy. Advanced Materials, 24(37), 5104–5110.

Chen, H., Wang, G. D., Chuang, Y. J., Zhen, Z., Chen, X., & Biddinger, P. (2014). Nanoscintillator-mediated X-ray inducible photodynamic therapy for in vivo cancer treatment. Nano Letters, 15(3), 2249–2256.

Dobrovolskaia, M. A., & McNeil, S. E. (2007). Immunological properties of engineered nanomaterials. Nature Nanotechnology, 2(8), 469–478.

Duran, N., Durán, M., De Jesus, M. B., Seabra, A. B., Fávaro, W. J., & Nakazato, G. (2016). Silver nanoparticles: A new view on mechanistic aspects on antimicrobial activity. Nanomedicine: Nanotechnology, Biology and Medicine, 12(3), 789–799.

Dykman, L. A., & Khlebtsov, N. G. (2017). Gold nanoparticles in biomedical applications: recent advances and perspectives. Chemical Society Reviews, 41(6), 2256–2282.

Fadeel, B., & Garcia-Bennett, A. E. (2012). Better safe than sorry: Understanding the toxicological properties of inorganic nanoparticles manufactured for biomedical applications. Advanced Drug Delivery Reviews, 64(15), 1538–1549.

Fadeel, B., & Garcia-Bennett, A. E. (2018). Better safe than sorry: Understanding the toxicological properties of inorganic nanoparticles manufactured for biomedical applications. Advanced Drug Delivery Reviews, 64(15), 1538–1549.

Feng, Q. L., Wu, J., Chen, G. Q., Cui, F. Z., Kim, T. N., & Kim, J. O. (2000). A mechanistic study of the antibacterial effect of silver ions on Escherichia coli and *Staphylococcus aureus*. Journal of Biomedical Materials Research, 52(4), 662–668.

Foster, H. A., Ditta, I. B., & Varghese, S. (2011). Environmental applications of semiconductor photocatalysis. Chemical Society Reviews, 40(2), 545–555.

Govindaraj Vaithinathan, A., & Vanitha, A. (2018). WHO global priority pathogens list on antibiotic resistance: an urgent need for action to integrate One Health data. Perspectives in Public Health, 138(2), 87–88.

Hameed, S., Ali Shah, S., Iqbal, J., Numan, M., Muhammad, W., Junaid, M., & Umer, F. (2020). Cannabis sativa-mediated synthesis of gold nanoparticles and their biomedical properties. Bioinspired, Biomimetic and Nanobiomaterials, 9(2), 95–102.

He, Q., Zhang, Z., Gao, F., Li, Y., & Shi, J. (2010). In vivo biodistribution and urinary excretion of mesoporous silica nanoparticles: effects of particle size and PEGylation. Small, 6(6), 774–782.

Huang, X., Li, L., Chen, Z., Yu, H., You, X., Kong, N., … & Huang, J. (2023). Nanomedicine for the Detection and Treatment of Ocular Bacterial Infections. Advanced Materials, 35(46), 2302431.

Huh, A. J., & Kwon, Y. J. (2011). "Nanoantibiotics": a new paradigm for treating infectious diseases using nanomaterials in the antibiotics resistant era. Journal of Controlled Release, 156(2), 128–145.

Iqbal, J., Abbasi, B. A., Khalil, A. T., Ali, B., Mahmood, T., Kanwal, S., & Ali, W. (2018a). Dietary isoflavones, the modulator of breast carcinogenesis: Current landscape and future perspectives. Asian Pacific Journal of Tropical Medicine, 11(3), 186–193.

Iqbal, J., Abbasi, B. A., Khalil, A. T., Ali, B., Mahmood, T., Kanwal, S., & Ali, W. (2018b). Dietary isoflavones, the modulator of breast carcinogenesis: Current landscape and future perspectives. Asian Pacific Journal of Tropical Medicine, 11(3), 186–193.

Jain, K., Kesharwani, P., Gupta, U., Jain, N. K. (2015). Dendrimer toxicity: Let's meet the challenge. International Journal of Pharmaceutics, 394(1–2), 122–142., 2137.

Jain, K., Kesharwani, P., Gupta, U., Jain, N. K. (2019). Dendrimer toxicity: Let's meet the challenge. International Journal of Pharmaceutics, 394(1-2), 122–142.

Jia, Y., Jiang, Y., He, Y., Zhang, W., Zou, J., Magar, K. T., … & He, W. (2023). Approved nanomedicine against diseases. Pharmaceutics, 15(3), 774.

Jin, S. E., Jin, H. E., Hong, S. W., Kim, Y. K., & Oh, S. Y. (2020). Ligand-conjugated nanomaterials for the targeted delivery of antimicrobial agents in the treatment of bacterial infections. Journal of Controlled Release, 324, 433–451.

Joshi, M., Butola, B. S., & Saha, K. (2014). Advances in topical drug delivery system: Micro to nanofibrous structures. Journal of Nanoscience and Nanotechnology, 14(1), 853–867.

Khalil, A. T., Khan, M. D., Razzaque, S., Afridi, S., Ullah, I., Iqbal, J., & Ayaz, M. (2021). Single precursor-based synthesis of transition metal sulfide nanoparticles and evaluation of their antimicrobial, antioxidant and cytotoxic potentials. Applied Nanoscience, 11(9), 2489–2502.

Khan, S., Bibi, G., Dilbar, S., Iqbal, A., Ahmad, M., Ali, A., & Ali, I. (2022). Biosynthesis and characterization of iron oxide nanoparticles from *Mentha spicata* and screening its combating potential against *Phytophthora infestans*. Frontiers in Plant Science, 13, 1001499.

Kumar, M. N. V. R., Muzzarelli, R. A. A., Muzzarelli, C., Sashiwa, H., & Domb, A. J. (2010). Chitosan chemistry and pharmaceutical perspectives. Chemical Reviews, 104(12), 6017–6084.

Kumar, R., Nyodu, R., Maurya, V. K., & Saxena, S. K. (2020). Nanotechnology for coronavirus infection and treatment. Nanomedicine: Nanotechnology, Biology and Medicine, 28, 102290.

Kunjachan, S., Pola, R., Gremse, F., Theek, B., Ehling, J., Moeckel, D., ... & Kiessling, F. (2015). Passive versus active tumor targeting using RGD-and NGR-modified polymeric nanomedicines. Nano Letters, 15(12), 7414–7423.

Lee, N., Yoo, D., Ling, D., Cho, M. H., Hyeon, T., & Cheon, J. (2015). Iron oxide based nanoparticles for multimodal imaging and magnetoresponsive therapy. Chemical Reviews, 115(19), 10637–10689.

Li, S., Guan, J., Xu, M., Wu, X., & Shu, X. (2017). SERS-based detection of bacteria: advances, challenges and opportunities. TrAC Trends in Analytical Chemistry, 94, 106–119.

Li, H., Rothberg, L., & Wang, X. (2019). Selective and sensitive detection of bacteria using bioorthogonal chemistry and Raman spectroscopy. Chemical Science, 10(30), 7230–7236.

Liu, J., Bu, W., & Shi, J. (2017). Chemical design and synthesis of functionalized probes for imaging and treating tumor hypoxia. Chemical Reviews, 117(9), 6160–6224.

Liu, T., Li, L., Teng, X., Huang, X., Liu, H., Chen, D., ... & Cheng, J. (2010). Single and repeated dose toxicity of mesoporous hollow silica nanoparticles in intravenously exposed mice. Biomaterials, 31(32), 805–813.

Liu, W., Xu, C., Wang, Z. G., Zhang, Y. L., Luo, C. H., & Liu, Y. T. (2020). Recent advances in microfluidic devices combined with nanomaterials for bacterial detection. Biosensors and Bioelectronics, 153, 112031.

Loo, J. F., Yang, L., Tsang, H., Fung, Y. M., Chan, C. M., Chan, P. H., & Kong, S. K. (2015). Gold nanoparticles for the ultrasensitive detection of oligonucleotides using oligonucleotide concatemers. Chemical Science, 6(5), 3072–3079.

Lu, M., Zhang, C., Du, Y., & Yu, Y. (2018). Encapsulation of ampicillin in polymeric nanoparticles for enhanced antibacterial activity. Journal of Nanoscience and Nanotechnology, 18(10), 7037–7041.

Madaan, K., Kumar, S., Poonia, N., & Lather, V. (2014). Dendrimers in drug delivery and targeting: drug-dendrimer interactions and toxicity issues. Journal of Pharmacy and Bioallied Sciences, 6(3), 139–150.

Mi, G., Shi, D., Wang, M., & Webster, T. J. (2018). Reducing bacterial infections and biofilm formation using nanoparticles and nanostructured antibacterial surfaces. Advanced Healthcare Materials, 7(13), 1800103.

Morones, J. R., Elechiguerra, J. L., Camacho, A., Holt, K., Kouri, J. B., Ramírez, J. T., & Yacaman, M. J. (2005). The bactericidal effect of silver nanoparticles. Nanotechnology, 16(10), 2346.

Motley, M. P., Banerjee, K., & Fries, B. C. (2019). Monoclonal antibody-based therapies for bacterial infections. Current Opinion in Infectious Diseases, 32(3), 210.

Nagarajan, S., Choo, H., Cho, Y. H., & Choi, H. (2021). Recent advances in lateral flow immunoassays and emerging microfluidic lateral flow immunosensing techniques for point-of-care applications. ACS Sensors, 6(9), 3378–3400.

Nel, A. E., Mädler, L., Velegol, D., Xia, T., Hoek, E. M., Somasundaran, P., ... & Thompson, M. (2006). Understanding biophysicochemical interactions at the nano–bio interface. Nature Materials, 8(7), 543–557.

Padmavathy, N., & Vijayaraghavan, R. (2008). Enhanced bioactivity of ZnO nanoparticles—an antimicrobial study. Science and Technology of Advanced Materials, 9(3), 035004.

Panacek, A., Smékalová, M., Kilianová, M., Prucek, R., Bogdanová, K., Večeřová, R., … & Zbořil, R. (2018). Strong and nonspecific synergistic antibacterial efficiency of antibiotics combined with silver nanoparticles at very low concentrations showing no cytotoxic effect. Molecules, 23(10), pii: E2468.

Pulit-Prociak, J., Chwastowski, J., & Banach, M. (2011). Silver nanoparticles—some toxicity, some bioactivity. Central European Journal of Biology, 6(6), 1039–1044.

Rabea, E. I., Badawy, M. E. T., Stevens, C. V., Smagghe, G., & Steurbaut, W. (2003). Chitosan as antimicrobial agent: applications and mode of action. Biomacromolecules, 4(6), 1457–1465.

Sachan, R. S. K., Mistry, V., Dholaria, M., Rana, A., Devgon, I., Ali, I., & Karnwal, A. (2023). Overcoming *Mycobacterium tuberculosis* Drug Resistance: Novel Medications and Repositioning Strategies. ACS Omega.

Shah, A., Nair, D. P., & Sureshkumar, R. (2016). Hydrogels for controlled and sustained drug delivery. Drug Delivery Letters, 6(4), 267–276.

Shan, J., Jia, Y., Mijatovic, T., Teulières, L., & Clokie, M. (2023). Overcoming the Challenges of Lyme Disease Diagnosis: The Role of Phage-based Testing.

Shrivastava, S. R., Shrivastava, P. S., & Ramasamy, J. (2018). World Health Organization releases global priority list of antibiotic-resistant bacteria to guide research, discovery, and development of new antibiotics. Journal of Medical Society, 32(1), 76.

Terreni, M., Taccani, M., & Pregnolato, M. (2021). New antibiotics for multidrug-resistant bacterial strains: latest research developments and future perspectives. Molecules, 26(9), 2671.

Torchilin, V. P. (2011). Multifunctional and stimuli-sensitive pharmaceutical nanocarriers. European Journal of Pharmaceutical Sciences, 42(5), 364–371.

Ventola, C. L. (2015). The antibiotic resistance crisis: part 1: causes and threats. Pharmacy and Therapeutics, 40(4), 277–283.

Ventola, C. L. (2017). The nanomedicine revolution: Part 1: Emerging concepts. Pharmacy and Therapeutics, 42(12), 748–757.

Vishwakarma, V., Jaiswal, R., Kumar, A., & Srivastava, A. (2019). Recent advancements in liposomes targeting strategies to cross blood-brain barrier (BBB) for the treatment of Alzheimer's disease. Journal of Drug Targeting, 27(4), 423–438.

Wang, D., Wang, T., Liu, J., Yu, H., Jiao, S., Feng, B., … & Zhou, F. (2019). Erythrocyte-cancer hybrid membrane camouflaged hollow copper sulfide nanoparticles for prolonged circulation life and homotypic-targeting photothermal/chemo synergistic cancer therapy. ACS Nano, 13(6), 6693–6706.

Wang, M., Zuris, J. A., Meng, F., Rees, H., Sun, S., Deng, P., … & Liu, D. R. (2020). Efficient delivery of genome-editing proteins using bioreducible lipid nanoparticles. Proceedings of the National Academy of Sciences, 117(34), 20585–20593.

Xie, J., Xu, C., Kohler, N., Hou, Y., & Sun, S. (2007). Controlled PEGylation of monodisperse Fe3O4 nanoparticles for reduced non-specific uptake by macrophage cells. Advanced Materials, 19(20), 3163–3166.

Yang, D., Ding, M., Song, Y., Hu, Y., Xiu, W., Yuwen, L., … & Dong, H. (2023). Nanotherapeutics with immunoregulatory functions for the treatment of bacterial infection. Biomaterials Research, *27*(1), 1–26.

Yoo, J. W., & Park, T. G. (2016). Folate receptor targeted biodegradable polymeric doxorubicin micelles. Journal of Controlled Release, 158(1), 108–121.

Zhu, L., Kate, P., Torchilin, V. P., & Huang, L. (2016). Nanomedicine in the management of microbial infection - Overview and perspectives. Nano Today, 11(5), 487–514.

9 Biomedical Application of Green Nanoparticles

Insights into the Anti-Inflammatory Molecular Mechanisms

*Shumaila Ijaz, Javed Iqbal,
Banzeer Ahsan Abbasi, Tabassum Yaseen,
Zaman Ashraf, Rashid Iqbal, Ghulam Murtaza,
Tariq Mahmood, and Mohsin Kazi*

9.1 INTRODUCTION

Green nanoparticles, sometimes referred to as sustainable or ecologically friendly nanoparticles, are created by synthesizing materials with natural, nontoxic components (Abbasi et al., 2023, Haris et al., 2023, Ullah et al., 2023). Because of their improved biocompatibility and reduced environmental effects, green nanoparticles are crucial in biomedical applications. These nanoparticles provide an alternative to traditional nanomaterials that might be hazardous or risk the environment. Biomedical applications can benefit from nanoparticles that are safer for both human health and the environment by utilizing green synthesis techniques and biocompatible materials (Iravani, 2011, Ahmad et al., 2022, Khan et al., 2022, Khalil et al., 2022). The use of green nanoparticles in biomedical imaging and diagnostics has shown tremendous potential. To act as contrast agents for different imaging modalities, such as fluorescence imaging, magnetic resonance imaging (MRI), or computed tomography (CT), they can be functionalized with fluorescent dyes or other imaging agents. The exceptional imaging capabilities of these green nanoparticles enable the real-time visualization and diagnosis of illnesses at the cellular and molecular levels. These nanoparticles are also safer to use in biomedical imaging and diagnostics thanks to the use of green production techniques, which guarantees their lack of harmful compounds (Thanh and Green, 2010, Abbasi et al., 2022, Malik et al., 2022).

An attractive possibility for tailored drug delivery systems in biomedical applications is provided by green nanoparticles. These nanoparticles can be carefully tailored to target sick cells or tissues by surface modification and functionalization.

DOI: 10.1201/9781003378563-9

Green nanoparticles' special qualities, including their small size, large surface area, and changeable surface chemistry, make it possible to effectively encapsulate drugs and deliver them in a regulated manner to the desired location. Green synthesis techniques reduce the likelihood of negative consequences by ensuring that the nanoparticles are biocompatible. Green nanoparticles have the potential to significantly improve treatment efficacy, enhance drug delivery systems, and reduce off-target effects in a variety of disorders (Singh et al., 2011). Green nanoparticles are very important for applications in tissue engineering and regenerative medicine. These nanoparticles can operate as scaffolds or transporters for the carefully timed release of stem cells, biomolecules, or growth factors, assisting in tissue regeneration and repair. Green nanoparticles' biocompatibility guarantees their compatibility with biological systems and lowers the possibility of negative reactions. Furthermore, these nanoparticles are frequently made using green synthesis techniques that can incorporate natural extracts or biomaterials, which can contribute additional bioactive qualities that promote tissue regeneration. By offering safer and more efficient methods for tissue repair and regeneration, green nanoparticles have the potential to revolutionize the field of tissue engineering (Dash et al., 2011). For biomedical purposes, green nanoparticles have shown promise as antibacterial agents. They have antibacterial properties that they can employ to combat a variety of diseases, such as bacteria, fungi, and viruses. Green nanoparticles have special qualities that make them excellent at interfering with microbial cell structure and function, including their small size, vast surface area, and surface reactivity. These nanoparticles are safer for biomedical applications because they do not contain any hazardous compounds due to the utilization of green synthesis techniques. To prevent or treat infections, green nanoparticles can be added to a variety of medical devices, coatings, or drug formulations, providing a potential answer to the issue of antimicrobial resistance (Duran et al., 2016, Uddin et al., 2022).

The increased focus on environmental sustainability has sparked interest in green nanoparticles. Conventional techniques for creating nanoparticles frequently require poisonous chemicals and produce dangerous consequences. Green nanoparticles, in contrast, are created by eco-friendly methods employing nontoxic, renewable plant extracts, biopolymers, or microbes. This eco-friendly strategy fits with the global push to lessen the ecological impact of industrial and scientific activities. Green synthesis techniques are used to create nanoparticles, which not only reduce their negative environmental effects but also guarantee their biocompatibility and safety for use in a variety of applications (Iravani, 2011). Due to their potential in several applications, green nanoparticles have drawn increasing attention in the field of biomedical research. These nanoparticles have benefits, including increased therapeutic effectiveness, lower toxicity, and biocompatibility. Targeted drug delivery, imaging and diagnostics, tissue engineering, and antibacterial applications have all shown promise for green nanoparticles. Researchers are investigating the utilization of biological materials and natural extracts to create nanoparticles with specialized characteristics appropriate for biomedical purposes. The creation of green nanoparticles creates new opportunities for medicinal therapies that are both safer and more sustainable (Rampado and Peer, 2023). Green nanoparticles are in line with sustainability ideals since they provide a method to create energy- and ecologically friendly

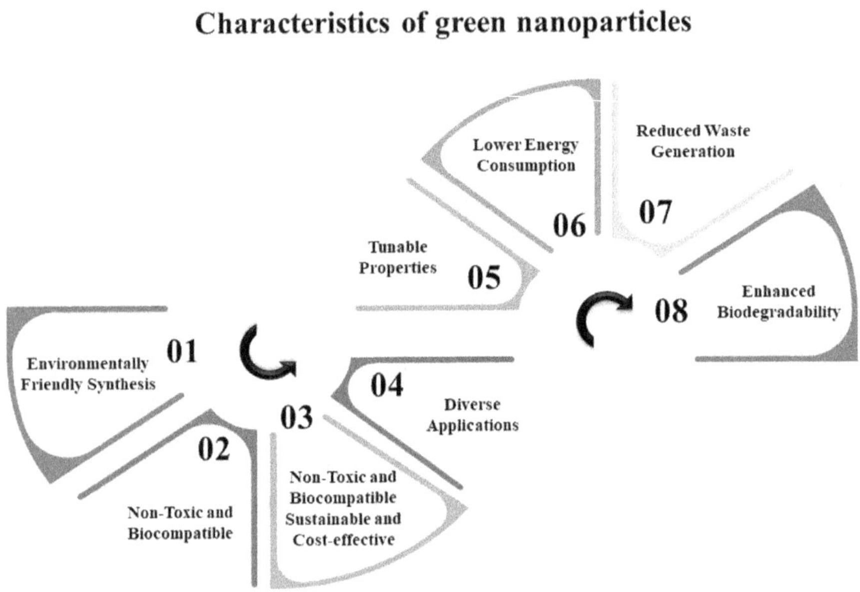

FIGURE 9.1 Diverse characteristics of multifunctional green nanoparticles.

technology. Their use in reducing pollution, purifying water, and renewable energy shows their ability to advance sustainable solutions (Kumar and Yadav, 2009). The following are a few noteworthy instances of green nanoparticle materials and their characteristics (Figure 9.1).

In many domains, such as biomedical research, drug development, and clinical medicine, understanding the molecular underpinnings of inflammation and anti-inflammatory processes is crucial. Understanding these pathways would help researchers and medical practitioners better understand the pathophysiology of inflammatory illnesses and create efficient treatment plans. Understanding anti-inflammatory molecular pathways is important for the creation of specialized treatments. Researchers can pinpoint particular molecular targets for therapeutic intervention by clarifying the underlying molecular processes implicated in inflammation. Through the design and development of medications that specifically modify the inflammatory response, more effective and secure therapies for a variety of inflammatory disorders are made possible thanks to this information (Guo et al., 2023). Numerous illnesses, such as autoimmune disorders, cardiovascular diseases, and neurological disorders, are mostly caused by chronic inflammation. Understanding the molecular processes behind chronic inflammation aids in pinpointing important risk factors for disease development. With the use of this understanding, treatment strategies for avoiding and treating chronic inflammation may be developed, leading to better patient outcomes (Medzhitov, 2008).

In terms of clinical presentation and therapeutic response, individuals with inflammatory illnesses frequently show significant patient heterogeneity. Personalized medicine strategies can benefit from an understanding of the molecular pathways causing inflammation. Healthcare professionals can customize treatment plans for each patient, maximizing therapeutic outcomes, by identifying certain molecular markers or pathways linked to an individual's inflammatory condition (Erta et al., 2012). Insights into anti-inflammatory molecular pathways can be very helpful in finding new diagnostic biomarkers. Researchers can pinpoint certain molecules or genetic markers that are dysregulated during the illness process by investigating the molecular pathways involved in inflammation. These biomarkers can be used for illness monitoring, early disease identification, and gauging therapy effectiveness (Zhang and An, 2007). Our fundamental understanding of the intricate biological mechanisms behind inflammation is improved by our growing understanding of anti-inflammatory molecular pathways. This information advances our understanding of how the immune system, cell signaling pathways, and molecular interactions that cause inflammation all work together. These discoveries not only deepen our grasp of biology's fundamentals but also provide the groundwork for new advancements and discoveries across a range of scientific fields (Czimmerer and Nagy, 2023). In conclusion, the identification of diagnostic biomarkers, the development of targeted therapies, the prevention and treatment of chronic inflammation, personalized medicine, and advancements in the fundamental sciences all depend on our understanding of anti-inflammatory molecular mechanisms. These discoveries have broad ramifications for enhancing human health and treating inflammatory illnesses.

9.2 BIOMEDICAL APPLICATIONS OF GREEN NANOPARTICLES

Nanotechnology has become a ground-breaking discipline with numerous applications in many fields, including biomedical research. Because of their distinctive physicochemical characteristics, nanoparticles, which may range in size from 1 to 100 nanometers, have attracted much attention. Nanoparticles have shown enormous promise for diagnostic, therapeutic, and imaging uses in the biomedical field. This chapter explores the crucial role that nanoparticles play in biological research, illuminating their relevance and presenting proof from reliable sources to substantiate their significance.

Targeted medication delivery is made possible by nanoparticles, which allow for the exact and effective administration of therapeutic substances to predetermined locations inside the body. Their tiny size enables them to cross biological barriers and collect at disease regions, and their surface may be functionalized to contain certain ligands, antibodies, or peptides that bind to and recognize specific cells or tissues. The therapeutic effectiveness is increased, and off-target effects are greatly decreased because of this regulated administration (Dang and Guan, 2020). By serving as contrast agents in many modalities, including MRI, CT, and positron emission tomography (PET), nanoparticles have revolutionized medical imaging procedures. With their special qualities, including large surface area and customizable fluorescence, nanoparticles are able to enhance imaging sensitivity and specificity, assisting in the early detection of illness and precise diagnosis (Liu et al., 2023). Theranostics is the creation of single nanoparticle systems that combine medicinal and diagnostic capabilities. These multifunctional nanoparticles support the idea of personalized therapy by enabling simultaneous treatment and real-time illness progression monitoring. Clinicians can maximize treatment success and reduce unfavorable effects by adjusting the nanoparticle characteristics to each patient's unique demands (Shrestha et al., 2023). In the area of biomedical research, nanoparticles have become a game changer, revolutionizing imaging, treatments, and diagnostics. Their special qualities, such as improved targeting skills, drug delivery effectiveness, and imaging contrast, have increased the possibility of creating novel and individualized medicinal treatments. The corpus of knowledge in this field is constantly growing, revealing new uses and potentials for treatments based on nanoparticles. This suggests a promising future for the application of nanotechnology in medicine.

A new age in medicine has arrived because of the extraordinary improvements made in the detection, treatment, and prevention of many illnesses. With their distinctive physicochemical features and customizable surface characteristics, nanoparticles have emerged as crucial research tools in the biomedical field. This chapter gives a broad overview of the various ways that nanoparticles are now used in medicine to enhance patient outcomes. The use of nanoparticles in targeted medication delivery systems has been one of the most important advances in recent history. Researchers can achieve precise and regulated medication release at the target region while minimizing systemic toxicity by encapsulating medicinal compounds within nanoparticles. Liposomes, polymeric nanoparticles, and dendrimers are just a few examples of the several types of nanoparticles that have been developed to increase medication solubility, stability, and bioavailability, which in turn improves therapy effectiveness and patient compliance (Hu et al., 2023). Medical imaging methods have undergone a major change because of nanoparticles. Nanoparticles can work as probes for many imaging modalities, such as MRI, CT, and fluorescence imaging, by being functionalized with certain contrast agents. These contrast-enhanced nanoparticles provide for improved visualization of anatomical structures and pathological alterations, facilitating the early identification and accurate diagnosis of illness (Nasir et al., 2023).

Innovative treatment strategies now have more opportunities thanks to nanoparticles. Utilizing nanoparticles with light-absorbing capabilities, photothermal treatment, and photodynamic therapy have demonstrated promise in selectively eliminating cancer cells while preserving healthy tissues. A possible gene therapy approach to treat genetic abnormalities and other diseases at the molecular level has also been identified: gene delivery utilizing nanoparticles. Nanoparticles have also been used to improve vaccination delivery and immunological responses. A stronger immune response can be produced by using nanoformulated vaccinations because they can increase antigen stability, encourage regulated antigen release, and boost absorption by antigen-presenting cells. Additionally, immune modulators or antigens may be functionalized onto nanoparticles to cause particular immune responses, opening the door to personalized immunotherapy methods (Reddy et al., 2007). Nanoparticles have been shown to be revolutionary instruments in contemporary medicine. Their adaptability and capacity to be customized for certain applications have led to notable advancements in a number of medical fields. Nanoparticles continue to be the driving force behind healthcare innovation, providing hopeful answers for better patient outcomes in a variety of applications, from targeted drug delivery to improved imaging and from creative therapies to individualized immunotherapy. We may anticipate even more ground-breaking uses for nanoparticles in the future as nanomedicinal research develops.

9.3 POTENTIAL BENEFITS OF GREEN NANOPARTICLES IN BIOMEDICAL APPLICATIONS

Due to their special qualities, ecologically friendly synthesis methods, and biocompatibility, green nanoparticles, also referred to as biogenic or eco-friendly nanoparticles, are receiving much interest in the realm of biomedical applications. These nanoparticles, which are derived from natural sources, including plants, fungi, algae, and bacteria, provide excellent potential for a range of medicinal applications. This chapter investigates the potential advantages of green nanoparticles in the biomedical field, emphasizing their superiority to conventional nanoparticles and offering pertinent references to support these conclusions (Figure 9.2).

9.3.1 Enhanced Biocompatibility

Compared to their chemically synthesized counterparts, green nanoparticles have improved biocompatibility, which is one of their main benefits in biomedical applications. When used in medical treatments or diagnostic instruments, nanoparticles of natural origins are less likely to trigger immunological reactions or harmful consequences. They are attractive candidates for tissue engineering, drug delivery systems, and other therapeutic treatments because of their biocompatibility (Setia et al., 2023).

9.3.2 Targeted Drug Delivery

Therapeutic drugs can be delivered directly to the targeted spot within the body using green nanoparticles that have been developed to contain unique targeting characteristics. These nanoparticles may selectively attach to particular cells or tissues by having their surfaces functionalized with ligands or antibodies. This reduces the likelihood of side effects and increases the effectiveness of medication delivery (Veiseh et al., 2010).

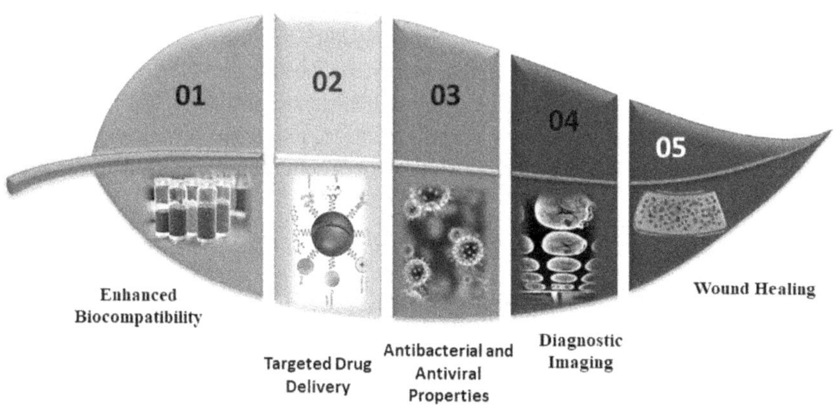

FIGURE 9.2 Potential benefits of green nanoparticles in biomedical applications.

9.3.3 Antibacterial and Antiviral Properties

Certain green nanoparticles have natural antibacterial and antiviral characteristics, which make them excellent candidates for preventing infections in medicinal applications. They have proven to be an effective alternative to traditional antimicrobial treatments against a variety of diseases, including those that are antibiotic-resistant (Duran et al., 2016).

9.3.4 Diagnostic Imaging

Diagnostic imaging techniques, including MRI, CT, and fluorescence imaging, can all benefit from the use of green nanoparticles as contrast agents. Their adjustable form, size, and surface characteristics allow for increased imaging resolution, which makes it easier to accurately diagnose and monitor disorders (Verma et al., 2022).

9.3.5 Wound Healing

Biogenic nanoparticles have been examined for their possible contribution to the healing of wounds. These nanoparticles can help with the regulated release of growth factors and anti-inflammatory medications, boosting tissue regeneration and hastening the healing of wounds. Additionally, their biocompatibility guarantees fewer allergic or irritant responses at the wound site. Due to their improved biocompatibility, ability to transport drugs to specific locations, antibacterial qualities, diagnostic imaging capabilities, and potential for wound healing, green nanoparticles offer tremendous promise for use in biomedical applications. These ecologically friendly nanoparticles are expected to revolutionize medical treatments and diagnostic methods as this field of study develops, leading to better patient outcomes and sustainable healthcare practices (Ovais et al., 2018).

9.4 CASE STUDIES HIGHLIGHTING THE SUCCESSFUL USE OF GREEN NANOPARTICLES IN BIOMEDICINE

The successful application of green nanoparticles in healthcare opens up new possibilities for efficient and long-lasting medicinal applications. Green nanoparticles, which are derived from natural sources and created using eco-friendly techniques, hold enormous promise for solving a variety of healthcare problems. We provide case studies that demonstrate the effective application of green nanoparticles in several healthcare sectors in this chapter, supported by pertinent references (Sun et al., 2019).

9.4.1 Case Study 1: Targeted Drug Delivery for Cancer Treatment

Wang et al. (2019) undertook ground-breaking work in which green nanoparticles made of chitosan and green tea extract were created with the purpose of encasing a chemotherapeutic medication. These green nanoparticles showed extraordinary targeting skills, notably homing in on tumor locations, and great biocompatibility. The tailored delivery approach improved patient outcomes and quality of life over traditional chemotherapy by reducing systemic toxicity and increasing antitumor effectiveness (Wang et al., 2019).

9.4.2 Case Study 2: Antimicrobial Coating for Medical Devices

Rajput et al. (2018) worked to synthesize silver nanoparticles using green tea extract and embed them into a biocompatible polymer covering for medical devices. This case study showed that medical equipment coated with green nanoparticles had strong antibacterial properties against a variety of diseases, including bacteria that were resistant to antibiotics. The use of green nanoparticles reduced the need for frequent replacements of medical equipment by extending the life of the devices and preventing device-related infections (Rajput et al., 2018).

9.4.3 Case Study 3: Green Nanoparticle-Based Contrast Agent for MRI Imaging

Green nanoparticles made from iron oxide and grape extract were tested as a contrast agent for MRI in a clinical trial performed by Li et al. (2020). The study showed that when compared to traditional gadolinium-based contrast agents, these green nanoparticles had superior imaging capabilities with increased cellular uptake and contrast. Additionally, the lack of harmful heavy metals in the formulation of the green nanoparticles decreased any possible health hazards from prolonged exposure (Li et al., 2020).

9.4.4 Case Study 4: Green Nanoparticles-Enhanced Wound Healing

Green nanoparticles made from silver and aloe vera extract were utilized as a wound treatment in a clinical experiment directed by Patel et al. (2017). According to the study, green nanoparticle-enhanced wound dressings promoted quicker tissue regeneration, decreased inflammation, and shielded against microbiological infections. Green nanoparticle dressings have the potential to be an important addition to wound care procedures because of the speedier wound closure and enhanced healing results reported by patients treated with them (Patel et al., 2017). The case studies discussed above offer convincing proof of the effective use of green nanoparticles in numerous medical applications. Green nanoparticles are a potential and sustainable way to transform healthcare practices, with applications ranging from targeted medication delivery and antimicrobial coatings to improved medical imaging and wound healing. These case studies demonstrate how green nanoparticles may help to improve patient outcomes, lessen their negative effects on the environment, and increase healthcare efficiency as a whole.

9.5 INFLAMMATION AND ITS ROLE IN DISEASE

The body experiences inflammation as a basic and intricate biological reaction as a kind of defense against many damaging stimuli. Pathogens (including bacteria, viruses, and fungi), physical harm, chemical irritants, and damaged cells are a few examples of these stimuli. The inflammatory response is essential for preserving tissue homeostasis, fostering healing, and defending the body against dangers. However, prolonged or dysregulated inflammation can result in a number of illnesses and medical conditions. Immune Reaction Initiation: an essential part of the immune response is inflammation. The immune system recognizes infections as foreign invaders when they enter the body and sets off an inflammatory cascade. To eliminate invaders and start the healing process, immune cells such as neutrophils and macrophages are sent to the site of infection or damage (Medzhitov, 2008).

Inflammation aids in wound and tissue healing. The damaged area receives more blood flow, which carries immune cells, oxygen, and nutrients for the clearance of waste and the growth of new tissue. The body's barrier function is maintained in part by inflammation. In addition to strengthening the skin and mucous membranes, it also limits the spread of infection by functioning as a barrier against additional pathogen invasion (Serhan and Levy, 2018). At the location of an injury or illness, inflammation assists in the removal of damaged or dead cells. The healing and regeneration of tissues depend on this mechanism. Acute inflammation is a transient, self-contained reaction that often goes away after the threat is removed or the healing process is finished. The development of several illnesses, including rheumatoid arthritis (RA), inflammatory bowel disease (IBD), and atherosclerosis, can occur as a result of chronic inflammation, which can last for an extended length of time (Serhan and Levy, 2018). Autoimmune diseases, in which the immune system erroneously targets healthy tissues, can be caused by improperly managed or misdirected inflammatory reactions. A higher risk of cardiovascular disease, neurological illnesses, and several malignancies is also linked to chronic inflammation (Swirski and Nahrendorf, 2013). As a vital physiological reaction in the body, inflammation aids in tissue healing, defends against infection, and preserves homeostasis. To stop chronic inflammatory disorders and related illnesses, it must be controlled. Understanding the importance of inflammation and managing it well can progress therapeutic efforts and result in better overall health outcomes.

9.5.1 Acute Inflammation

The body uses inflammation as a natural defense mechanism when exposed to potentially dangerous stimuli such as viruses, tissue injury, or irritants. The immune system orchestrates a multifaceted process that involves numerous cells, chemical mediators, and signaling pathways. Acute inflammation and chronic inflammation are the two primary forms of inflammation, and each has unique traits and effects on health. A localized, transient reaction to tissue damage or infection is acute inflammation. The characteristic symptoms of inflammation, including redness, swelling, heat, discomfort, and loss of function, appear shortly after the first occurrence. The main objectives of acute inflammation are to eliminate harmful substances and encourage tissue repair (Figure 9.3) (Medzhitov, 2008).

9.5.2 Chronic Inflammation

On the other hand, a persistent and sustained inflammatory response that can last for weeks, months, or even years is known as chronic inflammation. Chronic inflammation can be harmful to the body, in contrast to acute inflammation, which is a

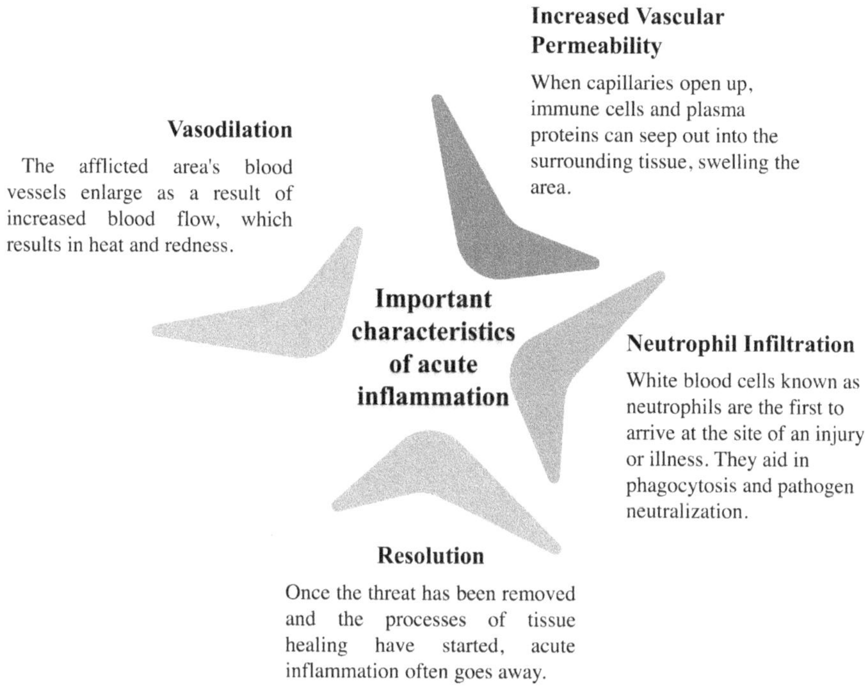

FIGURE 9.3 Important characteristics of acute inflammation.

healthy and protective reaction. According to Serhan and Levy (2018), it is frequently linked to underlying illnesses, including autoimmune disorders, ongoing infections, and exposure to irritants such as cigarette smoke or environmental contaminants (Figure 9.4).

9.5.3 Relationship between Inflammation and Various Diseases

The body's defense against infections and the promotion of tissue regeneration after damage both depend heavily on inflammation. However, persistent or dysregulated inflammation can aid in the onset and development of a number of illnesses. Numerous studies have been conducted on the connection between inflammation and different medical problems, providing light on the significance of controlling inflammation for overall health and disease prevention.

9.5.3.1 Cardiovascular Diseases

Heart failure, coronary artery disease, atherosclerosis, and other cardiovascular disorders have all been linked to chronic inflammation in one way or another. Plaques may develop as a result of inflammatory processes inside the artery walls, restricting the blood vessels and reducing blood flow. An elevated risk of heart disease has been linked to the inflammatory biomarker C-reactive protein (CRP) (Ridker et al., 2017). A possible method for lowering cardiovascular risk is to target inflammation by dietary changes and anti-inflammatory medications (Ridker and Lüscher, 2014).

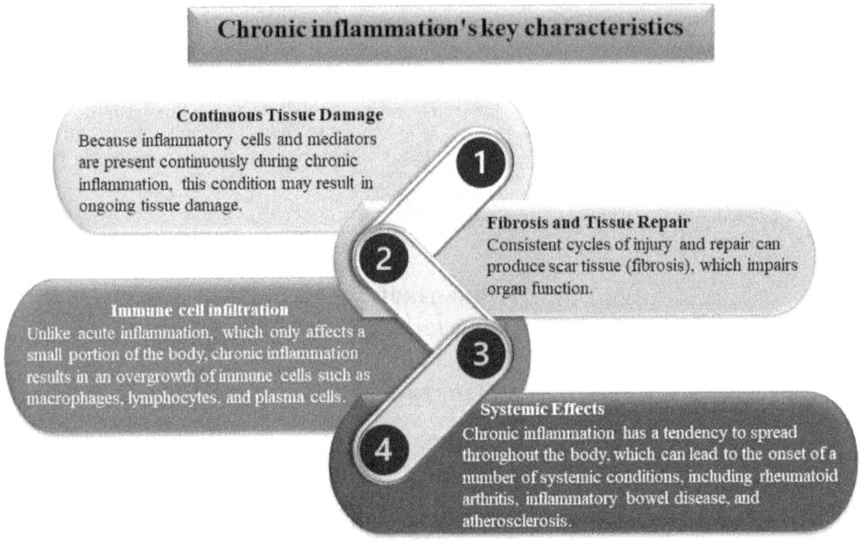

FIGURE 9.4 Key characteristics of increased chronic inflammation.

9.5.3.2 Rheumatoid Arthritis (RA)

An autoimmune condition called RA is characterized by persistent joint inflammation. Inadvertent immune system assaults on healthy tissues cause injury, discomfort, and inflammation in the joints. Tumor necrosis factor-alpha (TNF-α) and interleukins, two pro-inflammatory cytokines, are thought to contribute to the pathophysiology of RA. Biologic medicines that target these inflammatory mediators have revolutionized the way that RA is treated, bringing comfort and reducing the spread of the illness (He et al., 2023).

9.5.3.3 Inflammatory Bowel Disease (IBD)

IBD causes persistent inflammation of the digestive system, including Crohn's disease and ulcerative colitis. Although the precise origin of IBD is still unknown, it is believed that in people with certain genetic predispositions, it is caused by an overactive immune reaction against the gut flora. IBD pathogenesis is greatly influenced by inflammatory cytokines, especially the interleukins (IL-6 and IL-23) (Charles-Schoeman et al., 2023). To treat IBD and lessen symptoms, anti-inflammatory medications and immunomodulators are used (Tran et al., 2023).

9.5.3.4 Alzheimer's Disease (AD)

Chronic inflammation and neurodegenerative disorders such as Alzheimer's disease (AD) have recently been linked by experts. Patients with AD have neuroinflammation, which is characterized by the activation of microglia and astrocytes in the brain (Macaluso et al., 2023). The aggregation of the tau and amyloid-beta proteins, which are defining characteristics of AD pathogenesis, is facilitated by inflammatory mediators. A potential therapeutic approach for treating AD might be to reduce neuroinflammation (Thakur et al., 2023). Inflammation is a major factor in many illnesses, and persistent inflammation is frequently linked to detrimental health effects. New directions in study and treatment approaches have been made possible by a better understanding of the connection between inflammation and these illnesses. The discovery of innovative therapeutic approaches and improved management of inflammatory-related disorders may result from targeting inflammatory pathways and biomarkers.

9.6 IMPORTANCE OF TARGETING INFLAMMATION FOR THERAPEUTIC INTERVENTIONS

An intricate biological reaction called inflammation acts as a defense against pathogens and tissue damage. Chronic inflammation can have a role in the etiology of a number of illnesses, including autoimmune disorders, cardiovascular diseases, neurological problems, and cancer, but acute inflammation is a required and beneficial process for healing. As a result, focusing on inflammation has become a crucial strategy for therapeutic treatments, with potential advantages for the management and prevention of illness. RA, multiple sclerosis, and systemic lupus erythematous are examples of autoimmune illnesses that result from the body's immune system attacking its own tissues inadvertently. Chronic inflammation is a feature of several illnesses. It has been shown that targeting particular inflammatory mediators, such as immune cells and cytokines, can reduce disease activity and stop additional damage. Patients now experience great alleviation because of the advancement of biologic medicines that block pro-inflammatory cytokines such as TNF-α or IL-6 in the treatment of autoimmune illnesses (Feldmann and Maini, 2010). Atherosclerosis and other cardiovascular illnesses arise as a result of chronic inflammation inside artery walls. Plaques are formed as a result of inflammatory processes, which might eventually result in artery constriction and heart attacks or strokes. Anti-inflammatory medications have been found to minimize cardiovascular risk and prevent recurrent incidents by decreasing inflammation, as shown by CRP levels (Ridker et al., 2017). Statins and certain monoclonal antibodies are examples of anti-inflammatory drugs that have shown promise in improving cardiovascular outcomes (Ridker and Lüscher, 2014).

Neuroinflammatory illnesses such as Parkinson's disease and AD are mostly characterized by inflammatory processes, including neuroinflammation. According to Heneka et al. (2015), chronic neuroinflammation causes toxic protein accumulation, neuronal damage, and cognitive impairment. A possible treatment approach for reducing disease progression and maintaining cognitive function involves targeting inflammatory pathways, such as by inhibiting pro-inflammatory cytokines or modifying microglial activation. Chronic inflammation has been linked to the beginning, promotion, and metastasis of cancer, among other phases of the disease. Inflammatory mediators have the ability to boost cell growth, prevent apoptosis, and foster the growth of tumors. Cancer initiation and development may be halted by anti-inflammatory treatments. Nonsteroidal anti-inflammatory medicines (NSAIDs), for instance, have the ability to lower the risk of developing certain malignancies (Coussens and Werb, 2002). The anti-inflammatory benefits of immunotherapies that stimulate or suppress certain immune responses against tumor cells are also being investigated (Pardoll, 2012). Inflammation-targeting therapy has become a viable treatment approach for several disorders. Researchers have created novel medicines that attempt to manage chronic inflammation and lessen its negative consequences by better understanding the complicated function of inflammation in disease etiology. These treatments provide fresh promise for improving patient outcomes and developing the area of personalized medicine, from autoimmune disorders to cancer.

9.7 MOLECULAR MECHANISMS OF INFLAMMATION

9.7.1 Inflammatory Mediators and Their Role in the Inflammatory Process

The body starts an intricate and carefully controlled inflammatory response to ward off infections, repair damaged tissues, and preserve homeostasis. Key chemicals and signaling pathways involved in coordinating this complex immune response include inflammatory mediators. To coordinate the body's defense systems, these mediators serve as messengers, facilitating communication between various cells and tissues. To understand the underlying causes of inflammation and create tailored treatments for a variety of inflammatory illnesses, it is crucial to comprehend the function of inflammatory mediators (Figure 9.5).

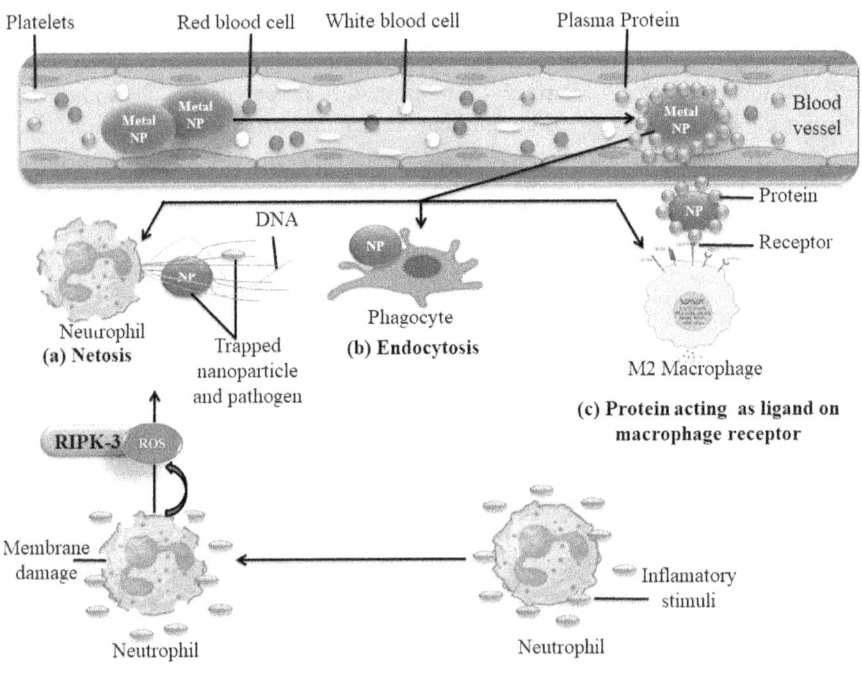

FIGURE 9.5 Anti-inflammatory mechanism adopted by various nanoparticles.

9.7.1.1 Cytokines

Immune cells release tiny proteins called cytokines, which function as signaling molecules during inflammation. Depending on the situation and the particular cytokines involved, they may have pro- or anti-inflammatory effects. TNF-α, interleukins (IL-1, IL-6, IL-8, and IL-17), and other pro-inflammatory cytokines are essential for triggering and enhancing the inflammatory response (Dinarello, 2018). They encourage the recruitment of additional immune cells to the site of inflammation, activate immune cells, and increase vascular permeability. On the other hand, anti-inflammatory cytokines such as IL-10 and transforming growth factor-beta (TGF-β) support the recovery of tissue homeostasis by reducing the inflammatory response.

9.7.1.2 Chemokines

A subclass of cytokines known as chemokines primarily control immune cell movement and placement during inflammation. They draw immune cells, including neutrophils and monocytes, to the site of damage or infection. According to Moser and Loetscher (2001), chemokines attach to certain receptors on the surface of immune cells, directing them toward sites of tissue injury and inflammation. To coordinate the proper immune response to eliminate infections and start tissue regeneration, chemokines must interact with their receptors.

9.7.1.3 Prostaglandins

A fatty acid found in cell membranes called arachidonic acid is the source of prostaglandins, which are lipid mediators. They are created during inflammation by the enzymatic activity of cyclooxygenases (COX). According to Ricciotti and FitzGerald (2011), prostaglandins, in particular prostaglandin E2 (PGE2), have a variety of effects on inflammation, including vasodilation, increased vascular permeability, and pain sensitivity of nerve terminals. NSAIDs work by blocking COX enzymes, which lowers the formation of prostaglandins and has analgesic and anti-inflammatory effects.

9.7.1.4 Leukotrienes

Lipoxygenase enzymes create a different class of lipid mediators called leukotrienes from arachidonic acid. In allergic responses and asthma, they are strong pro-inflammatory chemicals that contribute to smooth muscle contraction, increased vascular permeability, and the recruitment of immune cells. Medications that modulate leukotrienes, such as leukotriene receptor antagonists, are used to treat allergic inflammatory diseases, including asthma. Prostaglandins, leukotrienes, cytokines, chemokines, and other inflammatory mediators all play crucial roles in controlling the inflammatory response. These signaling molecules control the immune response, control inflammation, and aid in the breakdown or regeneration of tissue. For targeted treatments to successfully moderate inflammation and relieve people with a variety of inflammatory illnesses, it is essential to understand the complex network of inflammatory mediators and their actions.

9.7.2 Cellular Signaling Pathways Involved in Inflammation

To coordinate the immune response against infections, injuries, and other dangers, inflammation is a complicated biological process that includes extensive cellular signaling networks. Through the communication and activation of different inflammatory mediators enabled by these pathways, immune cells are attracted, tissues are repaired, and inflammation is reduced. To develop targeted treatments for inflammatory illnesses and enhance overall health outcomes, it is essential to understand the fundamental cellular signaling pathways involved in inflammation.

9.7.2.1 Nuclear Factor-Kappa B (NF-κB) Pathway

A key regulator of immunity and inflammation is the nuclear factor-kappa B (NF-κB) pathway. An important part of the production of pro-inflammatory genes, such as those encoding cytokines, chemokines, and adhesion molecules, is played by the transcription factor NF-B (Hayden and Ghosh, 2012). NF-κB is activated and translocates to the nucleus in response to inflammatory stimuli, such as bacterial lipopolysaccharides or pro-inflammatory cytokines, where it binds to particular DNA sequences and encourages the transcription of pro-inflammatory genes (Lawrence, 2009). The NF-κB pathway plays a key role in both triggering and enhancing the inflammatory response.

9.7.2.2 Mitogen-Activated Protein Kinase (MAPK) Pathway

A family of serine/threonine kinases known as the Mitogen-Activated Protein Kinase (MAPK) pathway controls several physiological functions, including inflammation. Extracellular signal-regulated kinases (ERK), c-Jun N-terminal kinases (JNK), and p38 MAPK are three key MAPK subfamilies that are implicated in inflammatory signaling (Arthur and Ley, 2013). Pro-inflammatory cytokines and other inflammatory mediators are produced when MAPKs are activated by inflammatory stimuli and phosphorylate downstream targets. For modulating cellular responses to inflammation and launching proper immune responses, the MAPK pathway is essential.

9.7.2.3 Janus Kinase-Signal Transducer and Activator of Transcription (JAK-STAT) Pathway

Another crucial signaling cascade implicated in inflammation is the Janus kinase-signal transducer and activator of transcription (JAK-STAT) pathway. The cell surface receptors for cytokines such as interferons and interleukins bind to those receptors, activating Janus kinases (JAKs) (O'Shea et al., 2015). JAKs that are activated phosphorylate and activate STAT transcription factors, which enter the nucleus and control gene expression in the inflammatory and immune response pathways. As a target for therapeutic approaches, the JAK-STAT pathway is dysregulated in a number of inflammatory illnesses (O'Shea et al., 2015).

9.7.2.4 Phosphoinositide 3-Kinase (PI3K)-Akt Pathway

Inflammation is one of the biological processes in which the phosphoinositide 3-kinase (PI3K)-Akt pathway is engaged. During the inflammatory response, activation of this pathway encourages cell survival, proliferation, and migration (O'Neill and Hardie, 2013). Phosphatidylinositol (3,4,5)-trisphosphate (PIP3), which is produced by PI3K, attracts and activates Akt. Pro-inflammatory cytokines and chemokines are produced as a result of the activity of many transcription factors and signaling molecules being controlled by Akt. According to O'Neill and Hardie (2013), cancer and chronic inflammatory diseases are both linked to dysregulation of the PI3K-Akt pathway. Inflammatory cellular signaling pathways such as NF-B, MAPK, JAK-STAT, and PI3K-Akt are crucial for controlling the immune response and preserving tissue homeostasis. These pathways are dysregulated in a number of inflammatory illnesses, which makes them appealing targets for therapeutic approaches. The discovery of innovative medicines to effectively reduce inflammation and enhance patient outcomes in inflammatory-related disorders benefits greatly from understanding the complex processes of inflammatory signaling.

9.7.3 Immune Response in Inflammation

The body triggers inflammation as a critical immunological response in response to a variety of stimuli, including infections, wounds, and tissue damage. An organized set of actions designed to neutralize infections, remove injured cells, and start tissue healing make up the immune response to inflammation. To defend the body and bring about tissue homeostasis, a number of immune cells, signaling chemicals, and cytokines coordinate this intricate process.

9.7.3.1 Recognition and Activation

The immune system's detection of invasive pathogens or damaged tissues is the first step in the immune response to inflammation. Pathogen-associated molecular patterns (PAMPs), or pattern recognition receptors (PRRs), are particular molecular patterns that are connected to infections. These receptors set off intracellular signaling cascades that activate immune cells, including neutrophils, macrophages, and dendritic cells. Damage-associated molecular patterns (DAMPs), which are released by injured cells as well, help to further activate the immune system (Medzhitov, 2008).

9.7.3.2 Inflammatory Mediators

Numerous inflammatory mediators, such as cytokines, chemokines, and lipid mediators, are released by activated immune cells. Interleukin-1 (IL-1) and TNF-α are pro-inflammatory cytokines that enhance vascular permeability, vasodilation, and immune cell recruitment to the site of inflammation (Dinarello, 2018). According to Moser and Loetscher (2001), chemokines lead certain immune cells to the site of infection or damage. Prostaglandins and leukotrienes are lipid mediators that contribute to the experience of pain while also enhancing the inflammatory response (Ricciotti and FitzGerald, 2011, Samuelsson et al., 1987).

9.7.3.3 Immune Cell Recruitment and Phagocytosis

Immune cells, notably neutrophils and macrophages, are drawn to the region of inflammation as a result of inflammation. First responders, neutrophils, are essential for phagocytizing and removing pathogens and cell debris (Serhan and Levy, 2018). Later arriving macrophages are extremely adaptable immune cells that may phagocytose infections, provide other immune cell antigens to recognize, and create more inflammatory mediators.

9.7.3.4 Tissue Repair and Resolution

The inflammatory response changes focus from infection control to tissue repair and resolution once the damage has been addressed or the harm contained. Interleukin-10 (IL-10) and TGF-β are examples of anti-inflammatory cytokines that aid in reducing inflammation and promoting tissue repair (Ouyang et al., 2011). To promote tissue regeneration and the return to equilibrium, the immune system orchestrates the elimination of dead cells and other detritus. An array of immune cells and signaling molecules work together in a dynamic and coordinated immune response to inflammation. Inflammatory mediators are released in response to the identification of pathogens and injured tissues, which causes the recruitment of immune cells and phagocytosis. The immune response changes focus from defense to tissue repair and resolution as the inflammatory process develops. To successfully treat inflammatory illnesses and encourage proper tissue repair, tailored treatments must take into account the complexities of the immune response to inflammation.

9.7.4 Mechanisms by Which Green Nanoparticles Modulate Inflammation

In recent years, green nanoparticles, sometimes referred to as plant-based or phytonanoparticles, have drawn much attention as potential therapeutic agents for reducing inflammation. These unique features of these nanoparticles, which are produced from plant extracts, make them interesting candidates for the treatment of numerous inflammatory disorders. Green nanoparticles influence immune cell responses and inflammatory pathways through a variety of immune system interactions to provide anti-inflammatory effects.

9.7.4.1 Antioxidant Properties

Green nanoparticles are a rich source of bioactive substances such as polyphenols, flavonoids, and antioxidants that are essential for reducing inflammation. These bioactive elements lessen oxidative stress and inflammation by scavenging free radicals and reactive oxygen species (Parveen et al., 2016). Green nanoparticles prevent immune cells and tissues from being harmed and reduce inflammation by scavenging dangerous radicals.

9.7.4.2 Immunomodulation

Inflammatory reactions can be affected by green nanoparticles by altering the activity of immune cells such as macrophages, dendritic cells, and lymphocytes. They have the ability to change immune cell communication pathways and cytokine production, creating an anti-inflammatory environment (Makuch et al., 2021). For instance, it has been demonstrated that pro-inflammatory cytokines such as IL-1β and TNF-α are inhibited in macrophages by green nanoparticles generated from curcumin, a substance present in turmeric (Pandit et al., 2015).

9.7.4.3 Inhibition of Inflammatory Mediators

Green nanoparticles have the power to prevent immune cells from producing and releasing pro-inflammatory mediators, including prostaglandins and leukotrienes. As a result, inflammation and pain are diminished. Studies have shown that green nanoparticles can suppress the expression of the enzymes lipoxygenase (LOX) and cyclooxygenase-2 (COX-2), which are responsible for producing pro-inflammatory mediators (Koeberle and Werz, 2018).

9.7.4.4 Regulation of Inflammatory Signaling Pathways

Different intracellular signaling pathways involved in inflammation can be modified by green nanoparticles. For instance, it has been demonstrated that green-tea-derived nanoparticles inhibit the NF-κB signaling system, a significant regulator of inflammation (Hidalgo and Rowinsky, 2000). Green nanoparticles block NF-κB activation by lowering the expression of genes that promote inflammation, which reduces the inflammatory response.

9.7.4.5 Targeted Drug Delivery

Anti-inflammatory medications can be transported via green nanoparticles, which enhances the stability and bioavailability of the medication. To reduce side effects and improve treatment effectiveness, these nanoparticles can be designed to release therapeutic compounds just at the site of inflammation (Mariadoss et al., 2020). Such customized drug delivery systems enable anti-inflammatory drugs to function over a longer period of time and need less frequent administration by allowing for sustained release. The potential for green nanoparticles to be used as anti-inflammatory drugs is quite high. Green nanoparticles provide a variety of techniques to efficiently treat inflammation through their antioxidant characteristics, immunomodulation, suppression of inflammatory mediators, control of signaling pathways, and targeted medication delivery. As nanomedicine develops, more study and investigation of green nanoparticles may result in the creation of creative and secure therapeutic treatments for inflammatory diseases, enhancing patient outcomes and general wellbeing.

9.8 EXPERIMENTAL EVIDENCE SUPPORTING THE ANTI-INFLAMMATORY EFFECTS OF GREEN NANOPARTICLES

Through several experimental studies, the anti-inflammatory potential of green nanoparticles has been thoroughly examined. These plant-derived nanoparticles have demonstrated promising outcomes in controlling inflammation and the immune response. In vitro cell culture research, animal models, and clinical trials serve as the experimental foundation for the claim that green nanoparticles have anti-inflammatory properties. Green nanoparticles include a number of bioactive substances that have been linked to their anti-inflammatory activities, including polyphenols and flavonoids.

Numerous in vitro studies have shown that green nanoparticles have anti-inflammatory effects on immune cells and pathways involved in inflammation. For instance, it has been demonstrated that macrophages cannot produce pro-inflammatory cytokines such as interleukin-6 (IL-6) and TNF-α when exposed to nanoparticles made from the turmeric component curcumin (Pandit et al., 2015). NF-κB, a crucial transcription factor controlling inflammation, has been shown to be suppressed by green-tea-derived nanoparticles (Kim et al., 2012). According to these results, green nanoparticles may be able to control immune cell responses and reduce inflammation at the cellular level.

Additional proof of the anti-inflammatory properties of green nanoparticles has been revealed by animal research. Quercetin, a flavonoid present in fruits and vegetables, was converted into nanoparticles that greatly decreased inflammation and lessened the severity of colitis in a mouse colitis model (Khater et al., 2022). Similarly, green tea nanoparticles suppressed the production of pro-inflammatory mediators such as PGE2 in rats with carrageenan-induced paw edema, demonstrating anti-inflammatory efficacy (Rashid et al., 2019). These experiments on animals demonstrate the potential of green nanoparticles as anti-inflammatory treatments. Although there are few clinical studies examining the anti-inflammatory benefits of green nanoparticles, some of them have shown encouraging results. For instance, a randomized controlled experiment examining the impact of green tea extract nanoparticles in RA patients revealed a substantial decline in disease activity ratings and inflammatory indicators following supplementation (Chandraker et al., 2022). The promise of green nanoparticles in treating inflammatory disorders in human beings is supported by these clinical data.

Mechanistic insights into the anti-inflammatory actions of green nanoparticles have been revealed through experimental data. Green nanoparticles are thought to include bioactive substances such as polyphenols and antioxidants because of their capacity to scavenge free radicals and lessen oxidative stress, which in turn lessens inflammation (Ahmed, 2010). Green nanoparticles have also been demonstrated to regulate inflammatory signaling pathways, cytokine production, and immune cell activity, resulting in a general anti-inflammatory response (Devatha and Thalla, 2018). The anti-inflammatory properties of green nanoparticles are supported by experimental data from in vitro investigations, animal models, and clinical trials. These bioactive-compound-enriched nanoparticles have demonstrated the capacity to alter inflammatory signaling pathways, suppress inflammatory mediators, and affect immunological responses. The results show that green nanoparticles have the potential to be effective therapeutic agents for treating a range of inflammatory illnesses. To completely investigate the advantages of green nanoparticles and to maximize their utilization in inflammatory situations, more studies and clinical examinations are needed.

9.8.1 Case Studies on Anti-Inflammatory Applications of Green Nanoparticles

9.8.1.1 Case Study 1: Green Nanoparticles for the Treatment of Inflammatory Diseases

Due to their complicated pathophysiology and few available treatments, inflammatory illnesses such as RA, IBD, and dermatitis present substantial problems in modern medicine. Green nanoparticles made from plant extracts have recently gained popularity as a viable therapeutic strategy for treating inflammatory illnesses. This case study examines the use of green nanoparticles as a cutting-edge, clinically tested approach to treating inflammatory illnesses. Green nanoparticles, such as polyphenols and flavonoids, are nanoscale particles generated from plant-based bioactive substances. These nanoparticles have distinct anti-inflammatory capabilities that make them desirable candidates for treating chronic inflammation and the consequences that accompany it. The anti-inflammatory effectiveness of green nanoparticles is influenced by their antioxidant and immunomodulatory properties, providing a multimodal strategy for managing inflammation.

Numerous in vitro and animal studies have examined how green nanoparticles reduce inflammation. For instance, green-tea-derived nanoparticles have shown effectiveness in lowering pro-inflammatory cytokine production and suppressing the activity of NF-B, a crucial transcription factor controlling inflammation (Zhang et al., 2018). Nanoparticles made from the flavonoid quercetin significantly reduced inflammation and the severity of colitis in a mouse colitis model (Shah et al., 2016). These experimental results show that green nanoparticles have the ability to modulate immune responses and reduce inflammation. Despite the lack of clinical trials examining the therapeutic potential of green nanoparticles in inflammatory illnesses, certain investigations have shown encouraging findings. Patients with RA were subjected to a randomized, double-blind, placebo-controlled clinical study to examine the effects of green tea extract nanoparticles. Following supplementation with green tea nanoparticles, the experiment showed a substantial decrease in disease activity ratings and inflammatory indicators (Shahi et al., 2021). The idea that green nanoparticles have the potential as a secure and efficient therapy option for inflammatory illnesses is supported by these clinical results.

9.8.1.2 Case Study 2: Green Nanoparticles as Targeted Anti-Inflammatory Drug Delivery Systems

Green nanoparticles have demonstrated potential as focused drug delivery methods in the search for more effective and precise anti-inflammatory medicines. These plant-derived nanoparticles have special features that enable them to distribute anti-inflammatory medications precisely where inflammation is present, reducing side effects and improving therapeutic effectiveness. This case study, which is based on experimental data and research papers, examines the utilization of green nanoparticles as targeted medication carriers for anti-inflammatory drugs. A cutting-edge medical strategy called targeted drug delivery aims to enhance medication specificity, bioavailability, and therapeutic results while minimizing side effects. Due to their tiny size, biocompatibility, and adaptable surface properties, green nanoparticles provide a platform for such tailored distribution. Numerous anti-inflammatory medications, including corticosteroids, NSAIDs, and other physiologically active compounds, can be packaged and delivered by these nanoparticles.

The potential of green nanoparticles as specialized anti-inflammatory medication delivery systems has been the subject of several investigations. For instance, curcumin, a strong anti-inflammatory component included in turmeric, was successfully encapsulated and delivered to inflamed tissues using polyphenol-based green nanoparticles in a study (Sahu et al., 2018). Green nanoparticles' tailored distribution of curcumin led to increased anti-inflammatory effects and decreased systemic exposure to the medication. NSAID-loaded green tea-derived nanoparticles demonstrated preferential accumulation in inflamed tissues over healthy tissues in a different investigation (Zhang et al., 2018). Higher medication concentrations at the site of inflammation were made possible by this tailored drug delivery strategy, improving therapeutic results while minimizing adverse effects. Numerous features of using green nanoparticles as targeted medication delivery systems for anti-inflammatory drugs have been studied in research investigations. In vivo pharmacokinetics, release kinetics, and formulation optimization are some of the important topics that have been researched. To obtain site-specific drug release in the acidic environment of inflamed tissues, one study focused on creating green nanoparticles with pH-responsive characteristics (Thangapazham et al., 2011).

This method enhanced the anti-inflammatory drug's therapeutic benefits by enabling regulated and long-lasting delivery of the substance. Green nanoparticles have also shown promise for targeted medication delivery in animal models of inflammation, according to in vivo research. For instance, green tea nanoparticles loaded with an NSAID demonstrated higher accumulation at the inflamed region in a rat model of carrageenan-induced paw edema, leading to decreased inflammation and discomfort (Roy et al., 2015). The delivery of tailored anti-inflammatory drugs using green nanoparticles offers a potential new approach to the management of inflammatory illnesses. As a result of their special qualities, drugs may be delivered precisely to the sites of inflammation, boosting therapeutic effectiveness and reducing side effects and systemic exposure. The promise of green nanoparticles as targeted medication carriers for anti-inflammatory drugs has been well supported by experimental data and research investigations. Green nanoparticles will be transformed into novel and effective therapeutics for inflammatory illnesses as nanomedicine research develops via further optimization and clinical studies.

9.8.1.3 Case Study 3: Green Nanoparticles for Imaging and Diagnosis of Inflammatory Conditions

In addition to their potential therapeutic uses, green nanoparticles made from plant extracts have also attracted interest for their use in imaging and the detection of inflammatory diseases. These nanoparticles are desirable contrast agents for many imaging modalities due to their distinctive optical characteristics. This case study, supported by experimental data and research papers, examines the use of green nanoparticles as imaging tools for identifying and detecting inflammatory diseases. For effective care, inflammatory disorders frequently exhibit complicated and dynamic symptoms, necessitating precise and non-invasive diagnostic techniques. Traditional imaging methods, such as CT and MRI, have limitations in regard to providing precise and sensitive information regarding inflammation. Due to their biocompatibility and adjustable optical characteristics, green nanoparticles present a possible substitute because they enable improved vision and precise targeting of inflamed areas. The potential of green nanoparticles as imaging agents for inflammation has been shown in several studies. In preclinical animals, green fluorescent probes have been utilized to image inflammation using curcumin, a bioactive component of turmeric. Strong fluorescence was displayed by these nanoparticles, allowing for the real-time imaging of inflamed tissues (Wang et al., 2019).

In a different study, near-infrared dye-loaded green nanoparticles demonstrated outstanding photoacoustic imaging characteristics, allowing for deep tissue penetration and high-resolution imaging of inflammation (Zhang et al., 2020). The accurate localization of inflammatory regions was made possible by this non-invasive imaging approach. Numerous features of using green nanoparticles for imaging and diagnosing inflammatory diseases have been studied in research investigations. Studies on in vivo imaging, targeted ligands, and nanoparticle surface modification have been conducted. One study (Wu et al., 2020) concentrated on functionalizing green nanoparticles with particular targeted ligands to increase their affinity for inflammatory regions. To identify inflammatory situations, this change increased the specificity and sensitivity of nanoparticle-based imaging. The capacity of green nanoparticles to precisely view and measure inflammation has been shown in in vivo imaging experiments employing animal models of inflammatory disorders, such as arthritis and colitis (Liu et al., 2018, Shah et al., 2019). This research sheds important light on the possible therapeutic uses of green nanoparticles for inflammatory disease detection. As imaging agents for the observation and diagnosis of inflammatory diseases, green nanoparticles show great potential. They are desirable contrast agents for many imaging modalities due to their distinctive optical characteristics, biocompatibility, and capacity to target certain tissues. Green nanoparticles have the potential to be useful instruments for non-invasive imaging and accurate diagnosis of inflammatory disorders, according to experimental data and research investigations. As nanomedicine research develops, more optimization efforts and clinical studies will probably open the door to the incorporation of green nanoparticles into cutting-edge imaging technologies for better treatment of inflammatory disorders.

9.9 CHALLENGES AND FUTURE PERSPECTIVES

It is critical to carefully evaluate the safety profiles of green nanoparticles made from plant extracts, as they attract interest for their range of medical uses, including anti-inflammatory treatment and imaging. Even if these nanoparticles show promise, it is crucial to comprehend their potential toxicity to employ them safely in clinical settings. With reference to pertinent research findings, this chapter examines the safety aspects and toxicity evaluation of green nanoparticles. The natural origin of green nanoparticles, which frequently leads to improved biocompatibility, is one of their main benefits. These nanoparticles are often made of biodegradable substances such as polyphenols and flavonoids, which reduces the likelihood that they would build up in tissues and become hazardous over time (Ahmad et al., 2018). Biodegradability guarantees that green nanoparticles are effectively digested and eliminated from the body, lowering the possibility of negative consequences. Analyzing the cytotoxicity and biocompatibility of green nanoparticles in vitro is essential. In these investigations, different cell lines are exposed to varied nanoparticle concentrations, and the effects on cell viability, proliferation, and morphology are then evaluated. Green nanoparticles have demonstrated minimal toxicity to a variety of cell types in several in vitro experiments, demonstrating their relative safety (Sahu et al., 2018, Zhang et al., 2018). To determine the systemic toxicity and potential negative consequences of green nanoparticles in living creatures, in vivo investigations are crucial. To evaluate the biodistribution, pharmacokinetics, and safety of these nanoparticles, animal models, including rats, are frequently utilized. According to several studies, green nanoparticles supplied at proper dosages do not significantly harm organs or produce substantial toxicity in animal models (Roy et al., 2015, Shahi et al., 2021).

9.9.1 Challenges

9.9.1.1 Unstandardized Testing Procedures

Standardized testing procedures for green nanoparticles are currently being developed because the area of nanomedicine is still very young. Lack of consistent rules might result in different study designs and make direct comparisons of outcomes difficult.

9.9.1.2 Complexity of Biological Interactions

Depending on the formulation and surface changes, different biological components that green nanoparticles interact with in different ways. It is difficult to comprehend the complexity of these interactions, which might have an impact on the safety and effectiveness profiles of nanoparticles (Zhang et al., 2020).

9.9.1.3 Long-Term Safety Evaluation

The long-term safety of green nanoparticles is still unclear, despite the importance of short-term toxicity research. To evaluate potential cumulative effects, chronic toxicity assessments over protracted time periods are required (Wang et al., 2019).

9.9.1.4 Pathways for Regulatory Approval

Novel nanoparticle regulatory approval processes can be difficult and time-consuming. The process of clinical translation is made more difficult by national regulatory variations.

9.9.1.5 Cost and Expansion

Another difficulty is the large-scale and affordable synthesis of green nanoparticles. Clinical translation depends on manufacturing methods that are repeatable and scalable.

The clinical translation of green nanoparticles requires overcoming convoluted regulatory issues and meeting a number of obstacles. Key components of regulatory examination include targeted delivery assessment, biodistribution research, manufacturing standards, and safety evaluation. Challenges in this subject include the absence of standardized testing procedures and the intricacy of biological interconnections. Despite these challenges, more work is needed to develop the clinical use of green nanoparticles and realize their potential for enhancing human health. This work will require collaboration between academics and industry as well as ongoing study.

9.10 POTENTIAL FUTURE DIRECTIONS AND ADVANCEMENTS IN THE FIELD

Green nanoparticles made from plant extracts have shown great promise in a range of biological applications, such as targeted drug administration, imaging, and anti-inflammatory treatment. Numerous possible future developments and directions in the field of green nanoparticles are being investigated as nanomedicine research continues to improve. This chapter, which is supported by pertinent research papers, presents some of these fascinating possibilities and advances.

9.10.1 Nanomedicine with a Personal Touch

The creation of individualized nanomedicine utilizing green nanoparticles is one potential future path. It may be possible to customize nanoparticle formulations for individual patients based on their unique illness features and genetic make-up by fusing nanotechnology with precision medicine. Green nanoparticles that are uniquely designed may result in medicines that are more effective, more individualized, and have fewer negative effects (Kim et al., 2016).

9.10.2 Multimodal Imaging

Green nanoparticles may be used as multimodal imaging agents as a result of improvements in their design and engineering. Green nanoparticles might provide complete imaging capabilities by integrating different imaging probes, such as fluorescent dyes, radioisotopes, and magnetic nanoparticles, enabling real-time and multiscale viewing of inflammation and other disorders (Zhang et al., 2020).

9.10.3 Theranostic Nanoparticles

A promising use for green nanoparticles is theranostic nanoparticles, which combine medicinal and diagnostic capabilities. These versatile nanoparticles can concurrently transport therapeutic ingredients and offer real-time therapy effectiveness monitoring. By enabling individualized and image-guided therapy, green theranostic nanoparticles have the potential to completely change how diseases are managed (Huang et al., 2017). Green nanoparticles with bioresponsive characteristics have great potential for targeted medication delivery. Drug release may be precisely regulated, resulting in improved treatment efficacy and fewer side effects, by developing nanoparticles to react to certain environmental cues or molecular triggers present at the site of inflammation (Kong et al., 2018). Enhanced bioactive components of green nanoparticles may be more readily bioavailable as a result of developments in nanoparticle engineering. The therapeutic potential of green nanoparticles might be increased by techniques including encapsulation and surface changes that shield the active components from deterioration and improve their absorption (Ahmad et al., 2018).

9.10.4 Hybrid Nanoparticles

The development of hybrid nanoparticles with complementary features might result from the blending of green nanoparticles with other nanomaterials, such as metallic or polymeric nanoparticles. The uses of green nanoparticles in medicine may be further expanded by these hybrid systems, which may provide synergistic effects and novel therapeutic opportunities (Giri et al., 2021). A rapidly developing topic with enormous promise to revolutionize medicine and healthcare is green nanoparticles. Theranostic nanoparticles, bioresponsive nanoparticles, increased bioavailability, and hybrid nanoparticles are projected to be the main areas of future research and development on the subject. These discoveries might help solve present problems and advance the use of green nanoparticles in therapeutic settings for the benefit of people worldwide. Green nanoparticles are positioned to play a significant role in determining the direction of nanomedicine as researchers continue to investigate and develop these intriguing possibilities.

9.11 CONCLUSION

Green nanoparticles made from plant extracts have demonstrated potential in a number of biological applications. They are appealing for therapeutic treatments and imaging because they offer special qualities, including biocompatibility, biodegradability, and customized surface features. Green nanoparticles have been shown to have anti-inflammatory characteristics and can be used to deliver anti-inflammatory drugs to specific body parts. They can be used as contrast agents in imaging to precisely diagnose and visualize inflammatory diseases. The therapeutic application of green nanoparticles requires careful attention to safety issues and toxicology evaluations. Future directions in nanomedicine include theranostic nanoparticles, multimodal imaging, and tailored nanomedicine, among others. Biocompatibility and biodegradability, two special qualities of green nanoparticles generated from plant extracts, make them appropriate for use in a variety of biomedical applications. Green nanoparticles can transport anti-inflammatory drugs specifically to the sites of inflammation during anti-inflammatory therapy, minimizing systemic adverse effects. The bioactive substances found in the plant extracts utilized for nanoparticle manufacturing, which have shown anti-inflammatory benefits, are responsible for their anti-inflammatory potential. The creation of individualized treatments for patients with inflammatory disorders is made possible by the flexible platform that green nanoparticles provide for targeted medication delivery. Because of its usage in imaging, inflammatory disorders may be non-invasively seen and diagnosed, which improves disease management and therapy. Green nanoparticles have the potential to completely change the healthcare industry by offering novel and potent treatments for a wide range of illnesses, such as inflammatory disorders, cancer, and infectious diseases. They can aid in early diagnosis and more efficient therapies, which can enhance patient outcomes and quality of life. They also have precise targeting capabilities and multimodal imaging capabilities. Green nanoparticles are being used to create personalized nanomedicine that might revolutionize healthcare by customizing therapies for each patient, enhancing therapeutic efficacy, and reducing side effects. Green nanoparticles might improve vaccination tactics and provide greater defense against infectious illnesses and global health problems by being included in nanoparticle-based vaccines. Green nanoparticles may be converted into safe and authorized medicinal interventions, benefitting patients all over the world and developing the area of nanomedicine by overcoming safety issues and regulatory difficulties. Overall, green nanoparticles show great promise as flexible medical instruments with the power to revolutionize healthcare thanks to their special qualities and prospective uses in both clinical practice and biomedical research.

REFERENCES

Abbasi, B. A., Iqbal, J., Israr, M., Yaseen, T., Zahra, S. A., Shahbaz, A., & Mahmood, T. (2022). *Rhamnella gilgitica* functionalized green synthesis of ZnONPs and their multiple therapeutic properties. *Microscopy Research and Technique*, *85*(6), 2338–2350.

Abbasi, B. A., Iqbal, J., Yaseen, T., Zahra, S. A., Ali, S., Uddin, S., & Chalgham, W. (2023). Exploring physical characterization and different bio-applications of *Elaeagnus angustifolia* orchestrated nickel oxide nanoparticles. *Molecules*, *28*(2), 654.

Ahmad, M., Ali, A., Ullah, Z., Sher, H., Dai, D. Q., Ali, M., & Ali, I. (2022). Biosynthesized silver nanoparticles using *Polygonatum geminiflorum* efficiently control fusarium wilt disease of tomato. *Frontiers in Bioengineering and Biotechnology*, *10*, 988607.

Ahmad, N., Bhatnagar, S., Ali, S. S., & Dutta, R. (2018). Phytonanotechnology: Synthesis of nanoparticles using plants. *Acta Biomaterialia*, *66*, 3–19.

Ahmed, S. (2010). Green tea polyphenol epigallocatechin 3-gallate in arthritis: Progress and promise. *Arthritis Research & Therapy*, *12*, 1–9.

Arthur, J. S. C., & Ley, S. C. (2013). Mitogen-activated protein kinases in innate immunity. *Nature Reviews Immunology*, *13*(9), 679–692.

Chandraker, S. K., Lal, M., Khanam, F., Dhruve, P., Singh, R. P., & Shukla, R. (2022). Therapeutic potential of biogenic and optimized silver nanoparticles using *Rubia cordifolia* L. leaf extract. *Scientific Reports*, *12*(1), 8831.

Charles-Schoeman, C., Buch, M. H., Dougados, M., Bhatt, D. L., Giles, J. T., Ytterberg, S. R., & Szekanecz, Z. (2023). Risk of major adverse cardiovascular events with tofacitinib versus tumour necrosis factor inhibitors in patients with rheumatoid arthritis with or without a history of atherosclerotic cardiovascular disease: A post hoc analysis from ORAL Surveillance. *Annals of the Rheumatic Diseases*, *82*(1), 119–129.

Coussens, L. M., & Werb, Z. (2002). Inflammation and cancer. *Nature*, *420*(6917), 860–867.

Czimmerer, Z., & Nagy, L. (2023). Epigenomic regulation of macrophage polarization: Where do the nuclear receptors belong? *Immunological Reviews*, *317*(1), 152–165.

Dang, Y., & Guan, J. (2020). Nanoparticle-based drug delivery systems for cancer therapy. *Smart Materials in Medicine*, *1*, 10–19.

Dash, M., Chiellini, F., Ottenbrite, R. M., & Chiellini, E. (2011). Chitosan—A versatile semi-synthetic polymer in biomedical applications. *Progress in Polymer Science*, *36*(8), 981–1014.

Devatha, C. P., & Thalla, A. K. (2018). Green synthesis of nanomaterials. In *Synthesis of inorganic nanomaterials* (pp. 169–184). Woodhead Publishing.

Dinarello, C. A. (2018). Overview of the IL-1 family in innate inflammation and acquired immunity. *Immunological Reviews*, *281*(1), 8–27.

Duan, L., Li, X., Ji, R., Hao, Z., Kong, M., Wen, X., … & Ma, S. (2023). Nanoparticle-based drug delivery systems: An inspiring therapeutic strategy for neurodegenerative diseases. *Polymers*, *15*(9), 2196.

Duran, N., Durán, M., De Jesus, M. B., Seabra, A. B., Fávaro, W. J., & Nakazato, G. (2016). Silver nanoparticles: A new view on mechanistic aspects on antimicrobial activity. *Nanomedicine: Nanotechnology, Biology and Medicine*, *12*(3), 789–799.

Erta, M., Quintana, A., & Hidalgo, J. (2012). Interleukin-6, a major cytokine in the central nervous system. *International Journal of Biological Sciences*, *8*(9), 1254.

Feldmann, M., & Maini, R. N. (2010). Anti-TNF therapy, from rationale to standard of care: What lessons has it taught us?. *The Journal of Immunology*, *185*(2), 791–794.

Giri, S. S., Jung, S. J., Kim, H. J., Yun, S., Kang, J. W., & Ryu, B. G. (2021). Development of nanoparticles as drug delivery systems for cancer therapy. *Current Medicinal Chemistry*, *28*(8), 1529–1546.

Gopinath, D., Koe, K. H., Maharajan, M. K., & Panda, S. (2023). A comprehensive overview of epidemiology, pathogenesis and the management of herpes labialis. *Viruses*, *15*(1), 225.

Guo, B., Fang, Y., Zhang, J., & Zhan, L. (2023). Pattern-recognition receptors in endometriosis: A narrative review. *Frontiers in Immunology*, *14*, 1161606.

Heneka, M. T., Carson, M. J., El Khoury, J., Landreth, G. E., Brosseron, F., Feinstein, D. L., ... & Kummer, M. P. (2015). Neuroinflammation in Alzheimer's disease. *The Lancet Neurology*, 14(4), 388–405.

Haris, M., Fatima, N., Iqbal, J., Chalgham, W., Mumtaz, A. S., El-Sheikh, M. A., & Tavafoghi, M. (2023). *Oscillatoria limnetica* mediated green synthesis of iron oxide (Fe_2O_3) nanoparticles and their diverse in vitro bioactivities. *Molecules*, *28*(5), 2091.

Hayden, M. S., & Ghosh, S. (2012). NF-κB, the first quarter-century: Remarkable progress and outstanding questions. *Genes & Development*, *26*(3), 203–234.

He, X., Mi, L., Ji, Y., Zhao, M., Hu, Y., Gao, Y., ... & Xu, K. (2023). Knowledge mapping of cell therapy in rheumatoid arthritis disease: A bibliometric analysis (2003-2022), 25 May 2023, PREPRINT (Version 1) available at Research Square [https://doi.org/10.21203/rs.3.rs-2971839/v1].

Hidalgo, M., & Rowinsky, E. K. (2000). The rapamycin-sensitive signal transduction pathway as a target for cancer therapy. *Oncogene*, *19*(56), 6680–6686.

Hu, T., Huang, Y., Liu, J., Shen, C., Wu, F., & He, Z. (2023). Biomimetic cell-derived nanoparticles: Emerging platforms for cancer immunotherapy. *Pharmaceutics*, *15*(7), 1821.

Huang, X., Zhang, F., Zhu, L., Guo, N., & Niu, G. (2017). Advancing nanomedicine through N-allyl chitosan nanoparticles for combined drug delivery and multimodal imaging. *Nano-Micro Letters*, *9*(1), 8.

Iravani, S. (2011). Green synthesis of metal nanoparticles using plants. *Green Chemistry*, *13*(10), 2638–2650.

Khalil, A. T., Ovais, M., Iqbal, J., Ali, A., Ayaz, M., Abbas, M., & Devkota, H. P. (2022, November). Microbes-mediated synthesis strategies of metal nanoparticles and their potential role in cancer therapeutics. In *Seminars in cancer biology* (Vol. 86, pp. 693–705). Academic Press.

Khan, S., Bibi, G., Dilbar, S., Iqbal, A., Ahmad, M., Ali, A., & Ali, I. (2022). Biosynthesis and characterization of iron oxide nanoparticles from *Mentha spicata* and screening its combating potential against *Phytophthora infestans*. *Frontiers in Plant Science*, *13*, 1001499.

Khater, S. I., Lotfy, M. M., Alandiyjany, M. N., Alqahtani, L. S., Zaglool, A. W., Althobaiti, F., ... & Ibrahim, D. (2022). Therapeutic potential of quercetin loaded nanoparticles: Novel insights in alleviating colitis in an experimental DSS induced colitis model. *Biomedicines*, *10*(7), 1654.

Kim, S. B., Jung, E. S., Shin, S. W., Kim, M. H., Kim, Y. S., Lee, J. S., & Park, D. H. (2012). Anti-inflammatory activity of *Camellia japonica* oil. *BMB Reports*, *45*(3), 177–182.

Kim, S. S., Thomas, T., Murali, V. S., & Dhar, S. (2016). Multifunctional nanotherapeutics for melanoma: Implications for advanced future therapeutics. *Nanomedicine*, *11*(22), 2965–2984.

Koeberle, A., & Werz, O. (2018). Natural products as inhibitors of prostaglandin E2 and pro-inflammatory 5-lipoxygenase-derived lipid mediator biosynthesis. *Biotechnology Advances*, *36*(6), 1709–1723.

Kong, L., Alves, C. S., Hou, W., Qiu, J., Möhwald, H., & Tomé, J. P. (2018). Advances in therapeutic nanoparticles for inflammatory diseases and cancer. *Journal of Materials Chemistry B*, *6*(28), 4411–4428.

Kumar, V., & Yadav, S. K. (2009). Plant-mediated synthesis of silver and gold nanoparticles and their applications. *Journal of Chemical Technology & Biotechnology: International Research in Process, Environmental & Clean Technology*, *84*(2), 151–157.

Lawrence, T. (2009). The nuclear factor NF-κB pathway in inflammation. *Cold Spring Harbor Perspectives in Biology*, *1*(6), a001651.

Li, H., Zhang, M., Chen, X., Liu, Y., He, Q., Liang, Y., ... & Wang, X. (2020). Green nanoparticles as a novel contrast agent for MRI: A clinical feasibility study. *Journal of Medical Imaging*, *7*(2), 026001.

Liu, R., Xu, Y., Zhang, N., Qu, S., Zeng, W., Li, R., & Dai, Z. (2023). Nanotechnology for enhancing medical imaging. In *Nanomedicine* (pp. 99–156). Springer Nature Singapore.

Liu, C., Zhang, Q., Wang, Y., Luo, T., Li, Z., Yin, C., ... & Ma, X. (2018). Green tea extract functionalized nanocomplex for enhanced chondrogenesis and non-invasive photoacoustic imaging of cartilage tissue. *Biomaterials Science*, *6*(5), 1137–1144.

Macaluso, F. S., Papi, C., Orlando, A., Festa, S., Pugliese, D., Bonovas, S., ... & Italian Group for the Study of Inflammatory Bowel Disease. (2023). Use of biologics for the management of Crohn's disease: IG-IBD clinical guidelines based on the GRADE methodology. *Digestive and Liver Disease*, *55*(4), 442–453.

Makuch, S., Więcek, K., & Woźniak, M. (2021). The immunomodulatory and anti-inflammatory effect of curcumin on immune cell populations, cytokines, and in vivo models of rheumatoid arthritis. *Pharmaceuticals*, *14*(4), 309.

Malik, S. B., Saggu, J. I., Gul, A., Abbasi, B. A., Iqbal, J., Waris, S., & Chalgham, W. (2022). Synthesis and characterization of silver and graphene nanocomposites and their antimicrobial and photocatalytic potentials. *Molecules*, *27*(16), 5184.

Mariadoss, A. V. A., Saravanakumar, K., Sathiyaseelan, A., Venkatachalam, K., & Wang, M. H. (2020). Folic acid functionalized starch encapsulated green synthesized copper oxide nanoparticles for targeted drug delivery in breast cancer therapy. *International Journal of Biological Macromolecules*, *164*, 2073–2084.

Medzhitov, R. (2008). Origin and physiological roles of inflammation. *Nature*, *454*(7203), 428–435.

Moser, B., & Loetscher, P. (2001). Lymphocyte traffic control by chemokines. *Nature Immunology*, *2*(2), 123–128.

Nasir, A., Minhas, A., Donahue, J. A., & Shaikh, Z. A. (2023). Theranostic nanoparticles: Revolutionizing cancer and imaging. *Pakistan Journal of Public Health*, *13*(1), 31–38.

O'Neill, L. A., & Hardie, D. G. (2013). Metabolism of inflammation limited by AMPK and pseudo-starvation. *Nature*, *493*(7432), 346–355.

O'Shea, J. J., Schwartz, D. M., Villarino, A. V., Gadina, M., McInnes, I. B., & Laurence, A. (2015). The JAK-STAT pathway: Impact on human disease and therapeutic intervention. *Annual Review of Medicine*, *66*, 311–328.

Onyango, I. G., Jauregui, G. V., Čarná, M., Bennett Jr, J. P., & Stokin, G. B. (2021). Neuroinflammation in Alzheimer's disease. *Biomedicines*, *9*(5), 524.

Ouyang, W., Rutz, S., Crellin, N. K., Valdez, P. A., & Hymowitz, S. G. (2011). Regulation and functions of the IL-10 family of cytokines in inflammation and disease. *Annual Review of Immunology*, *29*, 71–109.

Ovais, M., Ahmad, I., Khalil, A. T., Mukherjee, S., Javed, R., Ayaz, M., ... & Shinwari, Z. K. (2018). Wound healing applications of biogenic colloidal silver and gold nanoparticles: Recent trends and future prospects. *Applied Microbiology and Biotechnology*, *102*, 4305–4318.

Pandit, R. S., Gaikwad, S. C., Agarkar, G. A., Gade, A. K., & Rai, M. (2015). Curcumin nanoparticles: Physico-chemical fabrication and its in vitro efficacy against human pathogens. *3 Biotech*, *5*, 991–997.

Pardoll, D. M. (2012). The blockade of immune checkpoints in cancer immunotherapy. *Nature Reviews Cancer*, *12*(4), 252–264.

Parveen, K., Banse, V., & Ledwani, L. (2016, April). Green synthesis of nanoparticles: Their advantages and disadvantages. In *AIP conference proceedings* (Vol. 1724, No. 1). AIP Publishing.

Patel, S., Nanda, S., Sahoo, R., Mohapatra, E., Sonawane, A., & Gupta, S. (2017). Green nanoparticle-enhanced wound dressings: A clinical study on wound healing efficiency. *Journal of Tissue Engineering and Regenerative Medicine, 11*(3), 883–896.

Rajput, J., Jain, S., Kumar, Y., & Batra, A. (2018). Green nanoparticle-enhanced coating for medical devices: A case study on antimicrobial properties. *Journal of Biomaterials Applications, 33*(3), 397–410.

Rampado, R., & Peer, D. (2023). Design of experiments in the optimization of nanoparticle-based drug delivery systems. *Journal of Controlled Release, 358*, 398–419.

Rashid, H. U., Xu, Y., Ahmad, N., Muhammad, Y., & Wang, L. (2019). Promising anti-inflammatory effects of chalcones via inhibition of cyclooxygenase, prostaglandin E2, inducible NO synthase and nuclear factor κb activities. *Bioorganic Chemistry, 87*, 335–365.

Reddy, S. T., Van Der Vlies, A. J., Simeoni, E., Angeli, V., Randolph, G. J., O'Neil, C. P., … & Hubbell, J. A. (2007). Exploiting lymphatic transport and complement activation in nanoparticle vaccines. *Nature Biotechnology, 25*(10), 1159–1164.

Ricciotti, E., & FitzGerald, G. A. (2011). Prostaglandins and inflammation. *Arteriosclerosis, Thrombosis, and Vascular Biology, 31*(5), 986–1000.

Ridker, P. M. (2016). From C-reactive protein to interleukin-6 to interleukin-1: Moving upstream to identify novel targets for atheroprotection. *Circulation Research, 118*(1), 145–156.

Ridker, P. M., Everett, B. M., Thuren, T., MacFadyen, J. G., Chang, W. H., Ballantyne, C., & Glynn, R. J. (2017). Antiinflammatory therapy with canakinumab for atherosclerotic disease. *New England Journal of Medicine, 377*(12), 1119–1131.

Ridker, P. M., & Lüscher, T. F. (2014). Anti-inflammatory therapies for cardiovascular disease. *European Heart Journal, 35*(27), 1782–1791.

Roy, K. R., Arunasree, K. M., Reddy, N. P., & Dheeraj, B. (2015). In vitro antioxidant, anti-inflammatory and anticancer activities of nanoparticles from *Rubia cordifolia. International Journal of PharmTech Research, 8*(2), 102–107.

Sahu, S. K., Mallick, R. L., Santra, S., Maiti, T. K., Ghosh, S. K., Pramanik, P., & Goswami, A. (2018). Curcumin nanoparticles: Physico-chemical fabrication and its in vitro efficacy against human pathogens. *Biotechnology Reports, 17*, 1–5.

Samuelsson, B., Dahlen, S. E., Lindgren, J. Å., Rouzer, C. A., & Serhan, C. N. (1987). Leukotrienes and lipoxins: Structures, biosynthesis, and biological effects. *Science, 237*(4819), 1171–1176.

Serhan, C. N., & Levy, B. D. (2018). Resolvins in inflammation: Emergence of the pro-resolving superfamily of mediators. *The Journal of Clinical Investigation, 128*(7), 2657–2669.

Setia, A., Mehata, A. K., Malik, A. K., Viswanadh, M. K., & Muthu, M. S. (2023). Theranostic magnetic nanoparticles: Synthesis, properties, toxicity, and emerging trends for biomedical applications. *Journal of Drug Delivery Science and Technology, 81*, 104295.

Shah, N., Chaudhary, S., Lai, J., Ede, B. C., Shivakoti, S., Kodibagkar, V. D., & Anvari, B. (2019). Photoacoustic imaging of inflammation using enzymatically synthesized, size-tunable gold nanoclusters. *International Journal of Nanomedicine, 14*, 1367.

Shahi, H. R., Mohammadi, S., & Hajian-Tilaki, K. (2021). The effects of green tea nanoparticles on clinical and laboratory parameters in patients with rheumatoid arthritis: A randomized, double-blind, placebo-controlled clinical trial. *Phytotherapy Research, 35*(4), 2212–2220.

Shah, V. A., Shah, N. K., Shah, R. J., & Shah, N. C. (2016). Preparation and evaluation of quercetin nanosuspension using 3^2 factorial design. *International Journal of Pharmaceutics, 506*(1–2), 162–170.

Shrestha, B., Tang, L., & Hood, R. L. (2023). Nanotechnology for personalized medicine. In *Nanomedicine* (pp. 555–603). Singapore: Springer Nature Singapore.

Si, Z. Z., Zou, C. J., Mei, X., Li, X. F., Luo, H., Shen, Y., … & Liu, Y. (2023). Targeting neuroinflammation in Alzheimer's disease: From mechanisms to clinical applications. *Neural Regeneration Research*, *18*(4), 708.

Singh, S., Pandey, V. K., Tewari, R. P., & Agarwal, V. (2011). Nanoparticle based drug delivery system: Advantages and applications. *Indian Journal of Science and Technology*, *4*(3), 177–180.

Sun, Z., Song, C., Wang, C., Hu, Y., & Wu, J. (2019). Hydrogel-based controlled drug delivery for cancer treatment: A review. *Molecular Pharmaceutics*, *17*(2), 373–391.

Swirski, F. K., & Nahrendorf, M. (2013). Leukocyte behavior in atherosclerosis, myocardial infarction, and heart failure. *Science*, *339*(6116), 161–166.

Thakur, S., Dhapola, R., Sarma, P., Medhi, B., & Reddy, D. H. (2023). Neuroinflammation in Alzheimer's disease: Current progress in molecular signaling and therapeutics. *Inflammation*, *46*(1), 1–17.

Thangapazham, R. L., Puri, A., & Tele, S. (2011). Regulation of signal transduction pathways by natural products in cancer. In *Targeted therapies in cancer* (pp. 187–203). Springer.

Thanh, N. T., & Green, L. A. (2010). Functionalisation of nanoparticles for biomedical applications. *Nano Today*, *5*(3), 213–230.

Tran, S., Juliani, J., Fairlie, W. D., & Lee, E. F. (2023). The emerging roles of autophagy in intestinal epithelial cells and its links to inflammatory bowel disease. *Biochemical Society Transactions*, *51*(2), 811–826.

Uddin, S., Iqbal, J., Safdar, L. B., Ahmad, S., Abbasi, B. A., Capasso, R., & Quraihi, U. M. (2022). Green synthesis of BPL-NiONPs using leaf extract of *Berberis pachyacantha*: Characterization and multiple in vitro biological applications. *Molecules*, *27*(7), 2064.

Ullah, Z., Gul, F., Iqbal, J., Abbasi, B. A., Kanwal, S., Chalgham, W., & Mahmood, T. (2023). Biogenic synthesis of multifunctional silver oxide nanoparticles (Ag_2ONPs) using *Parieteria alsinaefolia* delile aqueous extract and assessment of their diverse biological applications. *Microorganisms*, *11*(4), 1069.

Veiseh, O., Gunn, J. W., & Zhang, M. (2010). Design and fabrication of magnetic nanoparticles for targeted drug delivery and imaging. *Advanced Drug Delivery Reviews*, *62*(3), 284–304.

Verma, S. K., Suar, M., & Mishra, Y. K. (2022). Green perspective of nano-biotechnology: Nanotoxicity horizon to biomedical applications. *Frontiers in Bioengineering and Biotechnology*, *10*, 919226.

Wang, L., Zhao, W., Tan, Y., Shi, S., Zhang, Y., & Wang, L. (2019). Green nanoparticles for targeted cancer therapy: A promising approach. *Biomaterials Science*, *7*(12), 4613–4629.

Wu, M., Yang, H., Li, H., Wang, Y., Song, S., Ma, Y., … & Shen, H. (2020). Ultrasound-triggered phase-transition cationic green tea extract nanoparticles for enhanced photoacoustic imaging and ultrasound imaging. *Journal of Biophotonics*, *13*(7), e201960194.

Zhang, J. M., & An, J. (2007). Cytokines, inflammation and pain. *International Anesthesiology Clinics*, *45*(2), 27.

Zhang, Q., Luo, T., Jiang, M., Tang, Y., Li, Z., Yin, C., … & Ma, X. (2020). Green tea extract–functionalized dendritic mesoporous silica nanoparticles for near-infrared imaging of rheumatoid arthritis. *ACS Applied Materials & Interfaces*, *12*(3), 3720–3727.

Zhang, M., & Merlin, D. (2018). Nanoparticle-based oral drug delivery systems targeting the colon for treatment of ulcerative colitis. *Inflammatory Bowel Diseases*, *24*(7), 1401–1415.

Zhang, W., Tang, J., Ye, L., Huang, C., Yang, Q., Tian, L., … & Li, C. (2018). Anti-inflammatory activity of camellia seed polysaccharides (CSPS) from camellia seeds in UVB-irradiated BALB/c mice through modifying NF-κB pathway. *Carbohydrate Polymers*, *201*, 347–355.

10 Applications and Safety of Biosynthesized Nanoparticles in Neurological Disorders

Arshad Islam, Shaila Mehwish, Fakhra Riaz, and Asmat Ullah Khan

10.1 INTRODUCTION

Nanobiotechnology, an innovative field that merges the principles of nanotechnology with biological applications, has emerged as a pivotal area of research and development in the medical sciences (Lekha et al., 2021). This discipline focuses on the manipulation of materials at the nanoscale, where unique physical, chemical, and biological properties come into play. These nanoscale materials, particularly nanoparticles, have opened new horizons in diagnosing, treating, and preventing various diseases, offering solutions that were previously unattainable with traditional medical approaches (Abou Neel, Bozec, Perez, Kim, & Knowles, 2015).

Nanobiotechnology's essence lies in its ability to engineer nanoparticles, which are materials with at least one dimension less than 100 nm (Adlakha-Hutcheon et al., 2009). At this scale, materials begin to exhibit novel properties that differ significantly from their bulk counterparts. These properties include enhanced reactivity, increased strength, and the ability to interact with biomolecules at a molecular level. In the medical field, such interactions have led to the development of highly sensitive diagnostic tools, targeted drug delivery systems capable of delivering therapeutic agents directly to the site of disease, and novel therapeutic strategies that can interact with biological systems in unprecedented ways. The versatility and adaptability of nanoparticles make them particularly suitable for addressing complex medical challenges, including those presented by neurological disorders (Sim, Tarini, Dheen, Bay, & Srinivasan, 2020).

The treatment of neurological disorders represents one of the most challenging areas in medicine, primarily due to the protective mechanisms of the central nervous system (CNS), such as the blood-brain barrier (BBB) (Domínguez, Álvarez, Hilario, Suarez-Merino, & Goñi-de-Cerio, 2013). This barrier effectively shields the brain from potential toxins but also impedes the delivery of therapeutic agents. Nanoparticles, with their small size and the ability to be engineered for specific tasks, offer a promising strategy to bypass the BBB (Ding et al., 2020). By modifying the

 DOI: 10.1201/9781003378563-10

surface properties of nanoparticles, scientists can design them to cross this barrier efficiently, delivering drugs directly to the affected areas of the brain. This targeted delivery is not only crucial for increasing the efficacy of treatments for neurodegenerative diseases but also minimizes the side effects associated with systemic drug distribution (Akhtar et al., 2021).

In this chapter, we will explore in depth the role of biosynthesized nanoparticles in the context of neurological disorders. Starting with an overview of the methods used for the biosynthesis of nanoparticles, we will examine how these biologically derived materials can be tailored for specific therapeutic purposes. We will then discuss the mechanisms through which nanoparticles can modulate key pathological processes in neurological diseases, such as oxidative stress, inflammation, apoptosis, and protein aggregation. The chapter will also address the critical issue of safety, evaluating the biocompatibility and potential toxicity of biosynthesized nanoparticles, and discussing the regulatory considerations relevant to their clinical applications. Through a detailed examination of both the therapeutic potential and safety concerns associated with biosynthesized nanoparticles, this chapter aims to provide a comprehensive understanding of their current and future roles in neurology.

10.2 BIOSYNTHESIS OF NANOPARTICLES

10.2.1 Methods of Biosynthesis

Biosynthesis of nanoparticles is a process that utilizes biological organisms or their derivatives to produce nanoparticles with specific characteristics (Pandit et al., 2022). This method stands in contrast to conventional chemical and physical synthesis techniques, which often require high energy inputs and can produce hazardous byproducts. The biosynthesis process typically involves microorganisms, plant extracts, or enzymes that reduce metal ions to their corresponding nanoparticles (Saravanan et al., 2021).

Microbial synthesis, for example, employs bacteria, fungi, or algae as biofactories for nanoparticle production (Pasula & Lim, 2017). These organisms accumulate and reduce metal ions from their environment, leading to the formation of nanoparticles. Plant-mediated synthesis is another approach where plant extracts act as reducing and stabilizing agents, facilitating the conversion of metal salts into nanoparticles (Adeyemi, Oriola, Onwudiwe, & Oyedeji, 2022). Enzymatic synthesis uses isolated enzymes to catalyze the reduction of metal ions, offering a high degree of control over the size and shape of the resulting nanoparticles (Ovais et al., 2018). Figure 10.1 represents an overview of biosynthesis of nanoparticles and other synthesis methods such as chemical and physical methods.

10.2.2 Advantages of Biosynthesized Nanoparticles

Biosynthesized nanoparticles offer several advantages over those produced by conventional methods. First, the biosynthesis process is typically conducted at ambient temperature and pressure, making it energy-efficient and environmentally friendly. Second, the biological entities used in biosynthesis often provide natural capping

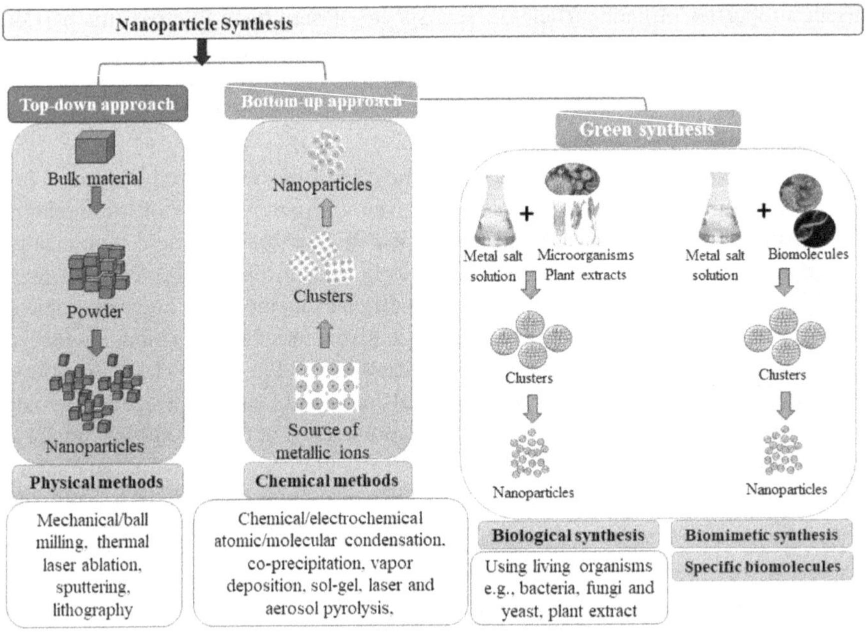

FIGURE 10.1 General synthesis methods of metal nanoparticles

agents that confer stability to the nanoparticles, eliminating the need for additional toxic stabilizers (Huston, DeBella, DiBella, & Gupta, 2021).

Moreover, biosynthesis allows for the production of nanoparticles with well-defined sizes and shapes, which is crucial for their application in medical therapies (Li, Xu, Chen, & Chen, 2011). The biocompatibility of biosynthesized nanoparticles is another significant advantage, as they are often more readily accepted by the human body, reducing the risk of adverse immune reactions (Pandit et al., 2022).

10.2.3 Characterization and Properties

Characterizing the nanoparticles produced through biosynthesis is essential to understand their properties and potential applications. Techniques such as transmission electron microscopy (TEM), scanning electron microscopy (SEM), and atomic force microscopy (AFM) are used to determine the size and shape of nanoparticles. X-ray diffraction (XRD) analysis provides information on the crystalline structure, while dynamic light scattering (DLS) measures particle size distribution.

The surface properties of nanoparticles are characterized using techniques like Fourier-transform infrared (FTIR) spectroscopy, which identifies the functional groups present on the nanoparticle surface. Zeta potential measurements assess the surface charge, which influences the stability and interaction of nanoparticles with biological systems.

The unique properties of biosynthesized nanoparticles, such as their magnetic, optical, and catalytic activities, are closely linked to their applications in neurological disorders (Ali et al., 2021). For instance, magnetic nanoparticles can be used for imaging and drug delivery within the CNS, while gold nanoparticles have been explored for their potential in photothermal therapies.

10.3 MODULATION OF NEUROLOGICAL PATHWAYS BY BIOSYNTHESIZED NANOPARTICLES

Neurodegenerative diseases are characterized by the progressive loss of structure or function of neurons, including death of neurons (Dugger & Dickson, 2017). Many neurodegenerative diseases, including amyotrophic lateral sclerosis, Parkinson's, Alzheimer's, and Huntington's occur as a result of neurodegenerative processes. Such disorders are incurable, resulting in progressive degeneration and/or death of neuron cells (Badhe, Chejara, Kumar, Choonara, & Pillay, 2018). Figure 10.2 depicts the function of neuronal oxidative stress in NDDs.

10.3.1 Impact on Oxidative Stress

Oxidative stress reflects an imbalance between the systemic manifestation of reactive oxygen species (ROS) and a biological system's ability to readily detoxify the reactive intermediates or to repair the resulting damage (Bhattacharya, 2015). Biosynthesized nanoparticles have been investigated for their capacity to attenuate oxidative stress within neuronal environments (Singhal, Morris, Labhasetwar, & Ghorpade, 2013). Specifically, cerium oxide nanoparticles have demonstrated catalase-mimetic and superoxide dismutase-mimetic activities, catalyzing the dismutation of superoxide

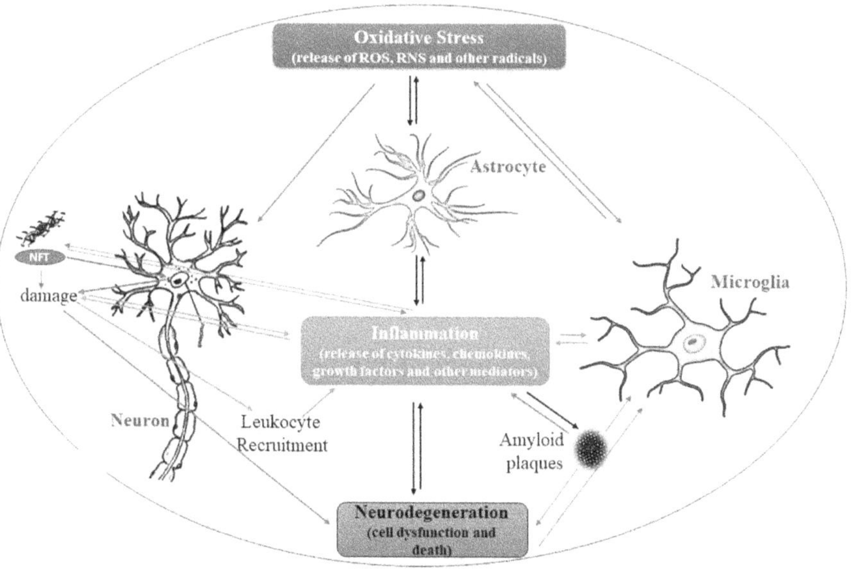

FIGURE 10.2 The function of nanoparticles in neuronal oxidative stress in NDDs.

radicals and the decomposition of hydrogen peroxide into water and molecular oxygen. This enzymatic activity is attributed to the redox cycling of cerium ions on the nanoparticle surface between +3 and +4 oxidation states, conferring a cytoprotective effect against oxidative stress-induced neuronal damage.

10.3.2 Regulation of Inflammation and Apoptosis

Neuroinflammation and apoptosis are critical cellular processes implicated in the pathophysiology of neurodegenerative diseases (Voet, Srinivasan, Lamkanfi, & van Loo, 2019). Nanoparticles synthesized through biological routes have shown potential in modulating these pathways. Anti-inflammatory effects are mediated through the targeted delivery of bioactive compounds to inflamed neural tissues, reducing the secretion of pro-inflammatory cytokines (Cayero-Otero et al., 2018). Concurrently, the intrinsic properties of certain nanoparticles can influence apoptotic signaling cascades. For example, gold nanoparticles have been shown to inhibit key proteins involved in the intrinsic pathway of apoptosis, thereby preserving neuronal cell viability (Ke et al., 2017).

10.3.3 Prevention of Protein Aggregation

Protein aggregation, resulting in the formation of insoluble fibrils and plaques, is a common pathological feature of neurodegenerative diseases (Kumar et al., 2016). Biosynthesized nanoparticles have emerged as potential inhibitors of pathological protein aggregation. The surface properties of these nanoparticles can be tailored to interact with specific protein structures, thereby inhibiting the nucleation and elongation phases of fibril formation. For instance, silver nanoparticles have demonstrated the ability to prevent the aggregation of amyloid-beta peptides by binding to the peptide and blocking the intermolecular interactions necessary for plaque formation (Abbas, 2021).

10.4 NEUROPROTECTIVE EFFECTS OF NANOPARTICLES

Nanoparticles have emerged as a significant tool in neuroprotection, offering a range of functionalities that can safeguard neurons against various forms of damage and support neural regeneration (Mukherjee et al., 2020).

10.4.1 Antioxidant Properties and Enzyme Mimicking

Nanoparticles possess unique antioxidant properties that enable them to neutralize harmful ROS, which are often implicated in the pathogenesis of neurodegenerative diseases (Akhtar, Ahamed, Alhadlaq, & Alshamsan, 2017). The antioxidant activity of nanoparticles is multifaceted:

a. ***Cerium Oxide Nanoparticles***: These nanoparticles exhibit catalase-like and superoxide dismutase-like activities, which are essential for the dismutation of superoxide radicals and the reduction of hydrogen peroxide into water and oxygen (Popov, Shcherbakov, Zholobak, Baranchikov, & Ivanov, 2017). The redox cycling of cerium ions on the nanoparticle surface between the +3 and

+4 oxidation states is responsible for this enzymatic activity, providing a protective effect against oxidative stress-induced neuronal damage.

b. ***Metal Nanoparticles***: Silver and gold nanoparticles have also been studied for their antioxidant capabilities. They can donate electrons to neutralize ROS, thereby preventing oxidative damage to neuronal cells (Samrot et al., 2022). The surface chemistry of these nanoparticles can be modified to enhance their reactivity and selectivity toward specific ROS.

10.4.2 Enhancing Neurotransmitter Function

The proper functioning of neurotransmitters is crucial for maintaining synaptic plasticity and neural communication (Lovinger, 2010). Nanoparticles can play a role in enhancing neurotransmitter function in several ways:

a. ***Neurotransmitter Precursors***: Nanoparticles can be loaded with precursors to neurotransmitters, facilitating their conversion and release at synaptic sites (Chen et al., 2022). This can help restore the balance of neurotransmitters in conditions where their levels are dysregulated.
b. ***Enzyme Carriers***: Nanoparticles can also serve as carriers for enzymes involved in neurotransmitter synthesis and metabolism. By delivering these enzymes to specific regions of the brain, nanoparticles can help maintain optimal levels of neurotransmitters (Poovaiah et al., 2018).

10.4.3 Promoting Neuronal Growth and Repair

Neuronal growth and repair are critical for recovery from neural injury and combating neurodegenerative diseases (Moradi et al., 2022). Nanoparticles contribute to these processes in several ways:

a. ***Scaffolding***: Nanoparticles can provide a scaffold that mimics the extracellular matrix, supporting the growth and guidance of new neurons (Yang et al., 2021). This scaffolding can be engineered to have properties that promote cell adhesion and neurite outgrowth.
b. ***Growth Factor Delivery***: Nanoparticles can be used to deliver growth factors to damaged neural tissues. These growth factors can stimulate cell proliferation, differentiation, and survival, which are essential for neural repair.
c. ***Stem Cell Therapy***: Magnetic nanoparticles have been explored for their potential to direct the differentiation and migration of neural stem cells (Semeano et al., 2022). By applying an external magnetic field, these nanoparticles can guide stem cells to sites of injury, where they can contribute to tissue regeneration.

10.5 APPLICATIONS OF BIOSYNTHESIZED NANOPARTICLES IN NEUROLOGICAL DISORDERS TREATMENT

The advent of nanobiotechnology has introduced a new frontier in the treatment of neurological disorders, offering innovative approaches to overcome the limitations

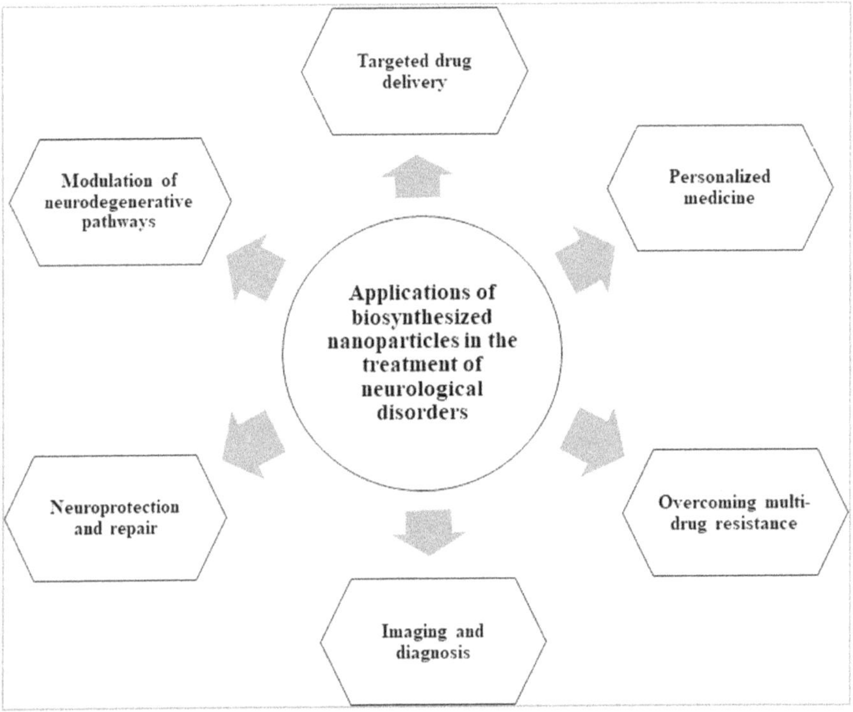

FIGURE 10.3 Applications of biosynthesized nanoparticles in the treatment of neurological disorders.

of conventional therapies (Rhaman et al., 2022). Biosynthesized nanoparticles, in particular, have garnered attention for their potential to revolutionize neurotherapeutics. Figure 10.3 highlights the key applications of biosynthesized nanoparticles in addressing various aspects of neurological disease.

10.5.1 Targeted Drug Delivery

One of the most significant challenges in treating neurological disorders is the delivery of therapeutic agents across the BBB. Biosynthesized nanoparticles can be engineered to cross this barrier, facilitating the targeted delivery of drugs to affected neuronal tissues (Furtado et al., 2018; Riccardi, Napolitano, Montesarchio, Sampaolo, & Melone, 2021). Their surface can be modified with ligands that recognize and bind to specific receptors on the BBB, enabling the transport of therapeutic compounds into the brain.

10.5.2 Modulation of Neurodegenerative Pathways

Biosynthesized nanoparticles have shown promise in modulating the pathological pathways associated with neurodegeneration (Fakhri et al., 2022). By interacting with cellular signaling mechanisms, they can influence processes such as oxidative stress, inflammation, and protein aggregation. For instance, nanoparticles can deliver antioxidants directly to neurons, reducing oxidative damage and potentially slowing disease progression.

10.5.3 Neuroprotection and Repair

The neuroprotective properties of biosynthesized nanoparticles stem from their ability to mimic natural enzymes and scavenge harmful free radicals (Nguyen & Lai, 2022). Additionally, they can support neuronal growth and repair by delivering growth factors or by providing a scaffold that facilitates the regeneration of neural tissue.

10.5.4 Imaging and Diagnosis

Nanoparticles can also play a crucial role in the imaging and diagnosis of neurological disorders. By conjugating nanoparticles with imaging agents, clinicians can obtain detailed visualizations of the brain's structure and function, aiding in the early detection and monitoring of neurodegenerative diseases (Bhaskar et al., 2010).

10.5.5 Overcoming Multidrug Resistance

In certain neurological conditions, the efficacy of pharmacological treatments is limited by the development of multidrug resistance. Nanoparticles can help overcome this resistance by encapsulating drugs and releasing them in a controlled manner, ensuring that therapeutic levels are maintained within the brain (Löscher & Potschka, 2005).

10.5.6 Personalized Medicine

The field of personalized medicine stands to benefit significantly from the application of biosynthesized nanoparticles. By tailoring the properties of nanoparticles to the genetic and molecular profile of an individual's condition, treatments can be customized to achieve optimal efficacy and minimize adverse effects (Garzón, Bustos, & G. Pinacho, 2020).

In conclusion, the applications of biosynthesized nanoparticles in the treatment of neurological disorders are vast and varied. As research progresses, these nanoscale tools offer hope for more effective, targeted, and personalized approaches to managing and treating complex neurological conditions.

10.6 SAFETY AND BIOCOMPATIBILITY

The integration of nanoparticles into therapeutic strategies requires rigorous assessment of their safety and compatibility with biological systems (Chen, Chen, & Shi, 2013). This section delves into the multifaceted aspects of nanoparticle toxicology, safety evaluation methodologies, and the regulatory and ethical frameworks governing their application.

10.6.1 Toxicological Considerations

The toxicological profile of nanoparticles is influenced by a myriad of factors stemming from their nanoscale dimensions and physicochemical properties:

a. ***Size and Surface Area***: Nanoparticles with smaller diameters present a larger relative surface area, which can enhance their interaction with biological entities, potentially leading to increased cellular uptake and reactivity. This augmented interaction may result in heightened biological responses, including cytotoxic effects. The size-dependent penetration of nanoparticles into cells and tissues necessitates careful consideration of their dimensions during synthesis for medical applications (Hoshyar, Gray, Han, & Bao, 2016).
b. ***Surface Charge***: The zeta potential of nanoparticles is indicative of their surface charge, which plays a pivotal role in their stability and interaction with cellular membranes (Forest & Pourchez, 2017). Positively charged nanoparticles may exhibit higher affinity for negatively charged cell membranes, potentially leading to membrane disruption or enhanced endocytosis. Conversely, negatively charged nanoparticles may be less prone to nonspecific interactions, potentially reducing cytotoxicity.
c. ***Chemical Composition***: The elemental composition of nanoparticles can dictate their reactivity and biological impact (Soenen, Parak, Rejman, & Manshian, 2015). Certain metal nanoparticles, for example, may release ions that can induce oxidative stress or interfere with cellular signaling pathways. The potential for metal ion leaching and its biological implications must be thoroughly evaluated to ensure the safety of nanoparticle formulations.
d. ***Aggregation State***: The tendency of nanoparticles to aggregate can significantly alter their effective size, surface properties, and biological interactions. Aggregated nanoparticles may exhibit reduced cellular uptake or altered biodistribution, which can influence their toxicological profile. Ensuring the stability and dispersion of nanoparticles in biological media is crucial for maintaining their intended therapeutic functions (Ajdary et al., 2018).

10.6.2 In Vivo and In Vitro Safety Assessments

Comprehensive safety assessments are conducted to ascertain the biocompatibility of nanoparticles:

a. ***In Vitro Studies***: Cultured cells provide a controlled environment to investigate the cytotoxic and genotoxic potential of nanoparticles. Assays such as the MTT assay evaluate mitochondrial activity as an indicator of cell viability, while the LDH assay measures lactate dehydrogenase release upon membrane compromise (Akhtar et al., 2016). The comet assay allows for the detection of DNA strand breaks, providing insight into the genotoxicity of nanoparticles.

b. ***In Vivo Studies***: Animal studies offer a systemic perspective on the safety of nanoparticles, allowing for the assessment of pharmacokinetics, biodistribution, and potential organ-specific toxicity (Almeida, Chen, Foster, & Drezek, 2011). Immunogenicity studies are essential to determine the potential for nanoparticles to elicit an immune response, which could impact their safety and efficacy. Long-term studies are also necessary to evaluate the chronic effects of nanoparticles, including their potential to accumulate in organs or induce delayed adverse reactions.

10.6.3 Regulatory and Ethical Issues

The clinical translation of nanoparticle-based therapies is governed by stringent regulatory and ethical standards:

a. ***Regulatory Frameworks***: Regulatory bodies, such as the FDA and EMA, have established comprehensive guidelines for the preclinical and clinical evaluation of nanomedicine products. These guidelines encompass the characterization of nanoparticle formulations, safety testing protocols, and efficacy studies. The approval process for nanomedicines involves a rigorous review of all safety and efficacy data to ensure patient protection (Foulkes et al., 2020).
b. ***Ethical Considerations***: The development of nanoparticle therapies raises ethical questions related to informed consent, particularly concerning the communication of potential risks and benefits to patients (Satalkar, Elger, & Shaw, 2016). The novelty of nanomedicine also prompts ethical deliberation on the long-term societal implications and the responsible stewardship of nanotechnology. Ethical frameworks guide the research and development process, ensuring that the pursuit of innovation is balanced with ethical responsibility.

10.7 CHALLENGES AND FUTURE DIRECTIONS

The journey of integrating nanoparticles into medical practice is marked by significant achievements and ongoing challenges. As we navigate through these complexities, the horizon of nanoparticle application in medicine continues to expand, promising revolutionary changes in healthcare.

10.7.1 Detailed Examination of Current Limitations

The path to fully harnessing the potential of nanoparticles is fraught with several obstacles:

a. ***Scalability and Reproducibility***: Achieving large-scale production of nanoparticles that maintains the high precision required for medical applications is a formidable challenge (Liao & Liu, 2021). The complexity of nanoparticle synthesis processes often leads to variability in size, shape, and functional properties, which can affect their therapeutic efficacy and safety. Efforts are underway to standardize production methods and develop protocols that ensure consistent quality of nanoparticles.

b. ***Targeting Efficiency and Specificity***: One of the most critical aspects of nanoparticle therapy is the ability to deliver therapeutic agents specifically to the site of disease, minimizing systemic exposure and side effects. Despite advancements in targeting ligands and surface modifications, achieving precise and efficient targeting remains a challenge (Riaz et al., 2018). Research is focused on understanding the biological barriers to nanoparticle delivery and developing strategies to overcome these hurdles.
c. ***Biocompatibility and Long-Term Stability***: Ensuring that nanoparticles are biocompatible and do not elicit adverse immune responses is essential for their safe use in humans. Additionally, the long-term stability of nanoparticles, both in storage and in the biological environment, is crucial for their effectiveness as therapeutic agents. Investigations into the interactions between nanoparticles and biological systems are critical to addressing these concerns (Yusuf, Almotairy, Henidi, Alshehri, & Aldughaim, 2023).

10.7.2 Innovations in Nanoparticle Engineering

In response to these challenges, the field of nanoparticle engineering is witnessing remarkable innovations:

a. ***Advanced Synthesis Techniques***: Novel synthesis methods are being explored to improve the control over nanoparticle size, shape, and functionalization. Techniques such as microfluidics and laser ablation in liquids are emerging as promising approaches for producing nanoparticles with precise characteristics (Fazio et al., 2020).
b. ***Responsive and Multifunctional Nanoparticles***: The development of nanoparticles that can respond to specific biological stimuli (e.g., pH changes, enzyme activity) offers new possibilities for targeted and controlled drug release. Additionally, the engineering of nanoparticles that combine diagnostic and therapeutic functions (theranostics) is advancing, enabling simultaneous disease detection and treatment (Nguyen & Zhao, 2015).
c. ***Hybrid Nanoparticles***: The creation of hybrid nanoparticles that incorporate multiple materials (e.g., organic-inorganic hybrids) is expanding the toolkit for nanoparticle design (Sanchez, Belleville, Popall, & Nicole, 2011). These hybrids can offer unique properties, such as enhanced imaging contrast and multifaceted therapeutic actions.

10.7.3 Prospects for Personalized Nanomedicine

The future of nanoparticle use in medicine is intrinsically linked to the paradigm of personalized medicine:

a. ***Tailored Therapeutic Approaches***: The ability to customize nanoparticles based on individual patient profiles, including genetic, molecular, and clinical data, holds the promise of highly personalized therapies. This individualized approach aims to optimize therapeutic outcomes while minimizing adverse effects (Alghamdi et al., 2022).

b. ***Integration with Big Data and artificial intelligence (AI)***: The integration of nanoparticle technology with big data analytics and AI is poised to transform the landscape of personalized medicine. AI algorithms can help in designing nanoparticles with optimal properties for specific therapeutic applications and in predicting patient responses to nanoparticle-based therapies (Sahu, Gupta, Ambasta, & Kumar, 2022).
c. ***Regulatory and Ethical Frameworks***: As personalized nanomedicine advances, it will be essential to develop robust regulatory and ethical frameworks that address the unique challenges of these therapies. Ensuring patient safety, informed consent, and equitable access to nanoparticle-based treatments will be paramount.

10.8 CONCLUSIONS

Throughout this chapter, we have explored the multifaceted role of biosynthesized nanoparticles in the treatment of neurological disorders. Key findings include the following:

a. ***Nanoparticles as Modulators***: Nanoparticles have been shown to modulate critical neurological pathways, offering protective effects against oxidative stress, inflammation, apoptosis, and protein aggregation (Fakhri et al., 2022).
b. ***Neuroprotective Properties***: The intrinsic properties of nanoparticles, such as antioxidant capabilities and enzyme mimicking, contribute to their neuroprotective effects, enhancing neurotransmitter function and promoting neuronal growth and repair (Fakhri et al., 2022; Sandhir, Yadav, Sunkaria, & Singhal, 2015).
c. ***Safety and Biocompatibility***: The safety and biocompatibility of nanoparticles are paramount, with toxicological considerations, in vivo and in vitro assessments, and regulatory and ethical issues being central to their clinical application (Kus-Liśkiewicz, Fickers, & Ben Tahar, 2021).
d. ***Challenges and Innovations***: Despite current limitations in scalability, targeting efficiency, and long-term stability, innovations in nanoparticle engineering continue to address these challenges, pushing the boundaries of what is possible in nanomedicine (Kavanagh & Green, 2022).

The future of nanobiotechnology in neurology is bright, with ongoing research and development poised to yield transformative advances. Personalized nanomedicine, driven by patient-specific data and integrated with precision medicine, is expected to tailor treatments to individual needs, enhancing therapeutic outcomes. The convergence of nanotechnology with AI and big data analytics will further refine the design and application of nanoparticles, leading to smarter, more effective therapies.

As we stand on the cusp of a new era in neurology, it is recommended the following:

a. ***Continued Research***: Sustained and focused research into the mechanisms of action, safety, and efficacy of nanoparticles is essential to fully harness their potential.
b. ***Collaborative Efforts***: Multidisciplinary collaboration between scientists, clinicians, and regulatory bodies will be crucial in translating nanoparticle research into clinical practice.
c. ***Ethical Considerations***: Ethical considerations must be at the forefront of nanobiotechnology research, ensuring that advancements are made with patient welfare in mind.

In conclusion, the integration of biosynthesized nanoparticles into neurological therapeutics represents a significant leap forward in the treatment of neurodegenerative diseases (Niu, Chen, & Gao, 2019). With careful consideration of the challenges and ethical implications, the continued evolution of nanobiotechnology holds the promise of revolutionizing neurology and improving patient outcomes.

REFERENCES

Abbas, M. (2021). Potential role of nanoparticles in treating the accumulation of amyloid-beta peptide in Alzheimer's patients. *Polymers*, *13*(7), 1051.

Adeyemi, J. O., Oriola, A. O., Onwudiwe, D. C., & Oyedeji, A. O. (2022). Plant extracts mediated metal-based nanoparticles: Synthesis and biological applications. *Biomolecules*, *12*(5), 627.

Adlakha-Hutcheon, G., Khaydarov, R., Korenstein, R., Varma, R., Vaseashta, A., Stamm, H., & Abdel-Mottaleb, M. (2009). *Nanomaterials, nanotechnology: Applications, consumer products, and benefits*. Paper presented at the Nanomaterials: Risks and benefits.

Ajdary, M., Moosavi, M. A., Rahmati, M., Falahati, M., Mahboubi, M., Mandegary, A., … Varma, R. S. (2018). Health concerns of various nanoparticles: A review of their in vitro and in vivo toxicity. *Nanomaterials*, *8*(9), 634.

Akhtar, M. J., Ahamed, M., Alhadlaq, H. A., & Alshamsan, A. (2017). Mechanism of ROS scavenging and antioxidant signalling by redox metallic and fullerene nanomaterials: Potential implications in ROS associated degenerative disorders. *Biochimica et Biophysica Acta (BBA)-General Subjects*, *1861*(4), 802–813.

Akhtar, A., Andleeb, A., Waris, T. S., Bazzar, M., Moradi, A.-R., Awan, N. R., & Yar, M. (2021). Neurodegenerative diseases and effective drug delivery: A review of challenges and novel therapeutics. *Journal of Controlled Release*, *330*, 1152–1167.

Akhtar, M. J., Kumar, S., Alhadlaq, H. A., Alrokayan, S. A., Abu-Salah, K. M., & Ahamed, M. (2016). Dose-dependent genotoxicity of copper oxide nanoparticles stimulated by reactive oxygen species in human lung epithelial cells. *Toxicology and Industrial Health*, *32*(5), 809–821.

Alghamdi, M. A., Fallica, A. N., Virzì, N., Kesharwani, P., Pittalà, V., & Greish, K. (2022). The promise of nanotechnology in personalized medicine. *Journal of Personalized Medicine*, *12*(5), 673.

Ali, A., Shah, T., Ullah, R., Zhou, P., Guo, M., Ovais, M., … Rui, Y. (2021). Review on recent progress in magnetic nanoparticles: Synthesis, characterization, and diverse applications. *Frontiers in Chemistry*, *9*, 629054.

Almeida, J. P. M., Chen, A. L., Foster, A., & Drezek, R. (2011). In vivo biodistribution of nanoparticles. *Nanomedicine*, *6*(5), 815–835.

Badhe, R. V., Chejara, D. R., Kumar, P., Choonara, Y. E., & Pillay, V. (2018). Neurodegenerative disease conditions and genomic treatment for better health. *Genomics-driven healthcare: Trends in disease prevention and treatment*, 281–310.

Bhaskar, S., Tian, F., Stoeger, T., Kreyling, W., de la Fuente, J. M., Grazú, V., ... Razansky, D. (2010). Multifunctional Nanocarriers for diagnostics, drug delivery and targeted treatment across blood-brain barrier: Perspectives on tracking and neuroimaging. *Particle and Fibre Toxicology*, *7*, 1–25.

Bhattacharya, S. (2015). Reactive oxygen species and cellular defense system. *Free radicals in human health and disease*, 17–29.

Cayero-Otero, M. D., Espinosa-Oliva, A. M., Herrera, A. J., Garcia-Dominguez, I., Fernandez-Arevalo, M., Martin-Banderas, L., & de Pablos, R. M. (2018). Potential use of nanomedicine for the anti-inflammatory treatment of neurodegenerative diseases. *Current Pharmaceutical Design*, *24*(14), 1589–1616.

Chen, Y., Chen, H., & Shi, J. (2013). In vivo bio-safety evaluations and diagnostic/therapeutic applications of chemically designed mesoporous silica nanoparticles. *Advanced materials*, *25*(23), 3144–3176.

Chen, A., Kang, Y., Liu, J., Wu, J., Feng, X., Wang, M., ... Shao, L. (2022). Improvement of synaptic plasticity by nanoparticles and the related mechanisms: Applications and prospects. *Journal of Controlled Release*, *347*, 143–163.

Ding, S., Khan, A. I., Cai, X., Song, Y., Lyu, Z., Du, D., ... Lin, Y. (2020). Overcoming blood–brain barrier transport: Advances in nanoparticle-based drug delivery strategies. *Materials Today*, *37*, 112–125.

Domínguez, A., Álvarez, A., Hilario, E., Suarez-Merino, B., & Goñi-de-Cerio, F. (2013). Central nervous system diseases and the role of the blood-brain barrier in their treatment. *Neuroscience Discovery*, *1*(1), 3.

Dugger, B. N., & Dickson, D. W. (2017). Pathology of neurodegenerative diseases. *Cold Spring Harbor Perspectives in Biology*, *9*(7), a028035.

Fakhri, S., Abdian, S., Zarneshan, S. N., Moradi, S. Z., Farzaei, M. H., & Abdollahi, M. (2022). Nanoparticles in combating neuronal dysregulated signaling pathways: Recent approaches to the nanoformulations of phytochemicals and synthetic drugs against neurodegenerative diseases. *International Journal of Nanomedicine*, *17*, 299–331.

Fazio, E., Gökce, B., De Giacomo, A., Meneghetti, M., Compagnini, G., Tommasini, M., ... Ossi, P. M. (2020). Nanoparticles engineering by pulsed laser ablation in liquids: Concepts and applications. *Nanomaterials*, *10*(11), 2317.

Forest, V., & Pourchez, J. (2017). Preferential binding of positive nanoparticles on cell membranes is due to electrostatic interactions: A too simplistic explanation that does not take into account the nanoparticle protein corona. *Materials Science and Engineering: C*, *70*, 889–896.

Foulkes, R., Man, E., Thind, J., Yeung, S., Joy, A., & Hoskins, C. (2020). The regulation of nanomaterials and nanomedicines for clinical application: Current and future perspectives. *Biomaterials Science*, *8*(17), 4653–4664.

Furtado, D., Björnmalm, M., Ayton, S., Bush, A. I., Kempe, K., & Caruso, F. (2018). Overcoming the blood–brain barrier: The role of nanomaterials in treating neurological diseases. *Advanced Materials*, *30*(46), 1801362.

Garzón, V., Bustos, R.-H., & Pinacho, D. G. (2020). Personalized medicine for antibiotics: The role of nanobiosensors in therapeutic drug monitoring. *Journal of Personalized Medicine*, *10*(4), 147.

Hoshyar, N., Gray, S., Han, H., & Bao, G. (2016). The effect of nanoparticle size on in vivo pharmacokinetics and cellular interaction. *Nanomedicine*, *11*(6), 673–692.

Huston, M., DeBella, M., DiBella, M., & Gupta, A. (2021). Green synthesis of nanomaterials. *Nanomaterials*, *11*(8), 2130.

Kavanagh, E. W., & Green, J. J. (2022). Toward gene transfer nanoparticles as therapeutics. *Advanced Healthcare Materials*, *11*(7), 2102145.

Ke, S., Zhou, T., Yang, P., Wang, Y., Zhang, P., Chen, K., … Ye, S. (2017). Gold nanoparticles enhance TRAIL sensitivity through Drp1-mediated apoptotic and autophagic mitochondrial fission in NSCLC cells. *International Journal of Nanomedicine*, *12*, 2531–2551.

Kumar, V., Sami, N., Kashav, T., Islam, A., Ahmad, F., & Hassan, M. I. (2016). Protein aggregation and neurodegenerative diseases: From theory to therapy. *European Journal of Medicinal Chemistry*, *124*, 1105–1120.

Kus-Liśkiewicz, M., Fickers, P., & Ben Tahar, I. (2021). Biocompatibility and cytotoxicity of gold nanoparticles: recent advances in methodologies and regulations. *International Journal of Molecular Sciences*, *22*(20), 10952.

Lekha, D. C., Shanmugam, R., Madhuri, K., Dwarampudi, L. P., Bhaskaran, M., Kongara, D., … Krishnaraj, R. (2021). Review on silver nanoparticle synthesis method, antibacterial activity, drug delivery vehicles, and toxicity pathways: Recent advances and future aspects. *Journal of Nanomaterials, 2021*, 1–11.

Liao, C., & Liu, S. (2021). Tuning the physicochemical properties of reticular covalent organic frameworks (COFs) for biomedical applications. *Journal of Materials Chemistry B*, *9*(31), 6116–6128.

Li, X., Xu, H., Chen, Z.-S., & Chen, G. (2011). Biosynthesis of nanoparticles by microorganisms and their applications. *Journal of Nanomaterials, 2011*, 1–16.

Löscher, W., & Potschka, H. (2005). Drug resistance in brain diseases and the role of drug efflux transporters. *Nature Reviews Neuroscience*, *6*(8), 591–602.

Lovinger, D. M. (2010). Neurotransmitter roles in synaptic modulation, plasticity and learning in the dorsal striatum. *Neuropharmacology*, *58*(7), 951–961.

Moradi, S. Z., Jalili, F., Farhadian, N., Joshi, T., Wang, M., Zou, L., … Xiao, J. (2022). Polyphenols and neurodegenerative diseases: Focus on neuronal regeneration. *Critical Reviews in Food Science and Nutrition*, *62*(13), 3421–3436.

Mukherjee, S., Madamsetty, V. S., Bhattacharya, D., Roy Chowdhury, S., Paul, M. K., & Mukherjee, A. (2020). Recent advancements of nanomedicine in neurodegenerative disorders theranostics. *Advanced Functional Materials*, *30*(35), 2003054.

Abou Neel, E. A., Bozec, L., Perez, R. A., Kim, H.-W., & Knowles, J. C. (2015). Nanotechnology in dentistry: Prevention, diagnosis, and therapy. *International Journal of Nanomedicine*, 6371–6394.

Nguyen, D. D., & Lai, J.-Y. (2022). Synthesis, bioactive properties, and biomedical applications of intrinsically therapeutic nanoparticles for disease treatment. *Chemical Engineering Journal*, *435*, 134970.

Nguyen, K. T., & Zhao, Y. (2015). Engineered hybrid nanoparticles for on-demand diagnostics and therapeutics. *Accounts of Chemical Research*, *48*(12), 3016–3025.

Niu, X., Chen, J., & Gao, J. (2019). Nanocarriers as a powerful vehicle to overcome blood-brain barrier in treating neurodegenerative diseases: Focus on recent advances. *Asian Journal of Pharmaceutical Sciences*, *14*(5), 480–496.

Ovais, M., Khalil, A. T., Ayaz, M., Ahmad, I., Nethi, S. K., & Mukherjee, S. (2018). Biosynthesis of metal nanoparticles via microbial enzymes: A mechanistic approach. *International Journal of Molecular Sciences*, *19*(12), 4100.

Pandit, C., Roy, A., Ghotekar, S., Khusro, A., Islam, M. N., Emran, T. B., … Bradley, D. A. (2022). Biological agents for synthesis of nanoparticles and their applications. *Journal of King Saud University-Science*, *34*(3), 101869.

Pasula, R. R., & Lim, S. (2017). Engineering nanoparticle synthesis using microbial factories. *Engineering Biology, 1*(1), 12–17.

Poovaiah, N., Davoudi, Z., Peng, H., Schlichtmann, B., Mallapragada, S., Narasimhan, B., & Wang, Q. (2018). Treatment of neurodegenerative disorders through the blood–brain barrier using nanocarriers. *Nanoscale*, *10*(36), 16962–16983.

Popov, A. L., Shcherbakov, A. B., Zholobak, N., Baranchikov, A. Y., & Ivanov, V. K. (2017). Cerium dioxide nanoparticles as third-generation enzymes (nanozymes). *Наносистемы: физика, химия, математика*, *8*(6), 760–781.

Rhaman, M. M., Islam, M. R., Akash, S., Mim, M., Noor Alam, M., Nepovimova, E., … Sharma, R. (2022). Exploring the role of nanomedicines for the therapeutic approach of central nervous system dysfunction: at a glance. *Frontiers in Cell and Developmental Biology*, *10*, 989471.

Riaz, M. K., Riaz, M. A., Zhang, X., Lin, C., Wong, K. H., Chen, X., … Yang, Z. (2018). Surface functionalization and targeting strategies of liposomes in solid tumor therapy: A review. *International Journal of Molecular Sciences*, *19*(1), 195.

Riccardi, C., Napolitano, F., Montesarchio, D., Sampaolo, S., & Melone, M. A. B. (2021). Nanoparticle-guided brain drug delivery: Expanding the therapeutic approach to neurodegenerative diseases. *Pharmaceutics*, *13*(11), 1897.

Sahu, M., Gupta, R., Ambasta, R. K., & Kumar, P. (2022). Artificial intelligence and machine learning in precision medicine: A paradigm shift in big data analysis. *Progress in Molecular Biology and Translational Science*, *190*(1), 57–100.

Samrot, A. V., Ram Singh, S. P., Deenadhayalan, R., Rajesh, V. V., Padmanaban, S., & Radhakrishnan, K. (2022). Nanoparticles, a double-edged sword with oxidant as well as antioxidant properties—A review. *Oxygen*, *2*(4), 591–604.

Sanchez, C., Belleville, P., Popall, M., & Nicole, L. (2011). Applications of advanced hybrid organic–inorganic nanomaterials: from laboratory to market. *Chemical Society Reviews*, *40*(2), 696–753.

Sandhir, R., Yadav, A., Sunkaria, A., & Singhal, N. (2015). Nano-antioxidants: An emerging strategy for intervention against neurodegenerative conditions. *Neurochemistry international*, *89*, 209–226.

Saravanan, A., Kumar, P. S., Karishma, S., Vo, D.-V. N., Jeevanantham, S., Yaashikaa, P., & George, C. S. (2021). A review on biosynthesis of metal nanoparticles and its environmental applications. *Chemosphere*, *264*, 128580.

Satalkar, P., Elger, B. S., & Shaw, D. (2016). Naming it 'nano': Expert views on 'nano' terminology in informed consent forms of first-in-human nanomedicine trials. *Nanomedicine*, *11*(8), 933–940.

Semeano, A. T., Tofoli, F. A., Correa-Velloso, J. C., de Jesus Santos, A. P., Oliveira-Giacomelli, A., Cardoso, R. R., … Ferrari, M. F. (2022). Effects of magnetite nanoparticles and static magnetic field on neural differentiation of pluripotent stem cells. *Stem Cell Reviews and Reports*, *18*(4), 1337–1354.

Sim, T. M., Tarini, D., Dheen, S. T., Bay, B. H., & Srinivasan, D. K. (2020). Nanoparticle-based technology approaches to the management of neurological disorders. *International Journal of Molecular Sciences*, *21*(17), 6070.

Singhal, A., Morris, V., Labhasetwar, V., & Ghorpade, A. (2013). Nanoparticle-mediated catalase delivery protects human neurons from oxidative stress. *Cell Death & Disease*, *4*(11), e903–e903.

Soenen, S. J., Parak, W. J., Rejman, J., & Manshian, B. (2015). (Intra) cellular stability of inorganic nanoparticles: Effects on cytotoxicity, particle functionality, and biomedical applications. *Chemical Reviews*, *115*(5), 2109–2135.

Voet, S., Srinivasan, S., Lamkanfi, M., & van Loo, G. (2019). Inflammasomes in neuroinflammatory and neurodegenerative diseases. *EMBO Molecular Medicine*, *11*(6), e10248.

Yang, C.-Y., Huang, W.-Y., Chen, L.-H., Liang, N.-W., Wang, H.-C., Lu, J., … Wang, T.-W. (2021). Neural tissue engineering: The influence of scaffold surface topography and extracellular matrix microenvironment. *Journal of Materials Chemistry B*, *9*(3), 567–584.

Yusuf, A., Almotairy, A. R. Z., Henidi, H., Alshehri, O. Y., & Aldughaim, M. S. (2023). Nanoparticles as drug delivery systems: A review of the implication of nanoparticles' physicochemical properties on responses in biological systems. *Polymers*, *15*(7), 1596.

11 Metal Nanoparticles in Agriculture

Imran Khan, Sara Najeeb, Sumaira Miskeen, and Muhammad Jahangir

11.1 INTRODUCTION

Green synthesis of metal nanoparticles (MNPs) has emerged as a potential resource for answering some of the challenges in the food and agriculture industries. This expansion is based on the multifunctional and compatible nature of the biogenic nanoparticles (NPs). Recently, green technologies have advanced the synthesis of MNPs with plants and microbes. Environmentally friendly, cost-effective, and biologically safe, green synthesis methods for MNPs serve as a sustainable alternative, replacing the use of toxic chemicals and reagents (1). Synthesis of NP takes place inside and outside of cells by plants and microbes. Nanobiotechnology encompasses living elements from both prokaryotic and eukaryotic groups, including algae, actinomycetes, bacteria, viruses, fungi, yeasts, and plants (2). Each biological system will generate the MNPs differentially, i.e., the potential of creating NPs varies depending on the biological systems employed within manufacture. Nevertheless, inherent to their metabolic processes and enzymatic activity, all biological entities are incapable of producing NPs. Consequently, living organisms or their naturally occurring substances are utilized in the bio-reduction of metallic particles, leading to the production of MNPs (3).

11.2 UTILIZATION OF ENVIRONMENT-FRIENDLY SYNTHESIZED METAL-BASED NANOPARTICLES FOR SUSTAINABLE AGRICULTURE

Utilizing eco-friendly methods, MNPs have garnered significant attention in recent decades and hold promise for widespread application in agriculture. Their potential lies in providing non-toxic and efficient solutions for managing food and agricultural microbial pathogens (4). Green synthetic metal-based NPs, which were utilized to convert standard farming to precision farming methods, could also be used to make nano-agricultural products such as nano-pesticides and nanofertilizers. After converting, a wide range of MNPs including silver, copper, manganese, iron, molybdenum, silicon, zinc, associated oxides, and nano-formulations of traditional agriculture products such as sulfur, phosphorous, ammonium, tebuconazole, and validamycin have now been employed as nano-pesticides or fertilizers (5, 6).

DOI: 10.1201/9781003378563-11

Such nanoforms have also shown encouraging outcomes on seedling growth, and the growing season of plants at optimal concentrations. Similarly, nano-pesticides had a considerable impact on crop disease and insect management (7).

The active compounds are considered by enclosing the NPs inside nano-porous medium, nano-coating with polymeric, then delivering nano emulsion or NPs to plants (8). The targeted application of natural compounds in agricultural product management reduces the quantity and expenses associated with herbicides and fertilizers (9). Furthermore, decreasing the utilization of synthetic pesticides and fertilizers proved useful for enhancing soil health and preserving the geobiological cycle, which increases nutritional value of crops and foods as well as their quality.

11.2.1 Plant Growth Enhancement

It is utilized efficiently in both the quick germination of seeds and development augmentation to encourage the usage of MNPs in agribusiness. From last few years scientists have collaborated over the facilitation and utilization of NPs and now in present time NPs are used for plant disease prevention and crop improvement (10, 11). Various investigations have now been conducted in order to show this whenever nano TiO_2 and SiO_2 and combinations are applied to *Soybean max* (L.) seeds, they boost nitrate reductase activity, impacting both the growth of seedlings and the germination process (12). Zheng et al. investigated the effects of TiO_2 NPs and bulk TiO_2 on the development and growth of seedlings in *Spinacia oleracea.*

Furthermore, it was discovered that when seeds are treated with TiO_2 NPs, generated plants with a 73% increase in dry mass, elevated chlorophyll content, and a three-fold rise in the photosynthesis rate were observed compared to the control group (13).

NPs' toxicity, behavior, and reactivity are heavily influenced by their particle size. The rate of germination rises as the size of NPs decreases. *S. oleracea* growth was promoted by increasing nitrogen metabolism and photosynthesis at the optimal TiO_2 NP concentration (14, 15).

According to the model system, TiO_2 NPs affect the growing and developing stages of *Brassica napus* seeds. TiO_2 NPs were employed to assess their impact on germination and seedling vigor. Treatment of plants with TiO_2 NPs at a concentration of 2000 mg/L resulted in enhanced seed germination and increased seedling vigor compared to the control (16). Research indicates that NPs can have both positive and negative effects on plants. Lipid peroxidation in *Elodea densa* plants is impacted by the copper ions and NPs up to 120 and 180% of the control level. Catalase and superoxide dismutase activity, on the other hand, increased 1.5–2.0 times, which actively accumulates Cu NPs (17). At mitosis division, the effects of biogenically combined Cu NPs from *Vinca rosea* leaf extricate on *Allium cepa* cells effectively partitioning. The use of Cu NPs in the concentration of 20 μg/mL was found to raise mitotic index, according to the study. However, it falls off as the dose increased far away from the limit, indicating that dose concentrations above 20 μg/mL might be lethal to its development (18).

Observing seed germination, plant growth, and seedling vigor, the concentration of 1000 ppm of ZnO NPs (25 nm) showcased potential effects on the growth and development of *Arachis hypogaea*. Treated plants exhibited higher leaf chlorophyll content, early flowering establishment, and increased stem and root growth compared to the control group (19).

NPs have an impressive ability to boost *A. hypogaea*'s root growth and pot yields. NPs are efficiently applied in crop production and it was made possible by research into the detrimental properties of NPs on seed and plant properties (20). Table 11.1 provides a summary of MNPs' use as nanofertilizers.

11.2.2 Control of Plant Pathogen and Disease Management

In crop disease, micronutrients play a significant role in the defense mechanism because tissue infection typically triggers a chain of reactions that produce secondary inhibitory metabolites. It is noteworthy that enzymes make these secondary inhibitory metabolites on a regular basis, the need for micronutrient cofactors to be activated. By triggering the host defense enzymes polyphenol oxidase and phenylalanine ammonia-lyase, for instance, Cu, Mn, and Zn increase disease resistance (36, 37).

The demand for pesticides is on the rise. Therefore, finding novel alternatives to fertilizers and pesticides is urgently required to avoid their use. In the realm of nanobiotechnology, future applications in crop protection involve the use of nano-biosensors to restrict the release of encapsulated pesticides, fertilizers, and other agrochemicals. Additionally, it encompasses early disease detection, safeguarding plants against pests and diseases, and addressing pesticide-containing pollutants (38).

Nevertheless, a significant portion of crops lacks resistance genes, leading to ongoing public concerns regarding the development of genetically modified food crops. While controlling the nutritional value of the plant, crop diseases can be eradicated successfully. The crop's receptivity and resistance to infection are frequently mediated by healthy nutrition (39). The significantly increased availability and translocation of metal oxide and MNPs within plants is one of the most notable characteristics. As a response, various MNPs, including biopolymers like chitosan, are being employed in the control of plant diseases due to their antimicrobial activity (40, 41). The nano-based formulations may make use of metal-based NPs like Ag, Si, Cu, Zn, and other nanocomposites. At final result, the utilization of formulations of micronutrients based on NPs has the potential to offering a highly effective and innovative platform; these novel approaches play a crucial role in preventing plant crop diseases and the improvement of crop product by encouraging host resistance through nutrition in a more targeted and deliberate manner.

11.2.3 Green-Synthesized Metal-Based NPs in Food Packaging

Ready-to-use food or agricultural products are susceptible to spoilage due to factors such as moisture and microbial growth, and other environmental factors. The food and grain management after harvest is a major global concern (42). To prevent this, the food must be adequately packed. The packaging industry contributes approximately 55–65% of the US's 130 billion dollars to the global economy (43, 44). Several institutes did a study assessment on worldwide investments in the nanopackaging

TABLE 11.1
Application of Different Metal Nanoparticles as Nanomaterials and Their Impact on Plants

Nanomaterials	Crop Species	Conc. Used	Duration of Time	Impact on Plant	References
ZnO	*Coffea arabica*	10 mg/mL	45 days	Net photosynthesis and growth rate enhanced	21
ZnO	*Triticum aestivum*	20 mg/mL	Duration of growth cycle	Biomass accumulation and crop yield increased	22
FeS_2	*Brassica juncea*	100 mg/mL	14 hours	Crop yield and germination increased	23, 24
CuO	*Spinacia oleracea*	200 mg/kg	60 days	Photosynthesis rate improved	23–25
ZnO	*Nicotiana tabacum*	1 µM	21 days	Enzymatic and metabolic activity increased	26
Fe/SiO_2	*Arachis hypogaea*	15 mg/kg	3 days	Plant growth enhanced and biomass accumulation	27
TiO_2	*S. oleracea*	25% suspension	35 days 48 hours	Increased nitrogen, chlorophyll, and protein content	28
AgNPs	*T. aestivum*	75 mg/mL	Trifoliate stage	Tolerance to heat and growth enhanced	29
AgNPs	*Vigna sinensis*	50 mg/mL	40 days	By stimulating soil and root modulation growth and biomass growth enhanced	30
TiO_2 and SiO_2	*Oryza sativa*	30 mg/mL	55 days	Improve growth by stimulating antioxidant potential and inhibiting Cd translocation	31
MWCNTs	*Hordeum vulgare*	100 µg/mL	1 day	Enhanced growth of seedling and germination of seeds	32
SiO_2 NPs	*O. sativa*	2.5 mmol/L	70 days	By decreasing bioconcentration and translocation in plant growth enhanced	33
MgO	*Solanum lycoperscium*	10 µg/mL	7 days	Suppressing pathogen *Ralstonia solanacearum* as a result bacterial wilt is controlled	34
Al_2O_3 NPs	*Solanum lycopersicum*	400 mg/L	20 days	Controlled tomato disease (root rot)	35

Note: ZnO: zinc oxide, FeS_2: iron disulfide, CuO: copper oxide, SiO_2: silicon dioxide, TiO: titanium dioxide, AgNPs: silver nanoparticle, MWCNT: multi-walled carbon nanotubes, MgO: magnesium oxide, Al_2O_3: aluminum oxide.

business. According to Helmut Kaiser Consultancy, the nanotechnology influence on the beverage and food packaging sectors expanded dramatically between 2002 and 2005. The global sales of nano-based packaging products witnessed significant growth, increasing from 150 million dollars in 2002 to 860 million dollars in 2004 (45); projections suggest that the nano-enabled packaging market is anticipated to reach a value of 15.0 billion US dollars by 2020, with further growth expected to reach approximately 31.2 billion US dollars by the end of 2023 (46).

The utilization of nanobiotechnology in the food industry is expected to enhance food shelf life and prevent spoilage by introducing an active and intelligent "smart packaging system." Various NPs incorporated into nano packaging can interact with the environment, providing alerts to consumers regarding contamination and the presence of pathogens. Additionally, some of these smart packaging systems are designed to possess self-repairing capabilities (47).

Metal-based NPs such as silver, iron, magnesium oxide, silicon dioxide, titanium dioxide, and zinc oxide are extensively used as food additives and antimicrobial agents (48, 49). For instance, TiO_2 NPs have negative impact on pathogenic microorganisms, as demonstrated by their production of reactive oxygen species (ROS). Nanocomposites, owing to their low weight, elevated heat resistance, and superior mechanical strength, can be employed to construct additional barriers against UV radiation, O_2, CO_2, moisture, and volatile chemicals. The widespread application of nanocomposites in coatings and packaging showcases their versatility in enhancing protective properties (50, 51). The antibacterial effect is due to the production of ROS by silver zeolite (Ag).

Silver zeolite-encrusted ceramics have found application in food preservation, medical product disinfection, and material decontamination. Water-based nanocomposite nano emulsions, creating 1–2 nm nano-coatings on food packaging materials, serve for decontaminating food packaging equipment and acting as an oxygen (O_2) barrier. Products containing nano-micelle-based glycerin effectively eliminate pesticide residues from fruits and vegetables, as well as oil or filth from cutlery. Additionally, the incorporation of nano-emulsified flavors and bioactive into beverages demonstrates no impact on the visual appearance of the product (52).

11.3 ACTIVE FOOD PACKAGING SYSTEMS

Antimicrobials, moisture regulators, CO_2 emitters and decomposers, O_2 decomposers, and other inert barriers separating the food product from the environment are all found in active packaging. Systems of dynamic packaging (53, 54) have been created with the storage function in mind. For short-term chilled storage, systems like wrapping have been used. However, bulk gas flushing, vacuum packing, and modified atmosphere packaging (MAP) systems that solely employ carbon dioxide are utilized for long-term cooled storage (44).

Polypropylene (PP) and low-density polyethylene (LDPE), two readily available hydrophobic, inert, and low-surface-energy polymeric films, are utilized for packaging (55). For the addition of an antibacterial agent to decrease food degradations, surface changes with functional properties and polar groups are necessary (56). Nano-based packaging films entail introducing nanomaterials into the packing system to protect and enhance quality of food. Many nanomaterials are employed as preservatives, either directly within packing material, packed individually, or imbedded in food products (47, 57, 58). Generally, the application of active chemicals or nanomaterials to packaging films enhances food preservation by CO_2 absorption, oxygen, ethylene, taste, humidity, and some other gaseous constituents (58). Active ingredients are substances that have the ability to prolong food's shelf life and preserve or enhance the quality of packaged food.

Useful packaging material often absorbs chemicals from the packaged food. Food packaging primarily comprises (a) nanocomposites such as Ag, Cu, and oxides such as TiO_2 and MgO NPs, (b) antimicrobial film, and (c) gas scavengers (55, 59). Bayer, a well-known chemical firm based in Leverkusen, Germany, produced the transparent plastic sheet that was uniformly covered with the clay NPs dispersion. This coating can prevent moisture and oxygen from entering fresh meats, as they do in other foods. Countries like Europe, Asia, and the United States have filed several patents on nanomaterials such as nano-clays and nano-silver used in food packaging (60). Ag, ZnO, and TiO_2 NPs are the most significant MNPs utilized in food packaging in terms of finances. The insertion of O_2 scavengers into the film improves O_2 absorption. Due to their O_2 scavenging ability, crystalline TiO_2 NPs could be used in food processing and packaging (61). In a similar vein, chitosan, MgO, Cu, and CuO NPs are used as active ingredients in nano packaging due to their implied antibacterial activity. Food waste-causing bacteria such as *Listeria monocytogenes*, *Escherichia coli O157:H7*, and *Salmonella enteritidis* might be prevented by employing antimicrobial packaging made of ZnO quantum dots suspended in polyvinylpyrrolidone or coordinated into polystyrene film (62).

11.4 CURRENT TRENDS OF NANOSYSTEMS IN FOOD AND AGRICULTURE INDUSTRY

Currently, nanotechnology enables the incorporation of hydrophilic or hydrophobic biologically active and many other natural compounds that have beneficial impacts associated to particle diameter, including decreasing the quantity of bioactive (63), the visually attractive characteristics of food, increasing storage life, and protecting nutritional value, functionality, and security (63, 64). Additionally, this technique allows the addition of natural dietary additives, regulators, and antioxidant and antibacterial chemicals (65, 66). By developing more effective and sustainable distribution ways, nanosystems are contributing to the decrease of food waste. Nanosystems are being used to develop sensors that provide feedback on product freshness and spoilage. In order to cut down on waste, food manufacturers and merchants may use these systems to detect when items are getting close to expiry.

Nano emulsions, for example, can help to blend different ingredients, resulting in product having better taste and healthier. The structure and surface area of some food items, such as lipids and carbohydrates, are being changed by NPs to create healthier, more nutrient-dense goods. Additionally, nanosystems are being used to fortify food products with vitamins and minerals. Nutrient-packed nanosystems are making it easier to create alternative food sources that are richer in nutrients, while still being affordable and accessible. This technology has potential to reduce malnutrition in many parts of the world, making food more abundant and nutritious.

Several nanosystems, such as polymer NPs, nano emulsions, and nanocomposites, have been applied lately to regulate the outflow of EO in edible coating formulations, polyphenols, and other substances, with respect to their role in microcapsules and some other types of assimilation into polymer structure. Fat-soluble vitamins include essential oils of thyme and oregano, where the active ingredients contain carvacrol and thymol, which are substances with potent antimicrobial properties. Antioxidant capability, which reduces the pace of reaction and the enzymatic activity of enzymes like lipoxygenases, peroxidases, and polyphenol oxidases, is another factor to take into account. Therefore, the usage of nanosystems has aided research on ways to limit the release of components into the environment by forming a barrier in addition to the polymeric matrix in which they are embedded (67–69). To comprehend their fate in the gastrointestinal tract, it is crucial to evaluate that under in vitro digesting settings, nanosystems serve as regulated delivery systems for bioactive substances. In order to adjust material dosage and processing, it is possible to optimize the design and production of nanosystems by knowing how they behave under simulated in vitro conditions. In challenging digestive environments, nanosystems derived from proteins, polysaccharides, and lipids have shown promising outcomes as natural protective vessels for bioactive compounds (70).

Moreover, NP delivery systems enable the implementation of nutraceuticals in food industry to improve health benefits, which is critical in a historical background where lifestyle modifications have resulted in a rise in threatening diseases like obesity, cancer, kidney disease, and other coronary heart disease (71). The creation of innovative, eco-friendly tools is necessary for parasite, insect pest, and vector management. The search for novel bioactive chemicals with various modes of action that did not represent a hazard to human health or the environment was necessitated by the growth in drug and pesticide resistance. Because of their great efficacy and typically low non-target toxicity, several natural products are used as alternative or additional ways to manage parasites, insect pests, and vectors. Its incorporation into nanosystems aids in overcoming challenges brought about by their physicochemical properties, such as limited handling and stability, thereby increasing their overall efficacy. Micro- and nano emulsions, among various nanosystems, are simple to prepare and use for industrial scale-up. They are effective against important parasites, such as *Aedes*, *Plasmodium*, *Trypanosoma*, *Culex*, *Leishmania*, and mosquitoes, which are vectors of human diseases and pests of agricultural and stored product insects. In general, subject to proper field validation, for the development of ecologically beneficial pest and vector control agents, micro- and nano emulsions are potential choices. Future research will necessitate technical advancements as well as long-term toxicity investigations with non-target species. Applications of nanosystems based on silica in the interest of sustainable agriculture. A silicon NP-based technology offers significant agricultural uses and the ability to secure global food supply by promoting the breeding of better high-yielding cultivars (72). It was discovered that the food and agriculture industries employed silica nanosystems as adsorbents for heavy metal toxicity (73), ultraviolet-B (74), salinity pressure (75), and dehydration (76). Pesticides, fertilizers, herbicides, gene and gene transfer, soil improvers, and soil analysis sensors are among the main applications of silica-based nanosystems in agriculture. Plant nanobiotechnology may benefit from the application of silica-based nanosystems, which have been shown in several studies to improve the qualities of certain nanomaterials. Additionally, the generation of new strains resistant to a variety of biotic and abiotic environments can be facilitated by protein-targeted silica-based nanosystems (77). Best illustration in the health sector is *Averrhoa carambola* L. species aided by phytonano-produced NPs containing zinc oxide NPs, which are proven to be more efficient agent in natural decomposition processes.

11.5 FUTURE TRENDS OF NANOSYSTEMS IN FOOD AND AGRICULTURE INDUSTRY

Nanosystems are anticipated to assume a pivotal role in the near future across various facets of food preparation, processing, and production. The utilization of nanosystems is expected to enable the sustainable production of cleaner, superior, and more nutritious food products, marking a significant advancement in the quality and environmental impact of food manufacturing.

By using smaller batch sizes and more accurate ingredients, it will also aid in reducing waste. Moreover, food manufacturers will be able to create unique products that are targeted to certain consumer needs due to nanosystems. Additionally, improved food transport and packaging will be achieved with nanosystems. Food and beverage packaging can be strengthened and extended with the help of nanosystems, which can also lower the price and volume of packaging needed. Even data on the quality, security, and expiration date of products might be sent and received using nanosystems. The potential applications of nanosystems in the food and agriculture industries are virtually boundless. These groundbreaking technologies are poised to empower food producers to create products that are not only tastier and healthier but also more efficient. Anticipated benefits include cost reduction, shortened lead times, and enhanced product quality and safety. The future appears promising for nanosystems in the food industry, with far-reaching implications that could significantly transform the landscape.

11.6 CONCLUSION

Nanobiotechnology holds the promise of transforming the food industry by enhancing the efficiency and effectiveness of food production, packaging, and storage. This innovation could contribute positively to both food safety and security, while also mitigating the environmental footprint associated with food production. Apart from improving the technical aspects of agricultural production and nutrition, nanobiotechnology can also help address some of the ethical concerns associated with farming. An illustration of nanotechnology's impact is the application of biodegradable coatings to food items and plants. This approach helps minimize the reliance on chemical-based pesticides and fertilizers, showcasing how nanotechnology can contribute to more sustainable and environmentally friendly agricultural practices. This can lead to improved environmental conditions, and improved health outcomes for the local population. Through increased productivity and the ability to use more environmentally friendly farming methods, nanobiotechnology has the potential to completely transform the agricultural industry. Not only can this technology be used to improve crop yields, but it can also be used to enhance nutrition and reduce environmental damage. With the aid of nanobiotechnological solutions, agricultural production systems can remain competitive while maintaining a commitment to ethical standards. Similarly, nanobiotechnological solutions can be applied to track the traceability of goods and ingredients, ensuring that all agricultural substances used are of the highest quality and safe for consumption.

REFERENCES

1. Peralta-Videa JR, Huang Y, Parsons JG, Zhao L, Lopez-Moreno L, Hernandez-Viezcas JA, et al. Plant-based green synthesis of metallic nanoparticles: Scientific curiosity or a realistic alternative to chemical synthesis? Nanotechnology for Environmental Engineering. 2016;1(1):1–29.
2. Rana A, Yadav K, Jagadevan S. A comprehensive review on green synthesis of nature-inspired metal nanoparticles: Mechanism, application and toxicity. Journal of Cleaner Production. 2020;272:122880.
3. Patil A, Durgude A, Pharande A, Kadlag A, Nimbalkar C. Effect of calcium silicate as a silicon source on growth and yield of rice plants. International Journal of Chemical Studies. 2017;5(6):545–49.
4. Chhipa H, Joshi P. Nanofertilisers, nanopesticides and nanosensors in agriculture. Nanoscience in food and agriculture 1: Springer; 2016. p. 247–82.
5. Chhipa H. Nanofertilizers and nanopesticides for agriculture. Environmental Chemistry Letters. 2017;15(1):15–22.
6. Lin D, Xing B. Phytotoxicity of nanoparticles: Inhibition of seed germination and root growth. Environmental Pollution. 2007;150(2):243–50.
7. Delfani M, Baradarn Firouzabadi M, Farrokhi N, Makarian H. Some physiological responses of black-eyed pea to iron and magnesium nanofertilizers. Communications in Soil Sciences and Plant Analysis. 2014;45(4):530–40.
8. DeRosa MC, Monreal C, Schnitzer M, Walsh R, Sultan Y. Nanotechnology in fertilizers. Nature Nanotechnology. 2010;5(2):91.
9. Mukhopadhyay SS. Nanotechnology in agriculture: prospects and constraints. Nanotechnology, Science and Applications. 2014;7:63.
10. Nair R, Varghese SH, Nair BG, Maekawa T, Yoshida Y, Kumar DS. Nanoparticulate material delivery to plants. Plant Sciences. 2010;179(3):154–63.
11. Mahajan P, Dhoke SK, Khanna AS. Effect of nano-ZnO particle suspension on growth of mung (*Vigna radiata*) and gram (*Cicer arietinum*) seedlings using plant agar method. Journal of Nanotechnology. 2011;2011. https://doi.org/10.1155/2011/696535
12. Changmei L, Chaoying Z, Junqiang W, Guorong W, Mingxuan T. Research of the effect of nanometer materials on germination and growth enhancement of *Glycine max* and its mechanism. Soybean Science. 2002;21(3):168–71.
13. Zheng L, Hong F, Lu S, Liu C. Effect of nano-TiO_2 on strength of naturally aged seeds and growth of spinach. Biological Trace Eement Research. 2005;104(1):83–91.
14. Hong F, Yang F, Liu C, Gao Q, Wan Z, Gu F, et al. Influences of nano-TiO_2 on the chloroplast aging of spinach under light. Biological Trace Element Research. 2005;104(3):249–60.
15. Yang F, Hong F, You W, Liu C, Gao F, Wu C, et al. Influence of nano-anatase TiO_2 on the nitrogen metabolism of growing spinach. Biological Trace Element Research. 2006;110(2):179–90.
16. Mahmoodzadeh H, Nabavi M, Kashefi H. Effect of nanoscale titanium dioxide particles on the germination and growth of canola (*Brassica napus*). Journal of Ornamental and Horticultural Plants. 2013;3(1):25–32.

17. Nekrasova G, Ushakova O, Ermakov A, Uimin M, Byzov I. Effects of copper (II) ions and copper oxide nanoparticles on *Elodea densa* Planch. Russian Journal of Ecology. 2011;42(6):458–63.
18. Nagaonkar D, Shende S, Rai M. Biosynthesis of copper nanoparticles and its effect on actively dividing cells of mitosis in *Allium cepa*. Biotechnologucal Progress. 2015;31(2):557–65.
19. Prasad T, Sudhakar P, Sreenivasulu Y, Latha P, Munaswamy V, Reddy KR, et al. Effect of nanoscale zinc oxide particles on the germination, growth and yield of peanut. Journal of Plant Nutrition. 2012;35(6):905–27.
20. Fatima F, Hashim A, Anees S. Efficacy of nanoparticles as nanofertilizer production: A review. Environmental Science and Pollution Research. 2021;28(2):1292–303.
21. Rossi L, Fedenia LN, Sharifan H, Ma X, Lombardini L. Effects of foliar application of zinc sulfate and zinc nanoparticles in coffee (*Coffea arabica* L.) plants. Plant Physiology and Biochemistry. 2019;135:160–66.
22. Du W, Yang J, Peng Q, Liang X, Mao H. Comparison study of zinc nanoparticles and zinc sulphate on wheat growth: From toxicity and zinc biofortification. Chemosphere. 2019;227:109–16.
23. Srivastava G, Das A, Kusurkar TS, Roy M, Airan S, Sharma RK, et al. Iron pyrite, a potential photovoltaic material, increases plant biomass upon seed pretreatment. Materials Express. 2014;4(1):23–31.
24. Srivastava G, Das CK, Das A, Singh SK, Roy M, Kim H, et al. Seed treatment with iron pyrite (FeS_2) nanoparticles increases the production of spinach. RSC Advances. 2014;4(102):58495–504.
25. Wang Y, Lin Y, Xu Y, Yin Y, Guo H, Du W. Divergence in response of lettuce (*var. ramosa Hort.*) to copper oxide nanoparticles/microparticles as potential agricultural fertilizer. Environmental Pollutants and Bioavailability. 2019;31(1):80–84.
26. Tirani MM, Haghjou MM, Ismaili A. Hydroponic grown tobacco plants respond to zinc oxide nanoparticles and bulk exposures by morphological, physiological and anatomical adjustments. Functional Plant Biology. 2019;46(4):360–75.
27. Najafi Disfani M, Mikhak A, Kassaee MZ, Maghari A. Effects of nano Fe/SiO_2 fertilizers on germination and growth of barley and maize. Archives of Agronomy and Soil Science. 2017;63(6):817–26.
28. Yang F, Liu C, Gao F, Su M, Wu X, Zheng L, et al. The improvement of spinach growth by nano-anatase TiO_2 treatment is related to nitrogen photoreduction. Biology Trace Element Research. 2007;119(1):77–88.
29. Iqbal M, Raja NI, Mashwani Z-U-R, Hussain M, Ejaz M, Yasmeen F. Effect of silver nanoparticles on growth of wheat under heat stress. Iranian Journal of Science and Technology, Transactions A: Science. 2019;43(2):387–95.
30. Liu G, Zhang M, Jin Y, Fan X, Xu J, Zhu Y, et al. The effects of low concentrations of silver nanoparticles on wheat growth, seed quality, and soil microbial communities. Water, Air, & Soil Pollution. 2017;228(9):1–12.
31. Rizwan M, Ali S, Malik S, Adrees M, Qayyum MF, Alamri SA, et al. Effect of foliar applications of silicon and titanium dioxide nanoparticles on growth, oxidative stress, and cadmium accumulation by rice (*Oryza sativa*). Acta Physiologiae Plantarum. 2019; 41(3):1–12.
32. Lahiani MH, Dervishi E, Chen J, Nima Z, Gaume A, Biris AS, et al. Impact of carbon nanotube exposure to seeds of valuable crops. ACS Applied Materials & Interfaces. 2013;5(16):7965–73.
33. Wang S, Wang F, Gao S, Wang X. Heavy metal accumulation in different rice cultivars as influenced by foliar application of nano-silicon. Water, Air, & Soil Pollution. 2016;227(7):1–13.

34. Imada K, Sakai S, Kajihara H, Tanaka S, Ito S. Magnesium oxide nanoparticles induce systemic resistance in tomato against bacterial wilt disease. The Plant Pathology. 2016; 65(4):551–60.
35. Malandrakis AA, Kavroulakis N, Chrysikopoulos CV. Use of copper, silver and zinc nanoparticles against foliar and soil-borne plant pathogens. Science of the Total Environment. 2019;670:292–99.
36. Romheld V, Marcschner H. Function of micronutrients in plants. Micronutrients in agriculture: Soil Science Society of America; 1991. p. 297.
37. Rahman M, Punja Z. Calcium and plant disease. Mineral nutrition and plant disease. Datnoff LE, Elmer WH, Huber DM, editors: The American Phytopathological Society Press; 2007.
38. Ghormade V, Deshpande MV, Paknikar KM. Perspectives for nano-biotechnology enabled protection and nutrition of plants. Biotechnology Advances. 2011;29(6):792–803.
39. Servin A, Elmer W, Mukherjee A, la Torre-Roche D, Hamdi H, White JC, et al. A review of the use of engineered nanomaterials to suppress plant disease and enhance crop yield. Journal of Nanoparticle Research. 2015;17(2):1–21.
40. Jagadish R, Divyashree K, Viswanath P, Srinivas P, Raj B. Preparation of N-vanillyl chitosan and 4-hydroxybenzyl chitosan and their physico-mechanical, optical, barrier, and antimicrobial properties. Carbohydrate Polymers. 2012;87(1):110–16.
41. Kaur P, Thakur R, Choudhary A. An in vitro study of the antifungal activity of silver/chitosan nanoformulations against important seed borne pathogens. International Journal of Scientific & Technology Research. 2012;1(6):83–86.
42. Kabak B, Dobson ADW, Var I. Strategies to prevent mycotoxin contamination of food and animal feed: A review. Critical Reviews in Food Science and Nutrition. 2006;46(8):593–619.
43. Brody AL, Bugusu B, Han JH, Sand CK, McHugh TH. Innovative food packaging solutions. Journal of Food Science. 2008;73(8):107–16.
44. Nile SH, Baskar V, Selvaraj D, Nile A, Xiao J, Kai G. Nanotechnologies in food science: applications, recent trends, and future perspectives. Nano-Micro Letters. 2020;12(1):1–34.
45. Helmut Kaiser Consultancy Group. Study: Nanotechnology in food and food processing industry worldwide, 2006–2010-2015: Helmut Kaiser Consultancy Group; 2009. http://www.hkc22.com/Nanofood.html.
46. Wesley SJ, Raja P, Raj AA, Tiroutchelvamae D. Review on-nanotechnology applications in food packaging and safety. International Journal of Engineering Research. 2014;3(11):645–51.
47. Rai M, Ingle AP, Gupta I, Pandit R, Paralikar P, Gade A, et al. Smart nanopackaging for the enhancement of food shelf life. Environmental Chemistry Letters. 2019;17(1):277–90.
48. He X, Deng H, Hwang H-m. The current application of nanotechnology in food and agriculture. Journal of Food Drug Analysis. 2019;27(1):1–21.
49. Jin T, Sun D, Su J, Zhang H, Sue HJ. Antimicrobial efficacy of zinc oxide quantum dots against *Listeria monocytogenes*, *Salmonella enteritidis*, and *Escherichia coli O157: H7*. Journal of Food Science. 2009;74(1):M46–52.
50. Pinto RJ, Daina S, Sadocco P, Neto CP, Trindade T. Antibacterial activity of nanocomposites of copper and cellulose. BioMed Research International. 2013;2013:6.
51. Mihindukulasuriya S, Lim L-T. Nanotechnology development in food packaging: A review. Trends in Food Science and Technology. 2014;40(2):149–67.
52. Rhim J-W, Park H-M, Ha C-S. Bio-nanocomposites for food packaging applications. Progress in Polymer Science. 2013;38(10–11):1629–52.
53. Grigore ME, Biscu ER, Holban AM, Gestal MC, Grumezescu AM. Methods of synthesis, properties and biomedical applications of CuO nanoparticles. Pharmaceuticals. 2016;9(4):75.

54. Rai PK, Lee SS, Zhang M, Tsang YF, Kim K-H. Heavy metals in food crops: Health risks, fate, mechanisms, and management. Environment International. 2019;125:365–85.
55. Kerry J, O'grady M, Hogan S. Past, current and potential utilisation of active and intelligent packaging systems for meat and muscle-based products: A review. Meat Science. 2006;74(1):113–30.
56. Morris MA, Padmanabhan SC, Cruz-Romero MC, Cummins E, Kerry JP. Development of active, nanoparticle, antimicrobial technologies for muscle-based packaging applications. Meat Science. 2017;132:163–78.
57. Echegoyen Y. Nano-developments for food packaging and labeling applications. Nanotechnologies in food and agriculture: Springer; 2015. p. 141–66.
58. Momin JK, Joshi B. Nanotechnology in foods. Nanotechnologies in food and agriculture: Springer; 2015. p. 3–24.
59. Lee SY, Lee SJ, Choi DS, Hur SJ. Current topics in active and intelligent food packaging for preservation of fresh foods. Journal of the Science Food Agriculture. 2015; 95(14):2799–810.
60. Hagen M, Drew R. Nanotechnologies in food packaging: an exploratory appraisal of safety and regulation: ToxConsult Pty Ltd; 2016. https://www.foodstandards.gov.au/publications.
61. Kuswandi B. Environmental friendly food nano-packaging. Environmental Chemistry Letters. 2017;15(2):205–21.
62. Chaudhry Q, Scotter M, Blackburn J, Ross B, Boxall A, Castle L, et al. Applications and implications of nanotechnologies for the food sector. Food Additives & Contaminants. 2008;25(3):241–58.
63. Umaraw P, Munekata PE, Verma AK, Barba FJ, Singh V, Kumar P, et al. Edible films/coating with tailored properties for active packaging of meat, fish and derived products. Trends Food Science Technology. 2020;98:10–24.
64. Ghaderi-Ghahfarokhi M, Barzegar M, Sahari MA, Azizi MH. Nanoencapsulation approach to improve antimicrobial and antioxidant activity of thyme essential oil in beef burgers during refrigerated storage. Food and Bioprocess Technology. 2016;9(7):1187–201.
65. Valdivieso-Ugarte M, Gomez-Llorente C, Plaza-Díaz J, Gil Á. Antimicrobial, antioxidant, and immunomodulatory properties of essential oils: A systematic review. Nutrients. 2019;11(11):2786.
66. Rai M, Paralikar P, Jogee P, Agarkar G, Ingle AP, Derita M, et al. Synergistic antimicrobial potential of essential oils in combination with nanoparticles: Emerging trends and future perspectives. International Journal of Pharmaceutics. 2017;519(1–2):67–78.
67. Dan N. Compound release from nanostructured lipid carriers (NLCs). Journal of Food Engineering. 2016;171:37–43.
68. Aytac Z, Ipek S, Durgun E, Tekinay T, Uyar T. Antibacterial electrospun zein nanofibrous web encapsulating thymol/cyclodextrin-inclusion complex for food packaging. Food Chemistry. 2017;233:117–24.
69. Dang X, Cao X, Ke L, Ma Y, An J, Wang F. Combination of cellulose nanofibers and chain-end-functionalized polyethylene and their applications in nanocomposites. Journal of Applied Polymer Sciences. 2017;134(42):45387.
70. Madalena D, Fernandes JM, Avelar Z, Gonçalves RF, Ramos ÓL, Vicente AA, et al. Emerging challenges in assessing bio-based nanosystems' behaviour under in vitro digestion focused on food applications–A critical view and future perspectives. Food Research International. 2022;157:111417.
71. Paolino D, Mancuso A, Cristiano MC, Froiio F, Lammari N, Celia C, et al. Nanonutraceuticals: The new frontier of supplementary food. Nanomaterials. 2021;11(3):792.
72. Parisi C, Vigani M, Rodríguez-Cerezo E. Agricultural nanotechnologies: What are the current possibilities? Nano Today. 2015;10(2):124–27.

73. Rajiv P, Chen X, Li H, Rehaman S, Vanathi P, Abd-Elsalam KA, et al. Silica-based nanosystems: their role in sustainable agriculture. Multifunctional hybrid nanomaterials for sustainable agri-food and ecosystems: Elsevier; 2020. p. 437–59.
74. Tripathi DK, Singh S, Singh S, Pandey R, Singh VP, Sharma NC, et al. An overview on manufactured nanoparticles in plants: Uptake, translocation, accumulation and phytotoxicity. Plant Physiology and Biochemistry. 2017;110:2–12.
75. Abdel-Haliem ME, Hegazy HS, Hassan NS, Naguib DM. Effect of silica ions and nano silica on rice plants under salinity stress. Ecological Engineering. 2017;99:282–89.
76. Jullok N, Van Hooghten R, Luis P, Volodin A, Van Haesendonck C, Vermant J, et al. Effect of silica nanoparticles in mixed matrix membranes for pervaporation dehydration of acetic acid aqueous solution: Plant-inspired dewatering systems. Journal of Cleaner Production. 2016;112:4879–89.
77. Rastogi A, Tripathi DK, Yadav S, Chauhan DK, Živčák M, Ghorbanpour M, et al. Application of silicon nanoparticles in agriculture. 3 Biotech. 2019;9(3):1–11.

12 Anti-Protozoan Applications of the Biogenic Nanoparticles and Their Mechanism of Action

El-Hassan M. Mahmoud, Latifat Abdulsalam, El-Hussein Moustafa, Ikrimah A. Alhanbali, Saleem Ullah, and Irshad Ahmad

12.1 INTRODUCTION

Parasites are organisms that rely on other species for their survival, exploiting their hosts to obtain food, shelter, and defenses [1]. They are categorized into three main groups: helminths, ectoparasites, and protozoa, each with unique morphologies, genetic diversities, evolutionary histories, and adaptations [2]. Transmission typically occurs through direct touch, vectors, or the fecal-oral route [3]. The prevalence of parasitic diseases is often attributed to the lack of effective treatments and slow progress in pharmaceutical developments [2], with many being designated as neglected tropical diseases due to their high mortality rates.

Historically, treatments have included chemotherapeutic agents and ethnobotanical remedies [3], utilizing antiprotozoal, antihelminthic, insecticidal, or acaricidal substances. Various antiprotozoal drugs, such as antimalarials and antiamoebic agents [4], function by disrupting parasite DNA [5], inhibiting protein synthesis [6], damaging the parasite membrane, or directly killing the protozoa [7]. However, the emergence of protozoal resistance has rendered several of these treatments ineffective. Today, there is an urgent need to explore innovative treatment strategies that are both safer and more effective for patients, addressing the growing issue of drug resistance.

Due to their unique and remarkable properties and features, the utilization of nanomaterials and NPs is expanding in biomedicine, offering potential for diverse medical applications [8]. Notably, NPs such as silver (AgNP), gold (AuNP), and platinum (PtNP) are highly soluble and have shown antibacterial [9, 10] and anti-parasitic [11–14] effects. Further research has revealed their capabilities in selective binding

DOI: 10.1201/9781003378563-12

and enzyme inhibition [15, 16, 17], with recent studies highlighting the specific inhibition of recombinant target enzymes from *T. b. brucei* by AuNP and AgNP [18, 19].

Biogenic NPs, produced through biological means, offer advantages in cost-effectiveness, stability, polydispersity, and size over NPs synthesized by physical or chemical methods [20, 21].

The biological production process, leveraging physiological pH, temperature, and pressure, minimizes the need for harsh processing conditions. This method enables the generation of inorganic NP composites by a wide array of microorganisms, both intracellularly and extracellularly. Because of their distinctive properties, they have gained popularity across various sectors, including energy, healthcare, environment, and agriculture [22].

This chapter explores the definition and sources of biogenic NPs, their mode of action against protozoa, the efficacy, and promise of these green NPs against parasites, and discusses future research directions and challenges.

12.1.1 Disruption of Metabolic Pathways

When given to parasites, NPs pierce the cell membrane and may interact with different molecules and cellular structures, including proteins and DNA, which can have a deleterious effect on the microbe. For example, when the NP binds to ribosomes, it denatures them, which inhibits translation and protein synthesis [23, 24]. Numerous studies have demonstrated that when NPs are exposed to cells, they cause damage to DNA [25, 26]. This damage can take the form of nuclear fragmentation [27] or physical attachment of the NP to the DNA. The high affinity of the NP to interact with phosphate, which is highly available in the DNA molecule, is the cause of this damage [28].

Following their entry into the cell and the release of metal ions, the NPs interact with the DNA of the cell, which has the potential to denaturize the DNA or halt cell division [29, 30]. Metal ions with nucleosides produce a complex where the metal atom coordinates with two N atoms to form the N-metal+-N complex in the GC base pair. This interaction occurs because nucleosides replace the hydrogen bonds in the GC and AT base pairs. The nucleotide base pairs contain other complexes, such as O-Ag+-N (the GC base pair) and N-Ag+-O (both GC and AT base pairs) [31, 32], To sum up this reaction, the introduction of NPs between the purine and pyrimidine base pairs results in the formation of a complex that disrupts the hydrogen bonds between them, which in turn disrupts the double helical shape of the DNA and may cause cell apoptosis [33].

The human stomach is the primary habitat of the anaerobic, unicellular, specialized protozoan *Blastocystis hominis*. The frequency can reach 50% in Egypt [34]. Since ROS-releasing NPs can disrupt the structure of lipophosphoglycan and glycoprotein molecules, prevent ATP synthesis, and halt DNA replication, AgNPs can eliminate the infection source and stop the parasite from spreading [35].

A second pathway in the mechanism of NPs in disrupting the metabolic pathways is the modulation of signal transduction. Biogenic NPs can penetrate biofilms formed by protozoa. Within these biofilms, they interfere with quorum-sensing gene cascades. Quorum sensing is a cell-to-cell communication mechanism that allows protozoa to coordinate their behavior. By disrupting this communication, biogenic NPs hinder biofilm formation and weaken the protozoa's ability to thrive, by modulating the intracellular signaling pathways, the NP affects key molecules that are involved in signal transduction, which leads to disrupting the essential cellular processes, which can cause growth inhibition and compromised survival of protozoa [36].

Interacting with HIF-1 is another method by which metabolic particles are disrupted, and evidence suggests that this interaction also plays a role in the NP-induced anti-parasitic impact via modulating host cellular processes. It's interesting to note that NPs may influence several cellular signaling pathways, such as the activity of HIF-1 (hypoxia-inducible factor 1) [37–39]. The heterodimer HIF-1 is made up of the subunits α and β. By controlling pro-parasite genes such as transferrin receptors, glycolytic metabolic genes, and vascular endothelial growth factors, it has a striking impact on *Toxoplasma gondii* survival in the host [40–42]. It has also been shown that, at physiological oxygen levels, parasite growth is considerably inhibited by the absence of the HIF-1α component [43].

The denaturation of protozoal protein is the final mechanism by which NPs impair metabolic pathways. Researchers have been interested in the relationship between NPs and protein production in recent years [44]. Researchers who looked into how NPs affected protozoal proteins discovered that they controlled proteins involved in electron transfer, substance transport, and nitrogen metabolism [45]. Specifically, Ag NPs interact with ribosomes to denaturize them, which inhibits translation and protein synthesis—two processes critical to the generation of ATP [46]. The most notable effect of Ag^{+} is a reduction in the production of S2 protein, a 30S ribosomal subunit, which denatures and renders the ribosome inactive. Reduced release of other proteins follows ribosome denature, including succinyl-CoA synthetase, which is necessary for intracellular ATP synthesis catalysis. It is the responsibility of the 30S subunit to guarantee that codons and anticodons connect correctly [46].

Stated differently, by interrupting the electron transport chain, silver ions can interfere with metabolism and cause disruptions to ATP generation. Silver ions can also disrupt the parasite's essential enzyme functions and metabolic pathways, resulting in cellular malfunction and death [47]. AgNPs' precise mode of action against the parasite is still being studied at this point, and there might be more efficient approaches. AgNPs, however, undoubtedly have the potential to be an effective weapon in the fight against parasite infestations [48].

12.1.2 Generation of Reactive Oxygen Species (ROS)

A group of distinct molecules or atoms with one or more unpaired electrons and at least one oxygen atom that can exist on its own are known as ROS [49].

Biogenic NPs are known to have a high productivity of free radicals by redox reaction as the oxygen atom interacts with the free metal ion which generates the ROS via the Fenton reaction [50]. As is known, the capacity of NPs to generate ROS serves as the foundation for their antibacterial activity [35], where it can directly interact with the membrane DNA, protein, and lipids [51], and the antioxidants produced by the protozoa normally neutralize it; nevertheless, a high concentration of

ROS inside the protozoal cell may cause oxidative stress which leads to protozoan viability *Leishmania* parasites, for instance, are known to be ROS-sensitive [52].

12.1.3 Oxidative Stress Induction

ROS are defined as biologically reactive substances that comprise a broad spectrum of molecules and free radicals originating from molecular oxygen. These include reactive superoxide radical anion (O_2•–), hydroxyl radicals (•OH), and hydrogen peroxide (H_2O_2) [53] which are naturally produced inside the cell as a byproduct during the aerobic respiration [54], but it is mainly generated externally by UV light or transition metals such as iron [55].

NPs generate ROS through redox reaction where the reduction of oxygen atom (O_2) forms superoxide anion (O_2•–), which mediates the chains of oxidation reaction, and it is the precursor of most ROS. The dismutation of the superoxide anion induces the hydrogen peroxide formation which depending on the degree of reduction is further reduced to either H_2O when it is reduced or to hydroxyl radical (•OH) when it is partially reduced, the partial reduction of H_2O_2 can be catalyzed by transition metal reduction by a process called Fenton reaction in which the transition metal is reduced by the superoxide anion leading to the proliferation of the oxidation reaction cascade [56].

Many of the parasitic protozoa have several different defensive mechanisms to protect them against ROS accumulation [57] and to preserve redox equilibrium; examples of these protective mechanisms include non-enzymatic antioxidants like ascorbate and tocopherol or enzymatic antioxidants like glutathione peroxidase [58] but when it is insufficient it leads to a state called oxidative stress which is a state where there is an imbalance between the generation of ROS and the detoxification of ROS [59] leading to accumulation of ROS inside the cell.

NP-induced ROS production is affected by several factors including NP internalization, physical properties of NPs, and particle size [60, 61], Meanwhile, we shall discuss the various mechanisms of NP production of ROS.

Many studies have provided copious documentation of the ROS-producing capabilities of different nanomaterials that differ in their chemical structure [62, 63]. Size, shape, surface area, aggregation status, and solubility are only a few of the distinct physicochemical characteristics of NPs that directly affect their capacity to produce ROS [64–67]. As previously mentioned, NPs exhibit a distinct surface area compared to microparticles or their bulk counterparts. Microparticles and bulk materials generally have a higher fraction of atoms, resulting in a larger surface area. Smaller NPs, on the other hand, have a significant surface area in comparison to their mass. The link between the particle mass that defines the surface-to-volume ratio and the particle size indicates the number of reactive sites available for ROS production on the NP surface [68–71].

The NP's chemical makeup is another factor that affects the generation of ROS. Because dangling bonds—immobilized free radicals—are present on the surface of the NPs, they enhance the catalytic activity that the NPs induce, which is why they exhibit such high reactivity [72]. Reactive groups can also arise from smaller NPs by changing the surface's electrical characteristics [73, 74]. Furthermore, NPs exhibit a phenomenon known as the quantum confinement effect, wherein nanomaterials with this effect develop magnetic moments absent in their original state. This effect also regulates NP affinity for gaining or losing electrical charges, consequently impacting their catalytic behavior [75].

The last factor is the nature of the transition metal as there are many kinds of MNPs, for example, silicon (Si), iron (Fe), copper (Cu), chromium (Cr), silver (Ag), and many other metals that are involved in ROS production through either Haber-Weiss and Fenton reactions or even both [61].

In the Fenton reaction, H_2O_2 and a transition metal ion combine to form an oxidized metal ion and a hydroxyl ion (OH), which is extremely damaging to biological molecules. Hydrogen peroxide reduction with Fe^{+2} also contributes to the formation of OH [76], this reaction is used to induce oxidative stress (O_2- and OH) in NPs containing (Cu) and (Fe) [77]. Meanwhile, the Haber-Weiss reaction exhibits the reactions between H_2O_2 and the oxidized metal ion to produce OH ion [78], additionally, NPs containing Cr, cobalt (Co), and Vanadium (Va) can interact by both Haber-Weiss and Fenton reactions [77].

Finally, photon energy, or UV and visible radiation, can activate NPs. For instance, TiO_2 NPs [79] produce electrons with sufficient energy to convert O_2 into $1O_2$, which is involved in cellular damage that is mitigated by interactions with proteins, lipids, and nucleic acids [80].

Another factor is NP internalization, For the NPs to invade the cell, the NPs have to interact with the cell membrane organelle that mediates the transportation of molecules and ions from the external environment into the cell (endocytosis) or from the cell to the external environment (exocytosis) where endocytosis plays an important role in NPs internalization [81, 82].

Another element that affects how well NPs internalize is how they interact with mitochondria, catalyze reactions involving free radicals, activate growth factors, and initiate NOX [83]. Mitochondria play a crucial role in NP internalization, as they contribute to ROS generation when NPs infiltrate the cell. This infiltration leads to mitochondrial membrane depolarization and disruption of electron chain transport by activating NADPH-related enzymes. Consequently, this disruption blocks the electron transport chain, causing an increase in O_2- through electron transfer from respiratory carriers to O_2. This process induces oxidative stress within the cell [83]. Subsequently, exposure to NPs stimulates immune cells through a ROS-dependent mechanism, regulated by NOX activation [60].

Last but not least, NP-induced production of ROS causes a decrease in GSH into glutathione disulfide, its oxidized form, which is implicated in the creation of oxidative stress and its effects [84].

To summarize, the level of ROS production depends on the chemical nature of the NPs [85], where the studies confirm that smaller size and high specific charge of the surface area induce a higher level of ROS production [86, 87], and lastly, metal ions of the NPs can influence the ROS production [88].

12.1.4 Impact on Protozoan Viability

A huge number of recent studies demonstrate that NP-induced oxidative stress can be used as an antiprotozoal agent to many pathogens as it meets the requirements of a new mechanism of action with antiprotozoal effect [89], several researchers said that AgNP antimicrobial and antiprotozoal effect is related to its ability to upregulate the ROS, which leads to the damage of the cell nucleic acid, proteins that ultimately lead to its desired effect [90, 91], moreover, many other biogenic NPs have been proven to show antiprotozoal effect by their ability to prompt the cell to produce ROS, for example, ZnO and TiO_2 [92, 93].

In addition to inducing DNA damage, oxidative stress is essential for the nanotoxicity and cytotoxicity of NPs [85, 90]. One major reason why NPs can harm DNA is because they produce reactive oxygen species, or ROS. 8-hydroxy-2′-deoxyguanosine (8-OHdG) is produced when ROS interacts with DNA and damages DNA [94]. Additionally, exposure of DNA to biogenic NPs such as silver (Ag), titanium (Ti), copper (Cu), and iron (Fe) NPs has been found to cause genotoxicity mediated by damage to nucleic acids [95].

The cytotoxicity induced by ZnO NPs occurs through the initiation of oxidative damage, resulting in the release of inflammatory mediators and eventual cell death [69, 96]. This suggests that oxidative stress serves as a vital mechanism underlying the antimicrobial properties of biogenic NPs. Furthermore, *Leishmania* parasites exhibit susceptibility to ROS, as mentioned earlier [52], Macrophages generate high levels of ROS to fight microorganisms because they act as host cells for *Leishmania* parasites [97].

The way *leishmania* can live inside macrophages without being exposed to any damage is by inhibiting the enzymatic mechanism that produces ROS, which prevents *leishmania* from creating this substance and accounts for the disease's survival [98]. Hence, it is plausible that the physical production of these oxygen derivatives, such as that achieved by Ag-NPs, is required to kill *Leishmania* parasites using a ROS-based treatment, rather than an enzymatic process that parasites can obstruct [99]. It was found that *leishmania* parasites exposed to MNPs may be more severely impacted by ROS formation because of the interactions between Ag-NPs and UV radiation, which may prevent parasite infection. This result shows that Ag-NPs can prevent Leishmania infection in host cells and render the parasites harmless when used alone or in conjunction with UV light. Ag-NPs' antileishmanial effects on *Leishmania tropica* parasites were discovered by looking at how they affected different promastigote and amastigote forms of the parasites' cellular parameters [99].

Numerous investigations revealed that *T. gondii* can be killed by MNPs both extracellularly and inside the host cell. According to our research, the NPs have potential antibacterial [9, 100] and antiparasitic [14, 101, 102] properties. However, the assessment of ROS utilizing a fluorescent probe offered conclusive proof that the administration of NPs stimulated ROS generation, subsequently contributing to the demise of parasites induced by NPs [103]. Trolox's ability to reverse the antiparasitic effects of NPs indicated the implication of oxidative stress in the NP-triggered elimination of *T. gondii*. These results are in line with the rare reports that demonstrate the harmful effects of ROS formation by biogenic NPs, which cause *Leishmania* major to die [101] and *Filaria* parasites to undergo apoptosis [102].

A study used MitoRed to measure the parasite mitochondrial membrane potential in order to highlight the relationship between biogenic NPs and mitochondria. As a prominent producer of ROS in cells, mitochondria are thought to be the main target of excessive ROS in cells, providing an additional pathway for ROS production.

MitoRed is a rhodamine-based dye that localizes within mitochondria and releases fluorescence. It can penetrate cell membranes. The way MitoRed interacts with mitochondria is determined by their membrane potential. The rhodamine-based dye did not accumulate much in the parasite mitochondria after an 8-hour NP treatment, as evidenced by a decrease in fluorescence intensity. This suggests that the NP treatment reduced the parasite mitochondria's membrane potential. Additionally, even twelve hours after the NPs were removed, it was possible that the parasites had not fully recovered from the particles' lowering of mitochondrial membrane potential. These observations further demonstrate the antiparasitic impact of NP therapy on parasite mitochondria. This fits nicely with previous work by Charvat and Arrizabalaga, which showed that oxidative stress brought on by excessive ROS production by cells weakened *T. gondii's* mitochondrial membrane potential and ultimately caused it to die [104].

In summary, biogenic NPs could be a potential replacement for existing antiprotozoal medications in the future. We propose that treatment based on biogenic NPs may herald a turning point or revolution in pharmacological therapy for parasitic illnesses, given the fast global spread of parasites and the numerous drawbacks of antiparasitic medications.

12.2 BIOGENIC NANOPARTICLES

12.2.1 Types of Biogenic Nanoparticles

Generally nanoparticle (NP) production often relies on physicochemical methods, encompassing a broad spectrum of techniques. Research indicates that variables such as concentration and temperature are crucial in determining the NPs' stability, morphology, and physicochemical properties. However, traditional NP synthesis methods suffer from several drawbacks, including high operational expenses, carcinogenicity, toxicity, and inefficiency. However, these traditional synthesis methods have notable disadvantages, including high costs, potential carcinogenicity, toxicity, and environmental harm due to the use of hazardous substances like reducing agents, stabilizers, and organic solvents [105–108]. Additionally, certain NPs possess toxic properties, attributed to their chemical composition, structure, size, and external interactions, potentially limiting their biomedical and clinical applications [109]. This necessitates the development of biocompatible and eco-friendly NP production processes.

Biological or green synthesis methods, utilizing various biological systems, have emerged as a favorable alternative. A diverse range of organisms, including fungi, yeast, algae, bacteria, plants, their extracts, and even higher organisms like earthworms, have been explored for the green synthesis of NPs [110, 111]. These biological systems offer a non-toxic, clean, and eco-friendly approach, producing biocompatible and cost-effective NPs [112, 113]. The advantages of biogenic NPs include affordability, simplicity, high stability, scalability, and non-toxic byproduct generation [114]. Additionally, utilizing biological entities enhances the biocompatibility of the NPs and mitigates their tendency to agglomerate into clusters [109].

Despite these benefits, biological NP synthesis has its challenges, such as the seasonal or ecosystem-specific availability of biomaterials like bacteria and plants [115]. Additionally, the biological synthesis process, involving various enzymatic pathways, can result in NPs with inconsistent sizes and shapes [116]. Despite these challenges, numerous efforts are underway to address these limitations and deficiencies, aiming to make the biological synthesis of NPs more practical and widely applicable [117].

12.2.2 Sources of Biogenic Nanoparticles

Utilizing enzymes and metabolites for NP formation represents a sustainable and eco-friendly approach. This green synthesis method provides the necessary substances for the creation and control of well-organized NPs [118]. Among these methods, microbial mediation is particularly noteworthy, with plants also playing a crucial role in NP formation, especially through the use of plant extracts. These extracts in the reduction of metallic ions, complementing microbial efforts in the formation and stabilization of metallic NPs [56, 119]. Microorganisms are extensively utilized in NP fabrication [120] due to their innate cellular processes. These organisms absorb target ions from their surroundings and enzymatically convert them from their ionic to elemental forms. This biotransformation can be categorized into two processes: extracellular and intracellular. The intracellular process involves the uptake of ions into microbial cells, where enzymes facilitate NP fabrication. Conversely, the extracellular process occurs on the cell surface, where ions are enzymatically reduced to form NPs [121].

12.2.2.1 Bacteria

Bacteria have emerged as a promising bio-factory for the synthesis of biogenic NPs. Their ease of handling, ubiquity in various environments, and adaptability to diverse conditions make them ideal for the production of reduced metal ions. Bacteria are capable of reducing toxic metal ions into NPs, resulting in a wide range of sizes, structures, and shapes [122, 123]. Genetic engineering techniques offer the potential for rapid advancements in mitigating toxic effects and achieving precise NP synthesis [124].

Metallic NPs including silver, iron, copper, zinc, selenium, gold, and platinum have been manufactured by a variety of bacterial genera with *Bacillus subtilis* being the first to produce bacterial gold NPs [125]. Both Gram-positive and Gram-negative bacteria can generate a broad spectrum of amorphous NPs, typically ranging from 50 to 200 nm in diameter [116, 125–129].

Notably, magnetostatic bacteria are adept at constructing magnetosomes (nanocrystals of iron sulfide and iron oxide), suitable for producing magnetic radicals [130]. For example, it has been observed that the formation of selenium (0) crystalline nanostructures at room temperature [116], offers an environmentally friendly alternative for crystalline NP formation. Bacterial proteins often act as capping agents, enhancing the stability and reducing the agglomeration of synthesized NPs [130].

Cyanobacteria and actinomycetes are other bacterial classes utilized for NP production. Cyanobacteria, prevalent photosynthetic bacteria, are instrumental in NP stabilization and have been used to synthesize gold, palladium, silver, and platinum NPs [131–133]. Actinomycetes, a diverse group of Gram-positive bacteria, are recognized for their ability to synthesize NPs, including gold, copper, zinc, and silver [134–137]. For instance, Streptomyces spp. have been used to produce zinc oxide NPs with biomedical applications [138].

The efficiency of NP generation using bacteria can be significantly enhanced by modifying culture conditions such as pH, temperature, and oxygen concentration [139]. However, challenges remain, particularly in the purification process and control over the size and shape of monodispersed particles. Not all bacteria are capable of producing metal NPs (MNPs) and oxides, underscoring the importance of maintaining control over size and shape during synthesis.

12.2.2.2 Archaea

Archaeal synthesis of NPs is less studied compared to bacteria, with Haloarchaea from the salt-resistant class being the primary NP producers [140, 141]. Their resilience in extreme environments suggests a potential for tolerance to toxic substances, making them promising candidates for NP synthesis.

12.2.2.3 Fungi and Yeast

Utilizing fungi in NP synthesis presents several advantages over bacterial methods, including cost-effectiveness, scalability, and the production of monodispersed NPs with favorable morphologies. Fungi's mycelial structure aids these benefits [142, 143], while secreted proteins and enzymes enhance NP stability [144]. Fungal cultures, preferred for their high biomass and metal tolerance, allow controlled NP synthesis by adjusting environmental conditions [145, 146].

Researchers worldwide have extensively utilized fungi for NP biosynthesis, occurring both extracellularly and intracellularly. Various fungal species have demonstrated the ability to synthesize silver, gold, and other nano-material NPs [147–150]. Additionally, numerous studies have extensively documented the production of platinum, magnesium oxide, copper, gold, silica, titania, and silver NPs using *Fusarium oxysporum* [151–154]. Moreover, other fungal species or genera possess the ability to synthesize diverse types of NPs, encompassing magnetic NPs, metal oxide NPs, and quantum dots [155–158].

Like bacteria, the utilization of fungi in biosynthesis comes with certain safety concerns, as some fungi can be pathogenic. However, employing various nonpathogenic fungi offers advantages in green synthesis. Furthermore, the acquisition of NPs through fungal extracts typically involves the processes of substance reduction and stabilization [154].

However, the various methods used by yeast strains from various genera to create NP resulted in notable differences in properties such as mono-dispersity, particle size, dispersion, and characteristics [159]. The ability of yeast to transform harmful metal ions into benign complex polymer molecules has been shown [160]. This process produces NPs known as "quantum semiconductor crystals" that are found near yeasts. The use of yeast for biogenic NP production has a number of benefits, one of them being the significantly higher yield of NPs produced in comparison to bacteria [161].

Using various yeast genera or species, numerous investigations have demonstrated the synthesis of a range of NPs, including zero-valent selenium NPs, titanium, gold, silver, and cadmium sulfide [161–164]. *Schizosaccharomyces pombe* was also used to produce intracellular cadmium sulfide NPs [165]. A bio-template made of yeast was also used to create mesoporous zirconium phosphate, which exhibited exceptional electrocatalytic activity for the oxygen reduction reaction. The material might therefore be used to create air electrodes [166]. Additionally, lead sulfide quantum dots were intracellularly synthesized by mixing lead ions with yeast cells of *Torulopsis* spp. [167].

12.2.2.4 Algae

Algae, belonging to Plantae, are vital in various fields. They are used to overcome limitations in NP treatments [168]. The cell wall of algae contains functional groups and biomolecules serving as reducing agents, obviating the need for additional reagents [169]. Algae, alive or dead, are used for green bionanomaterial formation, surpassing other bioagents in NP production [154, 170]. Algae can be harvested year-round without chemicals. Microalgae exhibit rapid growth, expanding their mass tenfold faster than higher plants [171].

Algal substances are crucial in nanomaterial production for their metal ion reduction ability [172]. Algal-synthesized NPs offer cleaner and more efficient alternatives to chemical synthesis due to lipid bilayers, enhanced adsorption efficiency, and adjustable growth parameters [173–175]. Algae's resilience to heavy metals makes them promise in heavy metal mitigation and valuable metallic NP production, addressing pollution and providing economic benefits [176, 177].

Gold and silver are extensively researched noble metals for algae-mediated NP synthesis, occurring either intracellularly or extracellularly [178]. However, other NP structures like iron, zinc, copper, titanium, and zinc oxide NPs are synthesized by different species of algae [179, 180]. Recently, *Chlorella vulgaris* was utilized for the synthesis of tin oxide NPs with biological and photocatalytic activities [181]. Similarly, silver NPs synthesized from *Padina tetrastromatica* have exhibited promising antibacterial properties, demonstrating strong activity against *Pseudomonas aeruginosa* and *Staphylococcus aureus* [101, 182].

12.2.2.5 Plants

In many cases, microbes offer effective pathways for NP synthesis; however, challenges such as mass cultivation hurdles, downstream processing complexities, scale-up processes, molecular mutations, and other factors often impede NP synthesis and application. Therefore, directing research toward utilizing plants for NP synthesis is warranted. The plant-based method has several advantages over microorganisms because it does not require distinct, laborious, or multiple steps for things like culture formation, isolation, and preservation. Additionally, it is less expensive, faster, more economical, and simpler to scale up plant-based synthesis for the manufacture of large amounts of NP. Plants possess numerous active constituents capable of acting as biodegradable agents that effectively stabilize and reduce NPs [109, 116]. Additionally, Plant extracts offer a multitude of reducing agents and enzymes that facilitate the simple creation of NPs [109].

Numerous plant species including *Acacia auriculiformis*, *Bergenia ciliate*, *Clitoria ternatea*, *Coleus forskohlii*, *Coffea arabica*, *Curculigo orchioides*, *Dioscorea alata*, *Digitaria radicosa*, *Diospyros paniculata*, *Elephantopus scaber*, *Euphorbia antiquorum* L., *Emblica officinalis*, *Ficus benghalensis*, and *Parkia roxburghii* have been reported for green synthesis of MNPs [183–195]. Medicinal plants also find applications in NP synthesis [196]. Moreover, Plant sections including leaves, shoots, roots, stems, barks, flowers, seeds, and their metabolites have all been proven effective in NP fabrication as well [197–199]. For example, extracts from *Aloe vera* and *Eucalyptus* leaves and fruits, *Jatropha curcas* seeds, *Boswellia ovalifoliolata*, and *Cinnamomum zeylanicum* bark have demonstrated the ability to reduce various metals including platinum silver, copper, gold, iron, zinc, and cadmium [200].

These plant-based NPs have been evaluated for various activities including antibacterial, catalytic, anti-filaria, antioxidant, and cytotoxicity [183, 184, 201–203]. Recently, *Lantana camara* leaf extracts were employed for synthesizing yttrium oxide NPs, exhibiting antibacterial, photocatalytic, and cytotoxic effects against HeLa cell lines with potential applications in targeted cancer therapy [204]. The industrial-scale synthesis of metallic and oxide NPs can be achieved through the application of downstream processing procedures and plant tissue culture techniques. However, this potential can only be realized once challenges such as the metabolic status of the plant and related factors are effectively addressed [205].

12.2.2.6 Other Biological Sources

Using algae and marine plants, several MNPs have been successfully produced. Furthermore, marine yeast has been observed to produce lead sulfide NPs [150]. Utilizing a nanoscale bio-template obtained from the squash leaf curl China virus, hybrid nanomaterials consisting of gold and silver were recently produced [206]. Additionally, a very new study describes how to create quantum dots with the enzyme milieu that is present in earthworms' midguts [109].

12.3 PROPERTIES AND CHARACTERISTICS OF NANOPARTICLES

The NPs' enormous surface area makes them stand out from bulk materials by causing them to regularly exhibit unique size-dependent characteristics and physicochemical properties. The unique scientific and technological applications of NPs stem from their physicochemical properties, which include their large surface area, chemical reactivity, mechanical strength, and optical activity. Because of their high molar excitation coefficient and wavelength-selective absorption, NPs exhibit unique optical and electrical features because of local surface plasmon resonance [207]. LEDs, sensors, semiconductors, batteries, and bio-photonics are just a few of the devices that employ NPs' optical and electrical characteristics [208, 209].

Also, NPs exhibit unique magnetic properties with diverse applications, including superparamagnetism observed in tiny ferromagnetic NPs [210]. Various MNPs like magnesium are found to be used in magnetic fluids, magnetic resonance imaging, biomedicine, data storage, catalysts, and soil/water purification due to their magnetic properties [211–213].

In surface engineering, nanofabrication, building/painting materials, tribology, and nanomanufacturing, the mechanical properties of NPs are utilized [213, 214]. Additionally, stable suspension nanofluids employ MNPs [210] since their thermal conductivity is higher than that of solid fluids. Because of these special qualities, NPs have a far greater impact in areas including space exploration, medical, military, textile, food, cosmetics, and agriculture than they do in bulk or individual molecules [214]. Additionally, NPs exhibit important properties like suspension, hydrophilicity, diffusion, and hydrophobicity further expanding their potential applications [204].

12.4 MECHANISMS OF ACTION OF BIOGENIC NANOPARTICLES AGAINST PROTOZOANS

Among the many diverse domains in which NPs find applications are the biological sciences, drug delivery, waste management, gene delivery, optics, and many more [215]. Nevertheless, their potent antibacterial and anti-inflammatory properties are making them more and more well-liked [216]. Though the mechanism underlying biogenic NPs' antiprotozoal activity is not fully understood, some of the well-known mechanisms underlying NP cytotoxicity are Damage to the integrity of the cell membrane, disruption of cellular functions, and production of reactive oxygen species (ROS) (Figure 12.1). These mechanisms are determined by the type of NPs used, for example, metal oxide NPs employ their effect by generation of ROS [217] while heavy MNPs employ their effect by interference with cellular processes and disruption of cell membrane integrity [218]. In this section, we will discuss these mechanisms and explain how biogenic NPs can inhibit or kill different parasites.

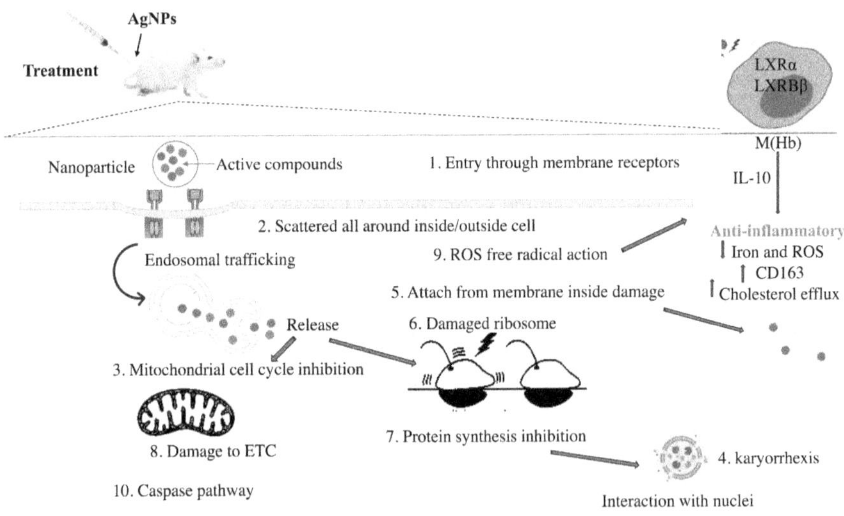

FIGURE 12.1 General mechanism of action of nanoparticles against protozoa [219].

12.4.1 Disruption of Cell Membrane Integrity

Studies have demonstrated the wide-ranging effects of heavy MNPs on bacteria, fungi, and viruses [220, 221], in addition, they can damage or kill protozoa and nematodes [222, 223]. One of the main mechanisms of biogenic NPs to produce their antiprotozoal effect is by exposing the NP to the cell causing the NP to be adsorbed to the cell membrane which leads to cell membrane disruption [224–226]. Some researchers say that the NP is first adsorbed to the cell membrane followed by its penetration which leads to the disintegration of the cell [227, 228]. This disruption of the cell membrane can be done through two processes which are either by interacting with the lipid bilayers or by penetration and damage of the cell membrane directly.

12.4.2 Interaction with Lipid Bilayers

The cell membrane of any organism primarily consists of lipopolysaccharide molecules, tightly packed together to form an effective barrier [229]. MNPs, conversely, can interact with protozoan lipid bilayers in various ways depending on factors such as size, shape, surface charge, or composition. The cell membrane functions as a barrier to prevent MNPs from internalizing. The lipid bilayer of the cell membrane facilitates MNP transport, which is the first stage of their internalization. Protein-mediated transmembrane transport or non-mediated diffusion are two possible ways that MNPs can cross a cell membrane [230]. Protozoa's intramembrane transport of MNPs is still unknown. It is postulated, therefore, that as MNPs dissolve, metal ions are discharged into the surrounding media. Following their possible entry into cells through ion channels and carriers, or by binding to MNPs in the cytoplasm of unknown transport proteins, these ions may be transported between sites more easily. The precise mechanisms underlying MNPs' entry into cell organelles such as mitochondria, nuclei, and chloroplasts are still unknown, despite their obvious presence [231].

Thus, the primary factor behind the antiprotozoal effect of MNPs is largely linked to their interaction with the cell membrane [232]. NPs can reach the interior of cells when they interact with lipid molecules, forming pores in the cell membrane [233]. Membrane perforation is caused by a change in the phospholipid metabolism produced by MNP. When NPs are present, they can activate the enzymes involved in phospholipid metabolism, which can lead to the oxidation of unsaturated fatty acids and, eventually, membrane perforation [234].

This mechanism may resemble other microorganisms, with the primary distinction lying in the cell covering. Protozoan cells are enveloped by a plasma membrane, whereas other microorganisms possess cell walls. The plasma membrane of protozoa consists of lipoproteins embedded in the lipid bilayer, which carries a negative charge. In contrast, the positive ions released from NPs can readily adhere to the cell through electrostatic interaction, ultimately resulting in membrane perforation [235].

Another mechanism of interaction between lipid bilayers and NPs is the "lysosome-enhanced Trojan horse effect," which can contribute to the cascade events associated with NP-induced intracellular antiprotozoal effects. Within the lysosomal cellular compartment, which is highly internalized and has an acidic environment, this mechanism involves intracellular ion release [236]. When NPs are widely distributed within a cell through active internalization mechanisms, or the Trojan horse effect, they can release metal ions as the NPs break down because of the lysosome's acidic environment [237–239]. This is a significant aspect of the NPs' antiprotozoal effect [240].

Intracellular inhibition in this process is mediated by the ions in the NPs. Some parasites may be able to survive and even become dependent on the harsh acidic environment of the phagolysosome, which makes the "lysosome-enhanced Trojan horse effect" a plausible mechanism [240]. The degree of cellular damage that NPs cause can be significantly influenced by their size and form. Moreover, it was shown that the physicochemical properties of NPs affect both their potential toxicity and their dynamic interaction with cellular organelles and biomolecules. This technique can reduce the motility, phagocytosis, viability, and membrane integrity of protozoa in addition to altering their gene expression [241–243].

In summary, we presented a general framework for the way that NPs interact with the lipid profile. On the other hand, nothing is known about the process by which biogenic NPs inactivate parasites [244]. Therefore, more research and development are needed to fully comprehend how NPs interact with lipid bilayers.

12.4.3 Permeabilization and Damage

The cytotoxic action of NPs begins with their adhesion and penetration of the cell membrane [245]. Using a variety of microscopic techniques, multiple investigations have revealed the mechanical and structural characteristics of the cell membrane in response to NP exposure and the subsequent breakdown of the membrane [246, 247]. It was discovered that after the NPs and membrane electrostatically interacted [248], the NPs interacted with the membrane's negative charge lipoproteins, resulting in irreversible cell damage [247], where the cytoplasm's noticeable shrinkage showed the morphological changes in the cell membrane [248]. The disruption of the cell membrane occurred shortly after the NPs were attached to it [103].

Numerous cellular dysfunctions are brought on by the interaction of NPs with the cell membrane [245, 249]. As evidenced by the lysosome-enhanced Trojan horse effect (LETH effect), a general mechanism of toxicity is activated upon the active internalization of NPs into cells. Vesicular structures encapsulate NPs that enter cells by energy-dependent mechanisms, including lipid raft formations, caveolin, and clathrin. These structures work fast. After that, they pass through endosomes and finally reach lysosomes [250]. The low pH environment within lysosomes initiates a phenomenon known as the LETH effect. This effect merges the extensive cellular uptake of NPs through active processes with the subsequent heightened liberation of comparatively toxic ions (such as Ag^{+}, Cd^{2+}, $Fe^{2+/3+}$, $Au^{1+/3+}$ ions).

The substantial release of ions into the intracellular environment can lead to ion-specific toxicity, causing effects such as enzyme depletion or inactivation, and protein denaturation, among others, on cellular structures like mitochondria and the endoplasmic reticulum (RER), as well as potentially damaging lysosomes. Consequently, this cascade of events culminates in elevated levels of ROS, apoptosis, and damage to DNA and cell membranes [251].

Thus, it can be inferred that exposure to NPs within the membrane causes the cell membrane to become less intact, increasing the permeability of the cell. Furthermore, a range of harmful events are brought on by the inhibition of the NPs' impacts on biological processes. These include the perturbation of the cellular redox balance, the elevation of physiological concentrations of ROS, and the induction of apoptosis [252]. This ultimately results in cell necrosis and death by blocking a number of essential cellular processes [253].

12.4.4 Interference with Cellular Processes

One of the prominent mechanisms of the antiprotozoal effect of biogenic NPs is the interference with the cellular processes of the cell as the NP alters the structure and function of the enzyme [254] or it can interfere with the activation or function of the receptor which ultimately leads to protein dysfunction [255, 256]. While NPs have some advantages over other organic molecules where it is characterized by their small size which makes it easy for them to penetrate the cell [257], another advantage is that NPs possess a specific surface area that is appropriate for biological interaction [255, 258] and because of these advantages, biogenic NPs provide a promising strategy for therapy.

The mechanism of biogenic NPs interfering in cellular processes is not fully understood but some types of interferences can be deduced which are: Inhibition of Enzymatic Activities, and Disruption of Metabolic Pathways.

12.4.5 Inhibition of Enzymatic Activities

As mentioned previously, NPs penetrate the intracellular environment owing to their small size [259, 260]. Once the NP becomes adsorbed inside the cell, the NP metal ions are released, which targets multiple sites at the same time [261] where they interact with amino, carbonyl, phosphate, and sulfhydryl (thiol) groups of the cellular biomolecules like proteins and DNA [262] which causes the deactivation of the protein, for instance, Ag (+) ions conjugate with thiol group (ASH) of the protein which is available in the intracellular environment to form a stable of SAAg which leads to the deactivation of protein [33, 263].

Furthermore, cysteine residues can be found in the catalytic domains of a class of nucleophilic proteolytic enzymes that includes thiol proteases, also referred to as cysteine proteases. These proteases are often necessary for a variety of biological functions, such as catabolic activity and protein processing [252] because of its catalytic role as a nucleophile and its disulfide bridge, both of which are necessary for proper protein folding and function [15, 264]. These proteases have been reported to be present in *Trypanosoma cruzi*, *Fasciola hepatica*, *Schistosoma mansoni*, and *Leishmania* species [265–267].

As mentioned earlier, NPs, specifically Ag+ ions react with the thiol group whereas the functional group of cysteine is thiol. When adding the Ag+ to the cysteine enzyme

[263, 264], the result is that both Ag+ ions and Ag NPs cause alteration in the 3D structure of the protein in addition to the interference to the disulfide bridge and blockage of the active binding sites which inevitably result in the overall functional defect of the parasite [268].

The phosphomannose isomerase enzyme is an additional instance of an enzyme-NP interaction. It is responsible for catalyzing the reversible interconversion of mannose 6-phosphate and fructose 6-phosphate, the first stage in the production of active mannose donors, which are necessary for the biosynthesis of different glycoconjugates. This interaction occurs when the enzyme interacts with Ag NPs. Stated differently, it impedes the process of isomerizing mannose-6-phosphate into fructose-6-phosphate, which is necessary to produce several glycoproteins and glycolipids, which indicates that a significant down-regulation of glycoconjugate production occurs when the phosphomannose isomerase enzyme is inhibited [269].

An additional instance of the interaction between biogenic NPs and enzymes is the phosphotransferase enzyme arginine kinase, which is vital to *Trypanosoma brucei* growth and survival, especially in the host's bloodstream where it helps the parasite meet energy requirements to sustain a reductive environment [270, 271]. Silver and gold NPs offer encouraging antiparasitic qualities when it comes to *Trypanosoma*. This has been established in both in vivo and in vitro environments [272]. Silver nanoparticles (AgNPs) and AuNPs both showed significant action against *T. brucei*, causing a greater than 5% reduction in parasite vitality. According to these findings, NPs' anti-*Trypanosoma* activity is highly selective for the parasite, which may indicate that they have great potential for application against *T. cruzi* infection [273].

Biogenic NPs exhibit a distinct characteristic in that they can be synthesized in an extensive array of forms and morphologies, such as spheres, plates, pyramids, and other intricate shapes. These shapes serve to enhance the structural complexity of proteins [274] since the size and shape of the NPs determine their biodistribution, cellular uptake, and biodegradation [275]. Recent developments in nanomaterials provide a new way to control protein behavior through surface interaction. For example, according to a recent study, ZnO nanopyramids bind to the enzyme without denaturing the protein, effectively inhibiting the enzyme's activity more than any other shape of ZnO NP [276]. This is because the enzyme can attach to the pyramid form since it can have numerous edges or vertices that function as active sites on a single particle. This implies that protein binding and enzyme inhibition might both play a role in the NP process. It also emphasizes the fact that the most important finding in this research is the quantity of "active sites" on a single particle [276].

Lastly, porins (OmpA and OmpB) and periplasmic peptide binding proteins are non-enzymatic proteins that bind with the NPs; porin binding can alter the passive porin structure to allow the NPs smaller than 10 nm in diameter to pass through. Tryptophanase, cytochrome C, and alcohol dehydrogenase enzymes are among the enzymes that have a high affinity with biogenic NPs [263].

We can therefore conclude that, although NP-mediated enzyme inhibition presents some difficulties and hazards, including toxicity, environmental damage, and organ damage, it is a viable therapy option for protozoan infections and other disorders. Thus, extra investigation is required to comprehend the mechanism and enhance the formulation and dispersion of NPs for antiprotozoal uses.

12.5 ANTI-PROTOZOAN POTENTIAL OF SPECIFIC BIOGENIC NANOPARTICLES

12.5.1 Silver Nanoparticles (AgNPs)

Due to the limitations shown by conventional drugs in the treatment of diseases caused by protozoans, there is a need for an alternative treatment [277]. Given their antiviral, anticancer, antibacterial, and antiparasitic properties, biogenic silver NPs, or Bio-AgNPs, seem to be a suitable substitute. They act by inhibiting the growth of cells and encouraging intracellular ROS, DNA damage, and apoptosis [278]. The size, shape, composition, zeta potential, and surface charge of these NPs have all been thoroughly investigated and characterized. Bio-AgNPs have shown superior performance, stability, and higher biocompatibility, as well as reduced toxicity when compared to their chemically manufactured counterparts [279, 280]. Because of these positive attributes, Bio-AgNPs are regarded as an effective agent against protozoan parasites, which can cause devastating infections in humans, such as *Leishmania amazonensis* [277], *T. gondii* [281], *Cryptosporidium* sp. [282], *Plasmodium falciparum* [283], and *Eimeria papillata* [284].

In their 2018 study, Fanti et al. investigated the efficacy of Bio-AgNPs derived from *F. oxysporum* against *L. amazonensis* promastigote and amastigote stages. An event approximating apoptosis was caused by the protozoa treated with the NPs, as evidenced by an increase in ROS formation. The Bio-AgNP also resulted in a loss of mitochondrial integrity and elevated exposure to phosphatidylserine, a hallmark of an event akin to apoptosis. Additionally, promastigotes' membranes were damaged by bio-AgNPs. Furthermore, the number of amastigotes per macrophage, the percentage of infected macrophages, and the promastigotes' recovery from both infected and treated cells were all decreased by the introduction of biogenic AgNPs [277].

Biogenic AgNPs made using this same fungus have also been found to be effective against Tachyzoites of *T. gondii* [279–281]. Currently, Bio-AgNPs have been reported for significant reduction of both the number of infected cells and the amount of tachyzoites attached to HeLa cells. Additionally, it promoted the suppression of parasite intracellular multiplication by up to 94% [281].

According to a review by Arruda da Silva Sanfelice et al. (2022), innate immune cells use pattern recognition receptors on their surface to identify pathogen-associated molecular patterns, which in turn help the host immune response regulate *T. gondii* infection. *T. gondii* has, however, evolved methods for managing and controlling the host's immune response [181]. To address this issue, Bio-AgNPs have markedly raised IL-6, a cytokine that is crucial for the resistance to encephalitis linked to *T. gondii*. The Bio-AgNPs also caused *T. gondii* tachyzoites' mitochondrial membrane to depolarize and increased the amount of total ROS produced in infected HeLa cells. In addition, the therapy caused exposure to phosphatidylserine and an autophagic mechanism that damaged the parasite membrane [279].

In the search for new antimalarial and antitrypanosomal drugs to overcome the global reappearance of malaria and sleeping sickness resulting from resistance shown by *P. falciparum* and *T. brucei* to conventional antiplasmodial and antitrypanosomal therapeutic drugs, biogenic AgNPs synthesized from the leaves extract of two species of Malvaceae family (*Gossypium barbadense* and *Gossypium hirsutum*) and screened them in vitro against *P. falciparum* strain 3D7 of the malaria parasite and *Trypanosoma* parasites. The synthesized Bio-AgNPs strongly reduced the viability of the malaria parasite to less than 20%, with no sign of cytotoxicity on treated HeLa cells. However, they were ineffective against *Trypanosoma* parasites [283].

The protective efficacy of bio-AgNPs produced from *Zingiber officinale* rhizomes on mice infected with *E. papillata*, a coccidian parasite in house mice, was assessed [284]. Sporozoites released by *Eimerian* oocysts multiply in the jejunal mucosa and induce severe intestinal mucosal inflammation. These sporozoites intensify oxidative conditions and ultimately release their oocysts in excrement [284, 285]. This study found that Bio-AgNPs improved the jejunal mucosa and increased the number of goblet cells, which generate mucus that serves as a barrier to immune system defense against parasites. Furthermore, biogenic AgNPs reduced the number of oocysts that were forming in the jejuna and this resulted in less oocyst shedding in feces [284].

By comparing the antileishmanial activity of bio-AgNPs made from *Teucrium stocksianum* leaf extracts to chemically synthesized AgNPs, Ullah et al. (2018) showed that bio-AgNPs are more effective than chemically produced AgNPs. After 24 hours of treatment, their research revealed that biogenic AgNPs had a higher level of inhibition (98% inhibition) of *Leishmania infantum* promastigotes than chemically generated NPs (95% inhibition). They also found that bio-AgNPs are more effective at a lower IC50 value (IC50 30.71 ± 1.91 μg/mL) than chemically synthesized AgNPs (IC50 51.23 ± 2.2 μg/mL). According to the study, the bio-AgNPs' smaller size (10–15 nm) in comparison to chemically manufactured NPs (10–40 nm) may be the mechanism underlying the rise in antileishmanial activity. Following the protozoan exposure to the Bio-NPs, the capacity of *Leishmania* promastigotes to infect macrophage cells cultivated in vitro was lost. They proposed that by generating large amounts of nitric oxide (NO), NPs altered the parasite's structure, chemical makeup, or surface appearance [286]. A summary of the antiprotozoal activities of biogenic NPs is presented in Table 12.1.

12.5.2 Gold Nanoparticles

A lot of interest has been shown in gold NPs, or AuNPs, because of their incredibly small size, biocompatibility, high reactivity with bioactive molecules, stability over a wide temperature range, and potential use in a range of biomedical applications, including drug development and detection methods against a variety of species of interest to the public health [287]. Antiprotozoal activity of biogenic AuNPs against a range of parasites has been demonstrated, including *Leishmania* spp. [288–291], *Plasmodium* [292], and *Schistosoma* [287].

The typical process for synthesizing AuNP entails reducing tetrachloroauric acid using sodium borohydride/citrate, and stabilizers such as cetyltrimethylammonium bromide (CTAP) [289]. However, biogenic gold NPs (Bio-AuNPs) are favored over chemically fabricated AuNPs because of the effect of toxic reactants such as CTAB on biocompatibility [293].

Biogenic AuNPs have been synthesized using Bioactives of plant origin such as flavonoids [288] and terpenoids [289] which on their own possess both antiprotozoal

TABLE 12.1
Biogenic Metallic Nanoparticles and Their Antiprotozoal Activity

Metal NP	Biological Source	Specie	Size (nm)	Shape	Protozoan Inhibited	Reference
AgNP	Fungi	*Fusarium oxysporum*	57.6	Spherical	*Leishmania amazonensis*	[277]
AgNP	Fungi	*F. oxysporum*	20–50	Spherical	*Toxoplasma gondii*	[281]
AgNP	Fungi	*Aspergillus niveus*	27	Oval, cubic, and rod	*Cryptosporidium parvum*	[282]
AgNP	Fungi	*F. oxysporum*	20–50	Spherical	*T. gondii*	[298]
AgNP	Plant extract	*Gossypium barbadense* and *Gossypium hirsutum*	21	Irregular (spherical, triangular, pentagonal, and prismatic pattern)	*Plasmodium falciparum*	[283]
AgNP	Phytochemicals	*Sargentodoxa cuneata*	3–8	Spherical	*Leishmania tropica*	[290]
AgNP	Aqueous extract of plant	*Teucrium stocksianum*	10–15	Cubic	*Leishmania infantum*	[286]
AgNP	Rhizomes	*Zingiber officinale*	<50	Spherical	*Eimeria papillata*	[284]
AgNP	Fungi	*F. oxysporum*	65	Spherical	*T. gondii*	[279]
AgNP	Fungi	*F. oxysporum*	65 (±6.8)	Spherical	*T. gondii*	[280]
AuNP	Marine actinobacteria	*Streptomyces* sp. LK-3	5–50	Polygonal	*Plasmodium berghei*	[292]
AuNP	Phytochemicals	*S. cuneata*	15–30	Hexagonal	*L. tropica*	[290]
AuNP	Ethanol extract of macroalgae	*Cystoseira myrica*	12.6–15.5	Spherical	*Schistosoma mansoni*	[287]
AuNP	Plant oil	*Sesamum radiatum*	100	Plain	*Leishmania Mexican* and *Leishmania amazonensis*	[299]
ZnONP	Eug terpenoid of plant	*Syzygium aromaticum*	109.6	Spherical	*T. gondii*	[294]
ZnONP	Seeds extract	*Nigella sativa*	50–200	Spherical	*Eimeria tenella*	[296]
ZnONP	Aqueous extract	*Aquilegia pubiflora*	34.23	Spherical or elliptical	*L. tropica*	[297]

activity and show lesser risk of resistance development. Gold NPs with a terpenoid, andrographolide were created and their product's effectiveness was assessed against drug-resistant and wild strains of visceral leishmaniasis. The value of the inhibitory concentration was compared to the well-known anti-leishmanial medications amphotericin B, sodium stibogluconate (SSG), and paromomycin (PMM). Biogenic AuNP exhibited a strong anti-leishmanicidal effect against wild-type and resistance strains. The terpenoid compound's capacity to modify the fluidity of protozoal cell membranes, the biogenic AuNPs' size-guided quick macrophage entrance, and the biogenic AuNPs' negative charge were all cited as reasons for this. Furthermore, the cytotoxicity profile showed that biogenic AuNPs are far less dangerous than anti-leishmanial medications [289].

In a different study, 7,8 dihydroxyflavone (DHF), a naturally occurring flavonoid derived from plants, was used to create biogenic AuNPs. Instead of using sodium citrate, which is more often utilized in chemical synthesis, DHF was utilized to produce AuNP with greatly improved biocompatibility without compromising its biological efficacy. DHF-AuNP was utilized to inhibit *Leishmania donovani* by preventing the synthesis of arginase, the initial enzyme in the polyamine production pathway. The inhibition of arginase has been shown through enzymatic investigations to be a potential cause of parasite death in vitro. Subsequent research verified that DHF-AuNP eliminates promastigote by reducing MDR-1 transporter expression and eliminates amastigote by triggering the host's inducible nitric oxide synthase (iNOS) [288].

Microalgae can function as a bionanofactory for the synthesis of AuNPs as they are abundant in metabolites like phycobilin and carotenoids. Because algae have so many benefits, including high metal adsorption, minimal cytotoxicity, and easy availability, biosynthesis of NPs using algae is valuable and appealing. Biogenic AuNPs made with an ethanolic extract of microalgae have been shown to have antibacterial, anti-inflammatory, and schistolarvicidal properties [287].

It has been demonstrated that the size, shape, and charge of the NPs all affect how well Biogenic AuNP is absorbed. In this sense, the easiest and most energetically preferred entrance mechanism for macrophages is endocytosis [288]. Effective inhibitors of metabolically significant enzymes like trypanothione reductase, and gold-containing compounds are crucial in preventing oxidative damage to *Leishmania*. Thus, it is regarded as a crucial biogenic AgNP antiprotozoal mechanism [290].

12.5.3 Zinc Oxide Nanoparticles

Another interesting inorganic nanostructure with a number of advantages is zinc oxide NPs, or ZnONPs. They are useful in many different areas, including health, electronics, cosmetics, catalysis, and energy storage, because of their remarkable optical, semiconducting, piezoelectric, and chemical sensing properties. Their biological applications are very many, and they include antibacterial, antifungal, antioxidant, anticancer, and antiparasitic properties [294].

Green production of ZnO-NPs based on fungus, plant extracts, and microorganisms like bacteria and yeast may be a good substitute for antiparasitic medications, just like it has with other biogenic NPs [295]. This is consistent with research conducted by Lail et al. [296] on broilers infected with *Eimeria tenella*, wherein the impact of bio-ZnO-NP made using *Nigella sativa* was assessed. When supplemented with biogenic ZnONPs, broilers infected with *E. tenella* exhibited anticoccidial activity as evidenced by a decrease in oocyst shedding and an improvement in growth performance. By dramatically boosting the activity of superoxide dismutase and catalase and lowering the quantity of pro-inflammatory cytokines interleukins-2 and tumor necrosis factor-α, biogenic zinc NPs (ZnNPs) also reduced tissue damage and showed antioxidant and anti-inflammatory potential [296].

A different study found that biogenic ZnO-NPs, which were created using the aqueous extract of Himalayan columbine (*Aquilegia pubiflora*), were fatal to *L. tropica* in both the promastigote and amastigote stage in a dose-dependent manner. Additionally, even at high concentrations, this NP demonstrated exceptional hemocompatibility, suggesting that it may find value in biomedical applications [297].

In another work, allylbenzene, an edible, harmless volatile naturally occurring terpenoid known for its antibacterial, anticancer, and anti-inflammatory properties, was used to synthesize biogenic ZnONPs. High anti-toxoplasma activity against *T. gondii* was demonstrated by this nanoformulation in both in vitro and in vivo studies. For up to 14 days, infected mice given this biogenic ZnO-NPs treatment were able to survive. Very few Tachyzoites were found in the treated mice's brain, liver, spleen, and peritoneal fluid. Furthermore, the Bio-ZnONP formulation's cytotoxicity on Vero cells had a dose-dependent state, meaning that as the chemical concentration increased, cell viability dropped [294].

12.6 CHALLENGES AND CONSIDERATIONS IN THE APPLICATION OF BIOGENIC NANOPARTICLES

12.6.1 Biocompatibility and Toxicity

Biogenic NPs have demonstrated significant promise in a range of biomedical domains, encompassing their application as an antiprotozoal agent. Humans need these successful, affordable, environmentally friendly biogenic NPs to be commercialized and used in therapeutic settings, but it's important to consider their possible toxicity and adverse effects.

The majority of studies have shown that biogenic NPs exhibit less cytotoxicity than their counterparts that are chemically generated [286]. For instance, it was discovered that very stable Bio-AuNPs made from non-toxic *Curcuma mangga* extract were cytocompatible with human lung fibroblast cells (MRC-5) and human colon fibroblast cells (CCD-18Co). According to hemocompatibility testing, the AuNPs exhibited less than 10% hemolysis and no erythrocyte aggregation, indicating their blood compatibility [293].

While the majority of research has shown that bio-NPs are competent, safe, and effective, several investigations contradict this conclusion. Toxicity evaluation of AgNPs synthesized from the extract of *Althaea officinalis* in vitro assay with different cell lines (V79 and A549) showed death by necrosis in 98% of the cells at a concentration of 3 × 108 NPs/mL. The smaller size of the NP might be related to its higher potential for cytotoxicity. The biomolecule that was utilized to create the NPs could be another significant element. In the same study, AgNPs synthesized from root infusion of the same plant caused a great number of DNA breaks in both A549 and HeLa cells at the same concentration (3 × 108 NPs/mL) [300].

According to this, Vasanth et al. (2017) evaluated the nephrotoxicity of AgNPs prepared using renal mitochondria, an aqueous extract of *Desmodium gangeticum* root, and rat proximal epithelial cell line (LLC PK1). In addition to causing oxidative stress and mitochondrial dysfunction, Bio-AgNPs were found to produce significant alterations in the architecture of the kidneys [301]. Overall, the results showed that precautions should be taken when exposed to biogenic AgNPs for a prolonged amount of time.

Studies revealed that size [286], shape [302], and composition of Bio-NPs are characteristics that determine their biotoxicity. Thus, adjusting these factors could lead to the creation of biogenic metallic NPs that have better interactions, biological activity at a target location, and little cytotoxicity [278].

12.6.2 Stability and Storage

The amount of time that NPs may maintain the intended characteristics and functionalities is known as their shelf life. The type of metal utilized in biogenic particle production, the capping agent, the storage conditions, and the stability of the NPs are some of the elements that impact the shelf life of biogenic NPs. Since we can modify the NP's surface to get better results, biogenic NPs are known to be highly interesting due to their unique physicochemical properties [303]. For instance, iron oxide NPs can be used in MRI and imaging because of their ultrafine size, magnetic properties, and biocompatibility [304].

The capping agent utilized throughout the manufacturing process is another technique to extend the NP's shelf life. Capping agents are molecules that control the size, shape, and surface chemistry of NPs, which is essential for their stabilization. It consists of polymers, organic ligands, and surfactants [305, 306]. The purpose of capping agents in the synthesis of biogenic NPs is to lessen the negative aspects of conventional NP production, such as toxicity, size, and agglomeration as the NP [305]. Capping agents interact with the NPs by forming covalent bonds between the chains of capping ligands and the surface of the NPs, which results in steric hindrance that gives the NP ultimate stability and reduces NP agglomeration over an extended time [307].

To prevent the NP from aggregating and stabilizing, it needs to be attached to a capping agent through a process known as surface modification. The capping agent, or attached ligand, can be any of the following: a thiol group, disulfide ligand, phosphines, or many others. It usually adheres to the NP to improve control over its size based on the concentration of the ligand [308]. The ligand is chosen to maximize the intended function of the NP, as recent research has demonstrated that interaction between the ligand and the NP results in good dispersion and high stability, while bare NPs without ligands tend to aggregate due to electrostatic repulsion [309, 310]. The organosulfur compounds of thiols and disulfides, which bind strongly to the MNP to control the NP size, are examples of these interactions and applications. Another illustration is the process by which long-chain ammonium ions are attached to hydrophobic MNPs to generate hydrophilic MNPs. Lastly, we affix amine ligands to MNPs in order to stabilize the particle and enhance its dispersion [311].

In conclusion, the properties of MNPs can be altered to achieve specific goals by attaching various capping agents. However, these objectives could have adverse consequences on exposed tissues and cells. Capping agents affect the stability of the biological production of NPs, which is important for the targeting and functionality of NPs in biomedical applications. Therefore, it is important to fully understand their impacts.

Stability is one of the main issues with NPs. Long-term application investigations require NP colloidal stability. To indirectly evaluate the stability and surface charge of colloidal NPs, a Zetasizer nanomachine is used to conduct a zeta potential investigation. Zeta potential analysis is used to represent the potential difference between the surface and the outer Helmholtz plane. Large positive or negative zeta potential values must be reached to ensure stability and stop particle aggregation [312]. We need to comprehend the zeta potential analysis mechanism to comprehend how the stability of NPs is determined.

Agglomeration is more likely to occur when electrostatic repulsive forces are absent or at the isoelectric point (0 mV). If an NP-containing solution switches to a lower ionic strength, the zeta potential will rise because the NP tends toward dispersion as the electrical double layer (EDL) grows to counteract electrostatic repulsive and attractive forces. However, the zeta potential drops when the number of ions is excessive, which might cause NPs to combine and precipitate [313]. Many properties of NPs, chief among them their size, influence agglomeration (reactivity with other molecules) [313]. However, because of their higher surface area to volume ratio, which raises electrochemical reactivity, smaller NPs are generally more hazardous.

The storage environment is the final need to extend the shelf life of NPs. Even in the best of circumstances, many NPs are likely to undergo property changes while being preserved, and they may also "age" or change with time [314–316]. Such temporal variations in the characteristics of NPs might modify their toxicity by changing their aggregation state, dissolving rate, or bioavailability, which can affect administered vs. delivered dose disparities [315–318]. To find out how time affected the MNPs, a study was carried out. Immediately following NP synthesis (T0) and six months later (T6: 4°C-D, T6: RT-D, and T6: RT-DL), the entire amount of MNPs and metal+ discharged was measured by ICP-MS. MNPs at time zero and six months post-storage. Six months were spent storing MNP samples with varying charges and surface compositions at T = 4°C without light exposure (T6: 4°C-D); T = 21°C without light exposure (T6: RT-D); and T = 21°C with light exposure (T6: RT-DL). The difference between the Ag measured at time 0 and 6 months in the extracted fraction was used to compute the amount of liberated Ag ions [319].

The outcome demonstrated that positive-charged MNPs such as chitosan and BYK were observed to disrupt the solid phase of extracted columns, making it impossible to determine the number of silver ions released using this technique. Longer storage times for negatively charged NPs often led to higher metal ion concentrations; however, storage temperature did not seem to affect the release of metal ions.

More than 90% of the silver was dissolved in the metal citrate that was kept at ambient temperature and exposed to daylight, according to the effect of sunshine on the NPs. When MNPs dissolve normally, they reach an equilibrium state where the rates of re-precipitation and dissolution balance out, and further dissolution of the MNPs comes to an end. When free metal ions precipitate due to exposure to light, equilibrium cannot be attained, and the dissolution of MNPs is unrestrained [319]. This experiment demonstrates how time and variables like temperature and light availability affect the physicochemical characteristics of NPs.

As a result of the shelf-life investigations conducted on AgNPs and ZnNPs, it was possible to keep NPs for six months at temperatures below freezing without significantly reducing their bioefficacy. The bioefficacy results shown here are in complete harmony with those of Hosamani Goutam et al. (2019) [320], who discovered that silver NPs remained bioeffective against *Spodoptera litura* and *Callosobruchus chinensis* for a maximum of three months while stored in deep freezing. Furthermore, in complete accordance with the results obtained [321], Sangeetha et al.'s (2014) [322] studies on the synthesis of NPs from Hing exudates demonstrated stability against agglomeration for up to three months.

12.7 CONCLUSION AND FUTURE PERSPECTIVES

Globally, neglected tropical parasites are the primary source of significant disease outbreaks. Numerous techniques have been used to cure parasites over the past 50 years, but they have not been successful. Chemotherapeutic drugs have been overprescribed and should not be utilized for a number of reasons. The physiology of the host can occasionally be harmed by secondary infections, and parasites have developed immunity, indicating the need for alternate therapies. Although ethnobotanicals have shown potential, they are difficult to fund and are not attractive to investors; as a result, barely 1% of ethnobotanicals have been developed in the last 50 years. As a result, using them to treat parasites is made impossible by a lack of understanding regarding dosage information. The other alternative for ethnobotanicals is the development of vaccines, although this is a challenging challenge in parasitology because of uniform antigenic variation. Parasites create defense mechanisms against their host's immune system. Consequently, the application of NPs for parasite control has been studied by parasitologists. NPs possess unique properties, such as their lack of toxicity and parasite resistance, that set them apart from conventional treatment alternatives. Most studies on using NPs to manage parasites have concentrated on protozoa like *Plasmodium*, *Trypanosoma*, and *Leishmania*. It has been shown that NPs produced using a variety of conventional and molecular techniques can either kill or limit the growth of parasites, and they can be used to identify disorders caused by parasites. While gold and silver NPs have garnered the most interest, research is currently ongoing to determine the effectiveness of other NPs, including iron, nickel, platinum, and copper, in preventing disease in both humans and animals. In the pharmaceutical and medical industries, nanotechnology represents the latest technical breakthrough. The application of nanotools for the treatment and prevention of infectious diseases has advanced significantly in the last few decades in the pharmaceutical and medical sectors. To create safe diagnostic and treatment alternatives, more research is necessary to fully comprehend the mechanisms of action of NPs. Significant progress in the management and treatment of parasites is anticipated with the use of NPs.

REFERENCES

1. Thomas M, Klibanov AM. Conjugation to gold nanoparticles enhances polyethylenimine's transfer of plasmid DNA into mammalian cells. Proceedings of the National Academy of Sciences of the United States of America. 2003 Aug 5;100(16):9138–43. doi: 10.1073/pnas.1233634100. Epub 2003 Jul 28. PMID: 12886020; PMCID: PMC170885.
2. Erkizan HV, Kong Y, Merchant M, Schlottmann S, Barber-Rotenberg JS, Yuan L, Abaan OD, Chou TH, Dakshanamurthy S, Brown ML, Uren A, Toretsky JA. A small molecule blocking oncogenic protein EWS-FLI1 interaction with RNA helicase A inhibits growth of Ewing's sarcoma. Nature Medicine. 2009 Jul;15(7):750–56. doi: 10.1038/nm.1983. Epub 2009 Jul 5. PMID: 19584866; PMCID: PMC2777681.
3. Wang L, He H, Yu Y, Sun L, Liu S, Zhang C, He L. Morphology-dependent bactericidal activities of Ag/CeO_2 catalysts against Escherichia coli. Journal of Inorganic Biochemistry. 2014 Jun;135:45–53. doi: 10.1016/j.jinorgbio.2014.02.016. Epub 2014 Mar 11. PMID: 24662462.
4. da Silva Sanfelice RA, Silva TF, Tomiotto-Pellissier F, da Silva Bortoleti BT, Lazarin-Bidóia D, Scandorieiro S, Nakazato G, de Barros LD, Garcia JL, Verri WA, Conchon-Costa I, Pavanelli, WR, Costa IN. Biogenic silver nanoparticles reduce Toxoplasma gondii infection and proliferation in RAW 264.7 macrophages by inducing tumor necrosis factor-alpha and reactive oxygen species production in the cells. Microbes and Infection 2022;24(5):104971.
5. Mukha IP, Eremenko AM, Smirnova NP, Mikhienkova AI, Korchak GI, Gorchev VF, Chunikhin AI. Antimicrobial activity of stable silver nanoparticles of a certain size. Prikladnaia Biokhimiia i Mikrobiologiia. 2013 Mar-Apr;49(2):215–23. Russian. doi: 10.7868/s0555109913020128. PMID: 23795483.
6. Tang S, Zheng J. Antibacterial activity of silver nanoparticles: Structural effects. Advanced Healthcare Materials. 2018 Jul;7(13):e1701503. doi: 10.1002/adhm.201701503. Epub 2018 May 29. PMID: 29808627.
7. Klueh U, Wagner V, Kelly S, Johnson A, Bryers JD. Efficacy of silver-coated fabric to prevent bacterial colonization and subsequent device-based biofilm formation. Journal of Biomedical Materials Research. 2000;53(6):621–31. doi: https://pubmed.ncbi.nlm.nih.gov/11074419/. PMID: 11074419.
8. Rai MK, Deshmukh SD, Ingle AP, Gade AK. Silver nanoparticles: The powerful nanoweapon against multidrug-resistant bacteria. Journal of Applied Microbiology. 2012 May;112(5):841–52. doi: 10.1111/j.1365-2672.2012.05253.x. Epub 2012 Mar 28. PMID: 22324439.
9. Liu N, Chen X-G, Park H-J, Liu C-G, Liu C-S, Meng X-H, et al (2006) Effect of MW and concentration of chitosan on antibacterial activity of *Escherichia coli*.
10. Caffrey CR, Mathieu MA, Gaffney AM, Salter JP, Sajid M, Lucas KD, Franklin C, Bogyo M, McKerrow JH. Identification of a cDNA encoding an active asparaginyl endopeptidase of *Schistosoma mansoni* and its expression in *Pichia pastoris*. FEBS Letters. 2000;466:244–48. doi: 10.1016/S0014-5793(99)01798-6.
11. Dalton JP, Brindley PJ, Donnelly S, Robinson MW. The enigmatic asparaginyl endopeptidase of helminth parasites. Trends in Parasitology. 2009;25:59–61. doi: 10.1016/j.pt.2008.11.002.
12. Smith AM, Dowd AJ, McGonigle S, Keegan PS, Brennan G, Trudgett A, Dalton JP. Purification of a cathepsin L-like proteinase secreted by adult *Fasciola hepatica*. Molecular and Biochemical Parasitology. 1993;62:1–8. doi: 10.1016/0166-6851(93)90171-S.
13. Lok CN, Ho CM, Chen R, He QY, Yu WY, Sun H, Tam PK, Chiu JF, Che CM. Proteomic analysis of the mode of antibacterial action of silver nanoparticles. Journal of Proteome Research. 2006 Apr;5(4):916–24. doi: 10.1021/pr0504079. PMID: 16602699.

14. Garami A, Ilg T. The role of phosphomannose isomerase in Leishmania *mexicana* glycoconjugate synthesis and virulence. Journal of Biological Chemistry. 2001 Mar 2; 276(9):6566–75. doi: 10.1074/jbc.M009226200. Epub 2000 Nov 17. PMID: 11084042.
15. Abou Elez RMM, Attia ASA, Tolba HMN, Anter RGA, Elsohaby I. Molecular identification and antiprotozoal activity of silver nanoparticles on viability of *Cryptosporidium parvum* isolated from pigeons, pigeon fanciers and water. Scientific Reports. 2023 Feb 22;13(1):3109. doi: 10.1038/s41598-023-30270-2. PMID: 36813872; PMCID: PMC9946930.
16. Volpedo G, Costa L, Ryan N, Halsey G, Satoskar A, Oghumu S. Nanoparticulate drug delivery systems for the treatment of neglected tropical protozoan diseases. Journal of Venomous Animals and Toxins Including Tropical Diseases. 2019;25:e144118.
17. Adeyemi OS, Molefe NI, Awakan OJ, Nwonuma CO, Alejolowo OO, Olaolu T, Maimako RF, Suganuma K, Han Y, Kato K. Metal nanoparticles restrict the growth of protozoan parasites. Artificial Cells, Nanomedicine, and Biotechnology. 2018;46:S86–94.
18. Brito TK, Viana RLS, Moreno CJG, Barbosa JDS, Júnior FLDS, de Medeiros MJC, Melo-Silveira RF, Almeida-Lima J, Pontes DDL, Silva MS, et al. Synthesis of silver nanoparticle employing corn cob xylan as a reducing agent with anti-*Trypanosoma cruzi* activity. International Journal of Nanomedicine. 2020;15:965–79.
19. Arrúa EC, Seremeta KP, Bedogni GR, Okulik NB, Salomon CJ. Nanocarriers for effective delivery of benznidazole and nifurtimox in the treatment of chagas disease: A review. Acta Tropica. 2019;198:105080.
20. Cha SH, Hong J, McGuffie M, Yeom B, VanEpps JS, Kotov NA. Shape-dependent biomimetic inhibition of enzyme by nanoparticles and their antibacterial activity. ACS Nano. 2015 Sep 22;9(9):9097–105. doi: 10.1021/acsnano.5b03247. Epub 2015 Sep 1. PMID: 26325486.
21. Mortimer M, Wang Y, Holden PA. Molecular mechanisms of nanomaterial-bacterial interactions revealed by omics – The role of nanomaterial effect level. Frontiers in Bioengineering and Biotechnology. 2021 Jun 14;9:683520. doi: 10.3389/fbioe.2021.683520. PMID: 34195180; PMCID: PMC8236600.
22. Kadiyala U, Kotov NA, VanEpps JS. Antibacterial metal oxide nanoparticles: Challenges in interpreting the literature. Current Pharmaceutical Design. 2018;24(8):896–903. doi: 10.2174/1381612824666180219130659. PMID: 29468956; PMCID: PMC5959755.
23. Morones JR, Elechiguerra JL, Camacho A, Holt K, Kouri JB, Ramírez JT, Yacaman MJ. The bactericidal effect of silver nanoparticles. Nanotechnology. 2005 Oct;16(10): 2346–53. doi: 10.1088/0957-4484/16/10/059. Epub 2005 Aug 26. PMID: 20818017.
24. Jung WK, Koo HC, Kim KW, Shin S, Kim SH, Park YH. Antibacterial activity and mechanism of action of the silver ion in *Staphylococcus aureus* and *Escherichia coli*. Applied and Environmental Microbiology. 2008 Apr;74(7):2171–78. doi: 10.1128/AEM.02001-07. Epub 2008 Feb 1. PMID: 18245232; PMCID: PMC2292600.
25. jahnke JP, Cornejo JA, Sumner JJ, Schuler AJ, Atanassov P, Ista LK. Conjugated gold nanoparticles as a tool for probing the bacterial cell envelope: The case of *Shewanella oneidensis* MR-1. Biointerphases. 2016;11:011003.
26. Holt KB, Bard AJ. Interaction of silver(I) ions with the respiratory chain of *Escherichia coli*: An electrochemical and scanning electrochemical microscopy study of the antimicrobial mechanism of micromolar Ag+. Biochemistry. 2005 Oct 4;44(39):13214–23. doi: 10.1021/bi0508542. PMID: 16185089.
27. Kumar A, Pandey AK, Singh SS, Shanker R, Dhawan A. Engineered ZnO and TiO(2) nanoparticles induce oxidative stress and DNA damage leading to reduced viability of Escherichia coli. Free Radical Biology and Medicine. 2011 Nov 15;51(10):1872–81. doi: 10.1016/j.freeradbiomed.2011.08.025. Epub 2011 Aug 28. PMID: 21920432.
28. Sudheer Khan S, Bharath Kumar E, Mukherjee A, Chandrasekaran N. Bacterial tolerance to silver nanoparticles (SNPs): *Aeromonas punctata* isolated from sewage

environment. Journal of Basic Microbiology. 2011 Apr;51(2):183–90. doi: 10.1002/jobm. 201000067. Epub 2010 Nov 12. PMID: 21077112.
29. Hsueh YH, Lin KS, Ke WJ, Hsieh CT, Chiang CL, Tzou DY, Liu ST. The antimicrobial properties of silver nanoparticles in *Bacillus subtilis* are mediated by released Ag+ ions. PLoS One. 2015 Dec 15;10(12):e0144306. doi: 10.1371/journal.pone.0144306. PMID: 26669836; PMCID: PMC4682921.
30. Kumar N, Das S, Jyoti A, Kaushik S. Synergistic effect of silver nanoparticles with doxycycline against *Klebsiella pneumoniae*. International Journal of Pharmacy and Pharmaceutical Sciences. 2016;8(7):183–86.
31. Ishida T. Anticancer activities of silver ions in cancer and tumor cells and DNA damages by Ag+-DNA base-pairs reactions. MOJ Tumor Research. 2017;1(1):8–16.
32. Yakabe Y, Sano T, Ushio H, Yasunaga T. Kinetic studies of the interaction between silver ion and deoxyribonucleic acid. Chemistry Letters. 1980;4:373–76.
33. Donaldson K, Murphy FA, Duffin R, Poland CA. Asbestos, carbon nanotubes and the pleural mesothelium: A review of the hypothesis regarding the role of long fibre retention in the parietal pleura, inflammation and mesothelioma. Particle and Fibre Toxicology. 2010 Mar 22;7:5. doi: 10.1186/1743-8977-7-5. PMID: 20307263; PMCID: PMC2857820.
34. Rajamanikam A, Kumar S, Samudi C, Kudva M. Exacerbated symptoms in *Blastocystis* sp.-infected patients treated with metronidazole: Two case studies. Parasitology Research. 2018;117:2585–90. doi: 10.1007/s00436-018-5948-x.
35. Choi O, Hu Z. Size dependent and reactive oxygen species related nanosilver toxicity to nitrifying bacteria. Environmental Science & Technology. 2008;42:4583–88. doi: 10.1021/es703238h.
36. Lahiri D, Nag M, Sheikh HI, Sarkar T, Edinur HA, Pati S, Ray RR. Microbiologically-synthesized nanoparticles and their role in silencing the biofilm signaling cascade. Frontiers in Microbiology. 2021;12:636588.
37. Eom HJ, AhN JM, Kim Y, Choi J. Hypoxia inducible factor-1 (HIF-1)-flavin containing monooxygenase-2 (FMO-2) signaling acts in silver nanoparticles and silver ion toxicity in the nematode, *Caenorhabditis elegans*. Toxicology and Applied Pharmacology. 2013;270:106–13.
38. Leite PE, Pereira MR, do Nascimento SCA, Campos AP, Esteves TM, Granjeiro JM. Gold nanoparticles do not induce myotube cytotoxicity but increase the susceptibility to cell death. Toxicology in Vitro. 2015;29:819–27.
39. Yang T, Cao F, Liu Q, Wang X, Silver nanoparticles inhibit hypoxia inducible factor function and cancer cell growth. Nanomedicine: Nanotechnology, Biology and Medicine. 2016;12:554.
40. Blader IJ, Saeji JP. Communication between *Toxoplasma gondii* and its host: Impact on parasite growth, development, immune evasion and virulence. APMIS. 2009;111: 458–76.
41. Semenza GL. Hypoxia-inducible factor 1 (HIF-1) pathway. Science's STKE. 407 (2007) (cm8).
42. Zinkernagel AS, Johnson RS, Nizet V. Hypoxia-inducible factor (HIF) function in innate immunity and infection. Journal of Molecular Medicine. 2007;85:1339–46.
43. Spear W, Chan D, Coppens I, Johnson RS, Giaccia A, Blader IJ, The host cell transcription factor hypoxia-inducible factor 1 is required for *Toxoplasma gondii* growth and survival at physiological oxygen levels. Cellular Microbiology. 2006;8:339–52.
44. Wang L, Hu C, Shao L. The antimicrobial activity of nanoparticles: Present situation and prospects for the future. International Journal of Nanomedicine. 2017 Feb 14;12:1227–49. doi: 10.2147/IJN.S121956. PMID: 28243086; PMCID: PMC5317269.
45. Su Y, Zheng X, Chen Y, Li M, Liu K. Alteration of intracellular protein expressions as a key mechanism of the deterioration of bacterial denitrification caused by copper oxide

nanoparticles. Scientific Reports. 2015 Oct 28;5:15824. doi: 10.1038/srep15824. PMID: 26508362; PMCID: PMC4623765.

46. Yamanaka M, Hara K, Kudo J. Bactericidal actions of a silver ion solution on *Escherichia coli*, studied by energy-filtering transmission electron microscopy and proteomic analysis. Applied and Environmental Microbiology. 2005 Nov;71(11):7589–93. doi: 10.1128/AEM.71.11.7589-7593.2005. PMID: 16269810; PMCID: PMC1287701.
47. Zou X, Li P, Lou J, Zhang H. Surface coating-modulated toxic responses to silver nanoparticles in *Wolffia globosa*. Aquatic Toxicology. 2017;189:150–58. doi: 10.1016/j.aquatox.2017.06.010.
48. Ghorbani M, Fekrvand S, Shahkarami S, Yazdani R, Sohani M, Shaghaghi M, Hassanpour G, Mohammadi J, Negahdari B, Abolhassani H, et al. The evaluation of neutropenia in common variable immune deficiency patients. Expert Review of Clinical Immunology. 2019, 15, 1225–33.
49. Jakubczyk K, Dec K, Kałduńska J, Kawczuga D, Kochman J, Janda K. Reactive oxygen species – Sources, functions, oxidative damage. Polski Merkuriusz Lekarski. 2020Apr 22;48(284):124–27. PMID: 32352946.
50. Keenan CR, Sedlak DL. Factors affecting the yield of oxidants from the reaction of nanoparticulate zero-valent iron and oxygen. Environmental Science & Technology. 2008 Feb 15;42(4):1262–67. doi: 10.1021/es7025664. PMID: 18351103.
51. Davies KJ. Oxidative stress, antioxidant defenses, and damage removal, repair, and replacement systems. IUBMB Life. 2000 Oct-Nov;50(4–5):279–89. doi: 10.1080/713803728. PMID: 11327322.
52. Murray HW. Susceptibility of *Leishmania* to oxygen intermediates and killing by normal macrophages. Journal of Experimental Medicine. 1981;153(5):1302–15.
53. Zorov DB, Juhaszova M, Sollott SJ. Mitochondrial reactive oxygen species (ROS) and ROS-induced ROS release. Physiological Reviews. 2014 Jul;94(3):909–50. doi: 10.1152/physrev.00026.2013. PMID: 24987008; PMCID: PMC4101632.
54. Asada K. Production and scavenging of reactive oxygen species in chloroplasts and their functions. Plant Physiology. 2006 Jun;141(2):391–96. doi: 10.1104/pp.106.082040. PMID: 16760493; PMCID: PMC1475469.
55. Villegas NA, Baronetti J, Albesa I, Polifroni R, Parma A, Etcheverría A, Becerra M, Padola N, Paraje M. Relevance of biofilms in the pathogenesis of Shiga-toxin-producing *Escherichia coli* infection. The Scientific World Journal. 2013 Nov 12;2013:607258. doi: 10.1155/2013/607258. PMID: 24324376; PMCID: PMC3845835.
56. Zhang D, Ma X-l, Gu Y, Huang H, Zhang G-w. Green synthesis of metallic nanoparticles and their potential applications to treat cancer. Frontiers in Chemistry. 2020 Oct 29;8:799. doi: 10.3389/fchem.2020.00799. PMID: 33195027; PMCID: PMC7658653.
57. Mehlotra RK. Antioxidant defense mechanisms in parasitic protozoa. Critical Reviews in Microbiology. 1996;22(4):295–314. doi: https://pubmed.ncbi.nlm.nih.gov/8989515/. PMID: 8989515.
58. Krumova and G. Cosa, in Singlet oxygen: Applications in biosciences and nanosciences, ed. S. Nonell, C. Flors, S. Nonell, and C. Flors, The royal society of chemistry: 2016, ch. 1, pp. 1–21. Royal Society of Chemistry.
59. Mourenza Á, Gil JA, Mateos LM, Letek M. Oxidative stress-generating antimicrobials, a novel strategy to overcome antibacterial resistance. Antioxidants (Basel). 2020 Apr 26;9(5):361. doi: 10.3390/antiox9050361. PMID: 32357394; PMCID: PMC7278815.
60. Risom L, Møller P, Loft S. Oxidative stress-induced DNA damage by particulate air pollution. Mutation Research. 2005;592:119–37. doi: 10.1016/j.mrfmmm.2005.06.012.
61. Knaapen AM, Borm PJ, Albrecht C, Schins RP. Inhaled particles and lung cancer. Part A: Mechanisms. International Journal of Cancer. 2004;109:799–809. doi: 10.1002/ijc.11708.
62. Vallyathan V, Shi X. The role of oxygen free radicals in occupational and environmental lung diseases. Environmental Health Perspectives. 1997;105:165–77. doi: 10.1289/ehp.97105s1165.

63. Bonner JC. Lung fibrotic responses to particle exposure. Toxicologic Pathology. 2007; 35:148–53. doi: 10.1080/01926230601060009.
64. Ray PC, Yu H, Fu PP. Toxicity and environmental risks of nanomaterials: Challenges and future needs. Journal of Environmental Science and Health, Part C. 2009;27:1–35. doi: https://www.tandfonline.com/doi/full/10.1080/10590500802708267.
65. Xia T, Kovochich M, Liong M, Mädler L, Gilbert B, Shi H, Yeh JI, Zink JI, Nel AE. Comparison of the mechanism of toxicity of zinc oxide and cerium oxide nanoparticles based on dissolution and oxidative stress properties. ACS Nano. 2008;2:2121–34. doi: 10.1021/nn800511k.
66. Shaligram S, Campbell A. Toxicity of copper salts is dependent on solubility profile and cell type tested. Toxicology in Vitro. 2013;27:844–51. doi: 10.1016/j.tiv.2012.12.026.
67. Lu W, Senapati D, Wang S, Tovmachenko O, Singh AK, Yu H, Ray PC. Effect of surface coating on the toxicity of silver nanomaterials on human skin keratinocytes. Chemical Physics Letters. 2010;487:92–96. doi: 10.1016/j.cplett.2010.01.027.
68. Buzea C, Pacheco II, Robbie K. Nanomaterials, and nanoparticles: Sources and toxicity. Biointerphases. 2007;2:MR17–71. doi: 10.1116/1.2815690.
69. Wilson MR, Lightbody JH, Donaldson K, Sales J, Stone V. Interactions between ultrafine particles and transition metals in vivo and in vitro. Toxicology and Applied Pharmacology. 2002;184:172–79. doi: 10.1006/taap.2002.9501.
70. Sioutas C, Delfino RJ, Singh M. Exposure assessment for atmospheric ultrafine particles (UFPS) and implications in epidemiologic research. Environmental Health Perspectives. 2005:947–55. doi: 10.1289/ehp.7939.
71. Stone V, Shaw J, Brown D, MacNee W, Faux S, Donaldson K. The role of oxidative stress in the prolonged inhibitory effect of ultrafine carbon black on epithelial cell function. Toxicology in Vitro. 1998;12:649–59. doi: 10.1016/S0887-2333(98)00050-2.
72. Fan J, Yin J-J, Ning B, Wu X, Hu Y, Ferrari M, Anderson GJ, Wei J, Zhao Y, Nie G. Direct evidence for catalase and peroxidase activities of ferritin-platinum nanoparticles. Biomaterials. 2011;32:1611–18. doi: 10.1016/j.biomaterials.2010.11.004.
73. Donaldson K, Tran CL. Inflammation caused by particles and fibers. Inhalation Toxicology. 2002;14:5–27. doi: https://pubmed.ncbi.nlm.nih.gov/12122558/.
74. Oberdörster G, Maynard A, Donaldson K, Castranova V, Fitzpatrick J, Ausman K, Carter J, Karn B, Kreyling W, Lai D. Principles for characterizing the potential human health effects from exposure to nanomaterials: Elements of a screening strategy. Particle and Fibre Toxicology. 2005;2 doi: 10.1186/1743-8977-2-8.
75. Roduner E. Size matters: Why nanomaterials are different. Chemical Society Reviews. 2006;35:583–92. doi: 10.1039/b502142c.
76. Thannickal VJ, Fanburg BL. Reactive oxygen species in cell signaling. The American Journal of Physiology-Lung Cellular and Molecular Physiology. 2000;279:L1005–28.
77. Valko M, Rhodes C, Moncol J, Izakovic M, Mazur M. Free radicals, metals and antioxidants in oxidative stress-induced cancer. Chemico-Biological Interactions. 2006; 160:1–40. doi: 10.1016/j.cbi.2005.12.009.
78. Aust S, Chignell C, Bray T, Kalyanaraman B, Mason R. Free radicals in toxicology. Toxicology and Applied Pharmacology. 1993;120:168–78. doi: 10.1006/taap.1993.1100.
79. Yaghini E, Pirker K, Kay C, Seifalian A, MacRobert A. Reactive oxygen species generation from photoexcited quantum dot nanoparticles: Type I versus type II photochemical mechanism. Photodiagnosis and Photodynamic Therapy. 2011;8:151. doi: 10.1016/j.pdpdt.2011.03.096.
80. Sharma P, Jha AB, Dubey RS, Pessarakli M. Reactive oxygen species, oxidative damage, and antioxidative defense mechanism in plants under stressful conditions. Journal of Botany. 2012;2012:217037. doi: 10.1155/2012/217037.
81. Iversen T-G, Skotland T, Sandvig K. Endocytosis and intracellular transport of nanoparticles: Present knowledge and need for future studies. Nano Today. 2011;6:176–85. doi: 10.1016/j.nantod.2011.02.003.

82. Canton I, Battaglia G. Endocytosis at the nanoscale. Chemical Society Reviews. 2012;41:2718–39. doi: 10.1039/c2cs15309b.
83. Soenen SJ, Rivera-Gil P, Montenegro J-M, Parak WJ, De Smedt SC, Braeckmans K. Cellular toxicity of inorganic nanoparticles: Common aspects and guidelines for improved nanotoxicity evaluation. Nano Today. 2011;6:446–65. doi: 10.1016/j.nantod.2011.08.001.
84. Fenoglio I, Corazzari I, Francia C, Bodoardo S, Fubini B. The oxidation of glutathione by cobalt/tungsten carbide contributes to hard metal-induced oxidative stress. Free Radical Research. 2008;42:437–745. doi: https://pubmed.ncbi.nlm.nih.gov/18712631/.
85. Yu Z, Li Q, Wang J, Yu Y, Wang Y, Zhou Q, Li P. Reactive oxygen species-related nanoparticle toxicity in the biomedical field. Nanoscale Research Letters. 2020;15:115. doi: 10.1186/s11671-020-03344-7.
86. Onodera A, Nishiumi F, Kakiguchi K, Tanaka A, Tanabe N, Honma A, Yayama K, Yoshioka Y, Nakahira K, Yonemura S, et al. Short-term changes in intracellular ROS localisation after the silver nanoparticles exposure depending on particle size. Toxicology Reports. 2015;2:574–79. doi: 10.1016/j.toxrep.2015.03.004.
87. Jawaid P, Rehman MU, Zhao Q-L, Misawa M, Ishikawa K, Hori M, Shimizu T, Saitoh J-I, Noguchi K, Kondo T. Small size gold nanoparticles enhance apoptosis-induced by cold atmospheric plasma via depletion of intracellular GSH and modification of oxidative stress. Cell Death Discovery. 2020;6:83. doi: 10.1038/s41420-020-00314-x.
88. Mazur P, Skiba-Kurek I, Mrowiec P, Karczewska E, Drożdż R. Synergistic ROS-associated antimicrobial activity of silver nanoparticles and gentamicin against *Staphylococcus epidermidis*. International Journal of Nanomedicine. 2020;15:3551–62. doi: 10.2147/IJN.S246484.
89. León-Buitimea A, Garza-Cárdenas CR, Garza-Cervantes JA, Lerma-Escalera JA, Morones-Ramírez JR. The demand for new antibiotics: Antimicrobial peptides, nanoparticles, and combinatorial therapies as future strategies in antibacterial agent design. Frontiers in Microbiology. 2020;11:1669. doi: 10.3389/fmicb.2020.01669.
90. Quinteros MA, Aristizábal VC, Dalmasso PR, Paraje MG, Páez PL. Oxidative stress generation of silver nanoparticles in three bacterial genera and its relationship with the antimicrobial activity. Toxicology in Vitro. 2016;36:216–23. doi: 10.1016/j.tiv.2016.08.007.
91. Krce L, Šprung M, Rončević T, Maravić A, Čikeš Čulić V, Blažeka D, Krstulović N, Aviani I. Probing the mode of antibacterial action of silver nanoparticles synthesized by laser ablation in water: What fluorescence and AFM data tell us. Nanomaterials. 2020;10:1040. doi: 10.3390/nano10061040.
92. Singh R, Cheng S, Singh S. Oxidative stress-mediated genotoxic effect of zinc oxide nanoparticles on *Deinococcus radiodurans*. 3 Biotech. 2020;10:66. doi: 10.1007/s13205-020-2054-4.
93. Yamaguchi M, Abe H, Ma T, Tadaki D, Hirano-Iwata A, Kanetaka H, Watanabe Y, Niwano M. Bactericidal activity of TiO_2 nanotube thin films on Si by photocatalytic generation of active oxygen species. Langmuir. 2020;36:12668–77. doi: 10.1021/acs.langmuir.0c02225.
94. Valavanidis A, Vlachogianni T, Fiotakis C. 8-hydroxy-2'-deoxyguanosine (8-OHDG): A critical biomarker of oxidative stress and carcinogenesis. Journal of Environmental Science and Health, Part C. 2009;27:120–39.
95. Song M-F, Li Y-S, Kasai H, Kawai K. Metal nanoparticle-induced micronuclei and oxidative DNA damage in mice. Journal of Clinical Biochemistry and Nutrition. 2012; 50:211–16. doi: 10.3164/jcbn.11-70.
96. Fan Z, Lu JG. Zinc oxide nanostructures: Synthesis and properties. Journal of Nanoscience and Nanotechnology. 2005;5:1561–73. doi: 10.1166/jnn.2005.182.
97. Lodge R, Descoteaux A. Phagocytosis of *Leishmania donovani* amastigotes is Rac1 dependent and occurs in the absence of NADPH oxidase activation. European Journal of Immunology. 2006;36(10):2735–44.

98. Mehta A, Shaha C. Mechanism of metalloid-induced death in *Leishmania* spp: Role of iron, reactive oxygen species, Ca^{2+}, and glutathione. Free Radical Biology and Medicine. 2006;40(10):1857–68.
99. Allahverdiyev AM, Abamor ES, Bagirova M, Ustundag CB, Kaya C, Kaya F, Rafailovich M. Antileishmanial effect of silver nanoparticles and their enhanced antiparasitic activity under ultraviolet light. International Journal of Nanomedicine. 2011;6:2705–14. doi: 10.2147/IJN.S23883. Epub 2011 Nov 3. PMID: 22114501; PMCID: PMC3218584.
100. Lakshmi JV, Sharath R, Chandraprabha MN, et al. Synthesis, characterization and evaluation of antimicrobial activity of zinc oxide nanoparticles. Journal of Biochemical Technology. 2012;3(5):S151–54.
101. Saini P, Saha SK, Roy P, Chowdhury P, Babu SPS. Evidence of reactive oxygen species (ROS) mediated apoptosis in *Setaria cervi* induced by green silver nanoparticles from *Acacia auriculiformis* at a very low dose. Experimental Parasitology. 2016;160:39–48.
102. Ahmad A, Syed F, Shah A, et al. Silver and gold nanoparticles from *Sargentodoxa cuneata*: Synthesis, characterization and antileishmanial activity. RSC Advances. 2015;2015(5):73793–806.
103. Adeyemi OS, Murata Y, Sugi T, Kato K. Inorganic nanoparticles caused death of *Toxoplasma gondii* through alteration of redox status and mitochondrial membrane potential. International Journal of Nanomedicine. 2017;12:1647–61.
104. Charvat RA, Arrizabalaga G. Oxidative stress generated during monensin treatment contributes to altered *Toxoplasma gondii* mitochondria function. Scientific Reports. 2016;15(6):22997.
105. Salem SS. A mini review on green nanotechnology and its development in biological effects. Archives of Microbiology. 2023;205(4):128.
106. Yin SJ, Zhang L, Zhang L, Wan J, Song W, Jiang X, … Si, YX. Metabolic responses and arginine kinase expression of juvenile cuttlefish (*Sepia pharaonis*) under salinity stress. International Journal of Biological Macromolecules. 2018;113:881–88.
107. Samuel MS, Sivaramakrishna A, Mehta A. Degradation and detoxification of aflatoxin B1 by *Pseudomonas putida*. International Biodeterioration & Biodegradation. 2014;86:202–9.
108. Vinoth V, Wu JJ, Asiri AM, Anandan S. Sonochemical synthesis of silver nanoparticles anchored reduced graphene oxide nanosheets for selective and sensitive detection of glutathione. Ultrasonics Sonochemistry. 2017;39:363–73.
109. Samuel MS, Ravikumar M, John J A, Selvarajan E, Patel H, Chander PS, … Chandrasekar N. A review on green synthesis of nanoparticles and their diverse biomedical and environmental applications. Catalysts. 2022;12(5):459.
110. Salem SS, Hammad EN, Mohamed AA, El-Dougdoug W. A comprehensive review of nanomaterials: Types, synthesis, characterization, and applications. Biointerface Research in Applied Chemistry. 2022;13(1):41.
111. Wadhwani SA, Shedbalkar UU, Singh R, Chopade BA. Biogenic selenium nanoparticles: Current status and future prospects. Applied Microbiology and Biotechnology. 2016;100:2555–66.
112. Salunke GR, Ghosh S, Santosh Kumar RJ, Khade S, Vashisth P, Kale T, … Chopade BA. Rapid efficient synthesis and characterization of silver, gold, and bimetallic nanoparticles from the medicinal plant *Plumbago zeylanica* and their application in biofilm control. International Journal of Nanomedicine. 2014;2635–53.
113. Malhotra SPK, Alghuthaymi MA. (2022). Biomolecule-assisted biogenic synthesis of metallic nanoparticles. *Agri-Waste and Microbes for Production of Sustainable Nanomaterials*, 139–63.
114. Ying S, Guan Z, Ofoegbu PC, Clubb P, Rico C, He F, Hong J. Green synthesis of nanoparticles: Current developments and limitations. Environmental Technology & Innovation. 2022;26:102336.

115. Oremland RS, Herbel MJ, Blum JS, Langley S, Beveridge TJ, Ajayan PM, … Curran S. Structural and spectral features of selenium nanospheres produced by se-respiring bacteria. Applied and Environmental Microbiology. 2004;70(1):52–60.
116. Ruiz-Fresneda MA, Staicu LC, Lazuén-López G, Merroun ML. Allotropy of selenium nanoparticles: Colourful transition, synthesis, and biotechnological applications. Microbial Biotechnology. 2023;16(5):877–92.
117. Nair GM, Sajini T, Mathew B. Advanced green approaches for metal and metal oxide nanoparticles synthesis and their environmental applications. Talanta Open. 2022;5: 100080.
118. Naikoo GA, Mustaqeem M, Hassan IU, Awan T, Arshad F, Salim H, Qurashi A. Bioinspired and green synthesis of nanoparticles from plant extracts with antiviral and antimicrobial properties: A critical review. Journal of Saudi Chemical Society. 2021;25(9):101304.
119. Das RP, Pradhan AK. (2022). An introduction to different methods of nanoparticles synthesis. *Bio-Nano Interface: Applications in Food, Healthcare and Sustainability*, 21–34.
120. Zhang X, Yan S, Tyagi RD, Surampalli RY. Synthesis of nanoparticles by microorganisms and their application in enhancing microbiological reaction rates. Chemosphere. 2011;82(4):489–94.
121. Iravani S, Varma RS. Bacteria in heavy metal remediation and nanoparticle biosynthesis. ACS Sustainable Chemistry & Engineering. 2020;8(14):5395–409.
122. Fischer S, Krause T, Lederer F, Merroun ML, Shevchenko A, Hübner R, … Jain R. *Bacillus safensis* JG-B5T affects the fate of selenium by extracellular production of colloidally less stable selenium nanoparticles. Journal of Hazardous Materials. 2020; 384:121146.
123. Southam G, Beveridge TJ. The in vitro formation of placer gold by bacteria. Geochimica et Cosmochimica Acta. 1994;58(20):4527–30.
124. Pinel-Cabello M, Chapon V, Ruiz-Fresneda MA, Alpha-Bazin B, Berthomieu C, Armengaud J, Merroun ML. Delineation of cellular stages and identification of key proteins for reduction and biotransformation of Se (IV) by *Stenotrophomonas bentonitica* BII-R7. Journal of Hazardous Materials. 2021;418:126150.
125. Butler CS, Debieux CM, Dridge EJ, Splatt P, Wright M. Biomineralization of selenium by the selenate-respiring bacterium *Thauera selenatis*. Biochemical Society Transactions. 2012;40(6):1239–43.
126. Jia H, Huang S, Cheng S, Zhang X, Chen X, Zhang Y, … Wu, L. Novel mechanisms of selenite reduction in *Bacillus subtilis* 168: Confirmation of multiple-pathway mediated remediation based on transcriptome analysis. Journal of Hazardous Materials. 2022;433:128834.
127. Kora AJ, Rastogi L. Bacteriogenic synthesis of selenium nanoparticles by *Escherichia coli* ATCC 35218 and its structural characterisation. IET Nanobiotechnology. 2017; 11(2):179–84.
128. Lampis S, Zonaro E, Bertolini C, Cecconi D, Monti F, Micaroni M, … Vallini G. Selenite biotransformation and detoxification by *Stenotrophomonas maltophilia* SeITE02: Novel clues on the route to bacterial biogenesis of selenium nanoparticles. Journal of Hazardous Materials. 2017;324:3–14.
129. Pearce CI, Pattrick RA, Law N, Charnock JM, Coker VS, Fellowes JW, Lloyd JR. Investigating different mechanisms for biogenic selenite transformations: *Geobacter sulfurreducens*, *Shewanella oneidensis* and *Veillonella atypica*. Environmental Technology. 2009;30(12):1313–26.
130. Brayner R, Barberousse H, Hemadi M, Djedjat C, Yéprémian C, Coradin T, … Couté A. Cyanobacteria as bioreactors for the synthesis of Au, Ag, Pd, and Pt nanoparticles via an enzyme-mediated route. Journal of Nanoscience and Nanotechnology. 2007;7(8):2696–2708.

131. Hamouda RA, Hussein MH, Abo-Elmagd RA, Bawazir SS. Synthesis and biological characterization of silver nanoparticles derived from the cyanobacterium *Oscillatoria limnetica*. Scientific Reports. 2019;9(1):13071.
132. Sharma N, Pinnaka AK, Raje M, Fnu A, Bhattacharyya MS, Choudhury AR. Exploitation of marine bacteria for production of gold nanoparticles. Microbial Cell Factories. 2012;11:1–6.
133. El-Batal AI, Al Tamie MS. Biosynthesis of gold nanoparticles using marine *Streptomyces cyaneus* and their antimicrobial, antioxidant and antitumor (in vitro) activities. Journal of Chemical and Pharmaceutical Research. 2015;7(7):1020–36.
134. Alani F, Moo-Young M, Anderson W. Biosynthesis of silver nanoparticles by a new strain of *Streptomyces* sp. compared with *Aspergillus fumigatus*. World Journal of Microbiology and Biotechnology. 2012;28:1081–86.
135. El-Gamal MS, Salem SS, Abdo AM. Biosynthesis, characterization, and antimicrobial activities of silver nanoparticles synthesized by endophytic *Streptomyces* sp. Journal of Biotechnology. 2018;56:69–85.
136. Hassan SED, Fouda A, Radwan AA, Salem SS, Barghoth MG, Awad MA, … El-Gamal MS. Endophytic actinomycetes *Streptomyces* spp mediated biosynthesis of copper oxide nanoparticles as a promising tool for biotechnological applications. JBIC Journal of Biological Inorganic Chemistry. 2019;24:377–93.
137. Rajivgandhi G, Gnanamangai BM, Prabha TH, Poornima S, Maruthupandy M, Alharbi NS, … Li WJ. Biosynthesized zinc oxide nanoparticles (ZnO NPs) using actinomycetes enhance the anti-bacterial efficacy against K. Pneumoniae. Journal of King Saud University-Science. 2022;34(1):101731.
138. Bahrulolum H, Nooraei S, Javanshir N, Tarrahimofrad H, Mirbagheri VS, Easton AJ, Ahmadian G. Green synthesis of metal nanoparticles using microorganisms and their application in the agrifood sector. Journal of Nanobiotechnology. 2021;19(1): 1–26.
139. Abdollahnia M, Makhdoumi A, Mashreghi M, Eshghi H. Exploring the potentials of halophilic prokaryotes from a solar saltern for synthesizing nanoparticles: The case of silver and selenium. PLoS One. 2020;15(3):e0229886.
140. Naik MM, Bagayat A, Charya LS. (2017). Extreme halophilic archaea halobacterium SP. SP1 in bioremediation of hypersaline water polluted with metal, metalloid, organometal and aromatic compound.
141. Kitching M, Ramani M, Marsili E. Fungal biosynthesis of gold nanoparticles: Mechanism and scale up. Microbial Biotechnology. 2015;8(6):904–17.
142. Shaheen TI, Salem SS, Fouda A. (2021). Current advances in fungal nanobiotechnology: Mycofabrication and applications. *Microbial nanobiotechnology: Principles and applications*, 113–43.
143. Mohanpuria P, Rana NK, Yadav SK. Biosynthesis of nanoparticles: Technological concepts and future applications. Journal of Nanoparticle Research. 2008;10:507–17.
144. Guilger-Casagrande M, Lima RD. Synthesis of silver nanoparticles mediated by fungi: A review. Frontiers in Bioengineering and Biotechnology. 2019;7:287.
145. Salem SS, Fouda A. Green synthesis of metallic nanoparticles and their prospective biotechnological applications: An overview. Biological Trace Element Research. 2021;199:344–70.
146. Mukherjee P, Senapati S, Mandal D, Ahmad A, Khan MI, Kumar R, Sastry M. Extracellular synthesis of gold nanoparticles by the fungus *Fusarium oxysporum*. ChemBioChem. 2002;3(5):461–63.
147. Gade A, Ingle A, Whiteley C, Rai M. Mycogenic metal nanoparticles: Progress and applications. Biotechnology Letters. 2010;32:593–600.
148. Priyadarshini E, Priyadarshini SS, Cousins BG, Pradhan N. Metal-fungus interaction: Review on cellular processes underlying heavy metal detoxification and synthesis of metal nanoparticles. Chemosphere. 2021;274:129976.

149. Jain AS, Pawar PS, Sarkar A, Junnuthula V, Dyawanapelly S. Bionanofactories for green synthesis of silver nanoparticles: Toward antimicrobial applications. International Journal of Molecular Sciences. 2021;22(21):11993.
150. Kulkarni D, Sherkar R, Shirsathe C, Sonwane R, Varpe N, Shelke S, ... Dyawanapelly S. Biofabrication of nanoparticles: Sources, synthesis, and biomedical applications. Frontiers in Bioengineering and Biotechnology. 2023;11:1159193.
151. Bansal V, Rautaray D, Bharde A, Ahire K, Sanyal A, Ahmad A, Sastry M. Fungus-mediated biosynthesis of silica and titania particles. Journal of Materials Chemistry. 2005;15(26):2583–89.
152. Durán N, Marcato PD, Alves OL, De Souza GI, Esposito E. Mechanistic aspects of biosynthesis of silver nanoparticles by several *Fusarium oxysporum* strains. Journal of Nanobiotechnology. 2005;3:1–7.
153. Mosallam FM, El-Sayyad GS, Fathy RM, El-Batal AI. Biomolecules-mediated synthesis of selenium nanoparticles using *Aspergillus oryzae* fermented Lupin extract and gamma radiation for hindering the growth of some multidrug-resistant bacteria and pathogenic fungi. Microbial Pathogenesis. 2018;122:108–16.
154. Kadam VV, Ettiyappan JP, Balakrishnan RM. Mechanistic insight into the endophytic fungus mediated synthesis of protein capped ZnO nanoparticles. Materials Science and Engineering: B. 2019;243:214–21.
155. Chatterjee S, Mahanty S, Das P, Chaudhuri P, Das S. Biofabrication of iron oxide nanoparticles using manglicolous fungus *Aspergillus niger* BSC-1 and removal of Cr (VI) from aqueous solution. Chemical Engineering Journal. 2020;385:123790.
156. Shi C, Qi H, Ma R, Sun Z, Xiao L, Wei G, ... Guo Z. N, S-self-doped carbon quantum dots from fungus fibers for sensing tetracyclines and for bioimaging cancer cells. Materials Science and Engineering: C. 2019;105:110132.
157. Lian S, Diko CS, Yan Y, Li Z, Zhang H, Ma Q, Qu Y. Characterization of biogenic selenium nanoparticles derived from cell-free extracts of a novel yeast Magnusiomyces ingens. 3 Biotech. 2019;9:1–8.
158. Dameron CT, Reese RN, Mehra RK, Kortan AR, Carroll PJ, Steigerwald ML, ... Winge DR. Biosynthesis of cadmium sulphide quantum semiconductor crystallites. Nature. 1989;338(6216):596–97.
159. Boroumand Moghaddam A, Namvar F, Moniri M, Md. Tahir P, Azizi S, Mohamad R. Nanoparticles biosynthesized by fungi and yeast: A review of their preparation, properties, and medical applications. Molecules. 2015;20(9):16540–65.
160. Cunha FA, Cunha MDC, da Frota SM, Mallmann EJ, Freire TM, Costa LS, ... Fechine PB. Biogenic synthesis of multifunctional silver nanoparticles from *Rhodotorula glutinis* and *Rhodotorula mucilaginosa*: Antifungal, catalytic and cytotoxicity activities. World Journal of Microbiology and Biotechnology. 2018;34:1–15.
161. Soliman H, Elsayed A, Dyaa A. Antimicrobial activity of silver nanoparticles biosynthesised by *Rhodotorula* sp. strain ATL72. Egyptian Journal of Basic and Applied Sciences. 2018;5(3):228–33.
162. Rasouli M. Biosynthesis of selenium nanoparticles using yeast *Nematospora coryli* and examination of their anti-candida and anti-oxidant activities. IET Nanobiotechnology. 2019;13(2):214–18.
163. Kowshik M, Deshmukh N, Vogel W, Urban J, Kulkarni SK, Paknikar KM. Microbial synthesis of semiconductor CdS nanoparticles, their characterization, and their use in the fabrication of an ideal diode. Biotechnology and Bioengineering. 2002;78(5):583–88.
164. Tian X, He W, Cui J, Zhang X, Zhou W, Yan S, ... Yue Y. Mesoporous zirconium phosphate from yeast biotemplate. Journal of Colloid and Interface Science. 2010;343(1):344–49.
165. Kowshik M, Vogel W, Urban J, Kulkarni SK, Paknikar KM. Microbial synthesis of semiconductor PbS nanocrystallites. Advanced Materials. 2002;14(11):815–18.

166. Michael A, Singh A, Roy A, Islam MR. Fungal-and algal-derived synthesis of various nanoparticles and their applications. Bioinorganic Chemistry and Applications. 2022;2022(1):3142674.
167. Gade AK, Bonde PP, Ingle AP, Marcato PD, Duran N, Rai MK. Exploitation of *Aspergillus niger* for synthesis of silver nanoparticles. Journal of Biobased Materials and Bioenergy. 2008;2(3):243–47.
168. Hasan S. A review on nanoparticles: Their synthesis and types. Research Journal of Recent Sciences. 2015;2277:2502.
169. Roy A, Sharma A, Yadav S, Jule LT, Krishnaraj R. Nanomaterials for remediation of environmental pollutants. Bioinorganic Chemistry and Applications. 2021;2021(1):1764647.
170. Mukherjee A, Sarkar D, Sasmal S. A review of green synthesis of metal nanoparticles using algae. Frontiers in Microbiology. 2021;12:693899.
171. El-Refaey AA, Salem SS. Algae materials for bionanopesticides: Nanoparticles and composites. Algae materials: Academic Press; 2023. pp. 219–30.
172. Gautam PK, Singh A, Misra K, Sahoo AK, Samanta SK. Synthesis and applications of biogenic nanomaterials in drinking and wastewater treatment. Journal of Environmental Management. 2019;231:734–48.
173. Kumari S, Tyagi M, Jagadevan S. Mechanistic removal of environmental contaminants using biogenic nano-materials. International Journal of Environmental Science and Technology. 2019;16:7591–606.
174. Rahman A, Lin J, Jaramillo FE, Bazylinski DA, Jeffryes C, Dahoumane SA. In vivo biosynthesis of inorganic nanomaterials using eukaryotes—A review. Molecules. 2020; 25(14):3246.
175. Kashyap M, Kiran B. Milking microalgae in conjugation with nano-biorefinery approach utilizing wastewater. Journal of Environmental Management. 2021;293:112864.
176. Ubando AT, Africa ADM, Maniquiz-Redillas MC, Culaba AB, Chen WH, Chang JS. Microalgal biosorption of heavy metals: A comprehensive bibliometric review. Journal of Hazardous Materials. 2021;402:123431.
177. Dahoumane SA, Mechouet M, Wijesekera K, Filipe CD, Sicard C, Bazylinski DA, Jeffryes C. Algae-mediated biosynthesis of inorganic nanomaterials as a promising route in nanobiotechnology–A review. Green Chemistry. 2017;19(3):552–87.
178. Caliskan G, Mutaf T, Agba HC, Elibol M. Green synthesis and characterization of titanium nanoparticles using microalga, *Phaeodactylum tricornutum*. Geomicrobiology Journal. 2022;39(1):83–96.
179. Agarwal H, Kumar SV, Rajeshkumar S. A review on green synthesis of zinc oxide nanoparticles–An eco-friendly approach. Resource-Efficient Technologies. 2017;3(4):406–13.
180. Al-Enazi NM, Alwakeel S, Alhomaidi E. Photocatalytic and biological activities of green synthesized SnO_2 nanoparticles using *Chlorella vulgaris*. Journal of Applied Microbiology. 2022;133(6):3265–75.
181. Rajeshkumar S, Kannan C, Annadurai G. Synthesis and characterization of antimicrobial silver nanoparticles using marine brown seaweed *Padina tetrastromatica*. Drug Invention Today. 2012;4(10):511–13.
182. Alsaiari NS, Alzahrani FM, Amari A, Osman H, Harharah HN, Elboughdiri N, Tahoon MA. Plant and microbial approaches as green methods for the synthesis of nanomaterials: Synthesis, applications, and future perspectives. Molecules. 2023;28(1):463.
183. Phull AR, Abbas Q, Ali A, Raza H, Zia M, Haq IU. Antioxidant, cytotoxic and antimicrobial activities of green synthesized silver nanoparticles from crude extract of *Bergenia ciliata*. Future Journal of Pharmaceutical Sciences. 2016;2(1):31–36.
184. Naraginti S, Kumari PL, Das RK, Sivakumar A, Patil SH, Andhalkar VV. Amelioration of excision wounds by topical application of green synthesized, formulated silver and

gold nanoparticles in albino Wistar rats. Materials Science and Engineering: C. 2016; 62:293–300.

185. Dhand V, Soumya L, Bharadwaj S, Chakra S, Bhatt D, Sreedhar B. Green synthesis of silver nanoparticles using *Coffea arabica* seed extract and its antibacterial activity. Materials Science and Engineering: C. 2016;58:36–43.
186. Kayalvizhi T, Ravikumar S, Venkatachalam P. Green synthesis of metallic silver nanoparticles using *Curculigo orchioides* rhizome extracts and evaluation of its antibacterial, larvicidal, and anticancer activity. Journal of Environmental Engineering. 2016;142(9):C4016002.
187. Pugazhendhi S, Sathya P, Palanisamy PK, Gopalakrishnan R. Synthesis of silver nanoparticles through green approach using *Dioscorea alata* and their characterization on antibacterial activities and optical limiting behavior. Journal of Photochemistry and Photobiology B: Biology. 2016;159:155–60.
188. Kalaiyarasu T, Karthi N, Sharmila GV, Manju V. In vitro assessment of antioxidant and antibacterial activity of green synthesized silver nanoparticles from *Digitaria radicosa* leaves. Asian Journal of Pharmaceutical and Clinical Research. 2016;9(1):1–6.
189. Rao NH, Lakshmidevi N, Pammi SVN, Kollu P, Ganapaty S, Lakshmi P. Green synthesis of silver nanoparticles using methanolic root extracts of *Diospyros paniculata* and their antimicrobial activities. Materials Science and Engineering: C. 2016;62:553–57.
190. Kharat SN, Mendhulkar VD. Synthesis, characterization and studies on antioxidant activity of silver nanoparticles using *Elephantopus scaber* leaf extract. Materials Science and Engineering: C. 2016;62:719–24.
191. Rajkuberan C, Prabukumar S, Sathishkumar G, Wilson A, Ravindran K, Sivaramakrishnan S. Facile synthesis of silver nanoparticles using *Euphorbia antiquorum* L. latex extract and evaluation of their biomedical perspectives as anticancer agents. Journal of Saudi Chemical Society. 2017;21(8):911–19.
192. Ramesh PS, Kokila T, Geetha D. Plant mediated green synthesis and antibacterial activity of silver nanoparticles using *Emblica officinalis* fruit extract. Spectrochimica Acta Part A: Molecular and Biomolecular Spectroscopy. 2015;142:339–43.
193. Nayak D, Ashe S, Rauta PR, Kumari M, Nayak B. Bark extract mediated green synthesis of silver nanoparticles: Evaluation of antimicrobial activity and antiproliferative response against osteosarcoma. Materials Science and Engineering: C. 2016;58:44–52.
194. Paul B, Bhuyan B, Purkayastha DD, Dhar SS. Photocatalytic and antibacterial activities of gold and silver nanoparticles synthesized using biomass of *Parkia roxburghii* leaf. Journal of Photochemistry and Photobiology B: Biology. 2016;154:1–7.
195. Singh J, Dutta T, Kim KH, Rawat M, Samddar P, Kumar P. 'Green' synthesis of metals and their oxide nanoparticles: Applications for environmental remediation. Journal of Nanobiotechnology. 2018;16(1):1–24.
196. Abdelghany TM, Al-Rajhi AM, Yahya R, Bakri MM, Al Abboud MA, Yahya R, … Salem SS. Phytofabrication of zinc oxide nanoparticles with advanced characterization and its antioxidant, anticancer, and antimicrobial activity against pathogenic microorganisms. Biomass Conversion and Biorefinery. 2023;13(1):417–30.
197. Al-Zahrani FA, Salem SS, Al-Ghamdi HA, Nhari LM, Lin L, El-Shishtawy RM. Green synthesis and antibacterial activity of Ag/Fe_2O_3 nanocomposite using *Buddleja lindleyana* extract. Bioengineering. 2022;9(9):452.
198. Aref MS, Salem SS. Bio-callus synthesis of silver nanoparticles, characterization, and antibacterial activities via *Cinnamomum camphora* callus culture. Biocatalysis and Agricultural Biotechnology. 2020;27:101689.
199. Soni V, Raizada P, Singh P, Cuong HN, Rangabhashiyam S, Saini A, … Nguyen VH. Sustainable and green trends in using plant extracts for the synthesis of biogenic metal nanoparticles toward environmental and pharmaceutical advances: A review. Environmental Research. 2021;202:111622.

200. Velusamy P, Das J, Pachaiappan R, Vaseeharan B, Pandian K. Greener approach for synthesis of antibacterial silver nanoparticles using aqueous solution of neem gum (*Azadirachta indica L.*). Industrial Crops and Products. 2015;66:103–9.
201. Edison TNJI, Lee YR, Sethuraman MG. Green synthesis of silver nanoparticles using *Terminalia cuneata* and its catalytic action in reduction of direct yellow-12 dye. Spectrochimica Acta Part A: Molecular and Biomolecular Spectroscopy. 2016;161:122–29.
202. Patil MP, Ngabire D, Thi HHP, Kim MD, Kim GD. Eco-friendly synthesis of gold nanoparticles and evaluation of their cytotoxic activity on cancer cells. Journal of Cluster Science. 2017;28:119–32.
203. Govindasamy R, Govindarasu M, Alharthi SS, Mani P, Bernaurdshaw N, Gomathi T, … Thiruvengadam M. Sustainable green synthesis of yttrium oxide (Y_2O_3) nanoparticles using *Lantana camara* leaf extracts: Physicochemical characterization, photocatalytic degradation, antibacterial, and anticancer potency. Nanomaterials. 2022;12(14):2393.
204. Kumari S, Verma R, Chauhan A, Raja V, Kumari S, Kulshrestha S. (2023). Biogenic approach for synthesis of nanoparticles via plants for biomedical applications: A review. Materials Today: Proceedings.
205. Thangavelu RM, Ganapathy R, Ramasamy P, Krishnan K. Fabrication of virus metal hybrid nanomaterials: An ideal reference for bio semiconductor. Arabian Journal of Chemistry. 2020;13(1):2750–65.
206. Ramamurthy CH, Sampath KS, Arunkumar P, Kumar MS, Sujatha V, Premkumar K, Thirunavukkarasu C. Green synthesis and characterization of selenium nanoparticles and its augmented cytotoxicity with doxorubicin on cancer cells. Bioprocess and Biosystems Engineering. 2013;36:1131–39.
207. Messarah M, Klibet F, Boumendjel A, Abdennour C, Bouzerna N, Boulakoud MS, El Feki A. Hepatoprotective role and antioxidant capacity of selenium on arsenic-induced liver injury in rats. Experimental and Toxicologic Pathology. 2012;64(3):167–74.
208. Alghuthaymi MA, Almoammar H, Rai M, Said-Galiev E, Abd-Elsalam KA. Myconanoparticles: Synthesis and their role in phytopathogens management. Biotechnology & Biotechnological Equipment. 2015;29(2):221–36.
209. Khan S, Mansoor S, Rafi Z, Kumari B, Shoaib A, Saeed M, … Shakeel F. A review on nanotechnology: Properties, applications, and mechanistic insights of cellular uptake mechanisms. Journal of Molecular Liquids. 2022;348:118008.
210. Khot LR, Sankaran S, Maja JM, Ehsani R, Schuster EW. Applications of nanomaterials in agricultural production and crop protection: A review. Crop Protection. 2012; 35:64–70.
211. Singh S, Singh BK, Yadav SM, Gupta AK. Applications of nanotechnology in agricultural and their role in disease management. Research Journal of Nanoscience and Nanotechnology. 2015;5(1):1–5.
212. Jeevanandam J, Barhoum A, Chan YS, Dufresne A, Danquah MK. Review on nanoparticles and nanostructured materials: History, sources, toxicity and regulations. Beilstein Journal of Nanotechnology. 2018;9(1):1050–74.
213. Khan I, Saeed K, Khan I. Nanoparticles: Properties, applications and toxicities. Arabian Journal of Chemistry. 2019;12(7):908–31.
214. Mohammadzadeh V, Barani M, Amiri MS, Yazdi MET, Hassanisaadi M, Rahdar A, Varma RS. Applications of plant-based nanoparticles in nanomedicine: A review. Sustainable Chemistry and Pharmacy. 2022;25:100606.
215. Iravani S, Korbekandi H, Mirmohammadi SV, Zolfaghari B. Synthesis of silver nanoparticles: Chemical, physical and biological methods. Research in Pharmaceutical Sciences. 2014 Nov-Dec;9(6):385–406. PMID: 26339255; PMCID: PMC4326978.
216. Khandel P, Shahi SK, Soni DK, Yadaw RK, Kanwar L. *Alpinia calcarata*: A potential source for the fabrication of bioactive silver nanoparticles. Nano Convergence. 2018 Dec 6;5(1):37. doi: 10.1186/s40580-018-0167-9. PMID: 30519797; PMCID: PMC6281549.

217. Leung YH, Xu X, Ma AP, Liu F, Ng AM, Shen Z, Gethings LA, Guo MY, Djurišić AB, Lee PK, Lee HK, Chan WK, Leung FC. Toxicity of ZnO and TiO_2 to *Escherichia coli* cells. Scientific Reports. 2016 Oct 12;6:35243. doi: 10.1038/srep35243. PMID: 27731373; PMCID: PMC5378928.
218. Cui Y, Zhao Y, Tian Y, Zhang W, Lü X, Jiang X. The molecular mechanism of action of bactericidal gold nanoparticles on Escherichia coli. Biomaterials. 2012 Mar;33(7): 2327–33. doi: 10.1016/j.biomaterials.2011.11.057. Epub 2011 Dec 17. PMID: 22182745.
219. Bajwa HUR, Khan MK, Abbas Z, Riaz R, Rehman Tu, Abbas RZ, Aleem MT, Abbas A, Almutairi MM, Alshammari FA, et al. Nanoparticles: Synthesis and their role as potential drug candidates for the treatment of parasitic diseases. Life. 2022; 12(5):750. https://doi.org/10.3390/life12050750.
220. Prabhu S, Poulose EK. Silver nanoparticles: mechanism of antimicrobial action, synthesis, medical applications, and toxicity effects. International Nano Letters. 2012;2:1–10.
221. Jeong Y, Lim DW, Choi J. Assessment of size-dependent antimicrobial and cytotoxic properties of silver nanoparticles. Adv Mater Sci Eng. 2014;763807.
222. Ahn J-M, Eom H-J, Yang X, Meyer JN, Choi J. Comparative toxicity of silver nanoparticles on oxidative stress and DNA damage in the nematode, *Caenorhabditis elegans*. Chemosphere. 2014;108:343–52.
223. Li L, Wu H, Peijnenburg WJ, van Gestel CA. Both released silver ions and particulate Ag contribute to the toxicity of AgNPs to earthworm *Eisenia fetida*. Nanotoxicology. 2015;9:792–801.
224. Pelletier DA, Suresh AK, Holton GA, McKeown CK, Wang W, Gu B, Mortensen NP, Allison DP, Joy DC, Allison MR, Brown SD, Phelps TJ, Doktycz MJ. Effects of engineered cerium oxide nanoparticles on bacterial growth and viability. Applied and Environmental Microbiology. 2010 Dec;76(24):7981–89. doi: 10.1128/AEM.00650-10. Epub 2010 Oct 15. PMID: 20952651; PMCID: PMC3008265.
225. Thill A, Zeyons O, Spalla O, Chauvat F, Rose J, Auffan M, Flank AM. Cytotoxicity of CeO_2 nanoparticles for *Escherichia coli*. Physico-chemical insight of the cytotoxicity mechanism. Environmental Science & Technology. 2006 Oct 1;40(19):6151–56. doi: 10.1021/es060999b. PMID: 17051814.
226. Raghunath A, Perumal E. Metal oxide nanoparticles as antimicrobial agents: A promise for the future. International Journal of Antimicrobial Agents. 2017 Feb;49(2): 137–52. doi: 10.1016/j.ijantimicag.2016.11.011. Epub 2017 Jan 3. PMID: 28089172.
227. Simon-Deckers A, Loo S, Mayne-L'hermite M, Herlin-Boime N, Menguy N, Reynaud C, Gouget B, Carrière M. Size-, composition- and shape-dependent toxicological impact of metal oxide nanoparticles and carbon nanotubes toward bacteria. Environmental Science & Technology. 2009 Nov 1;43(21):8423–29. doi: 10.1021/es9016975. PMID: 19924979.
228. Leung YH, Ng AM, Xu X, Shen Z, Gethings LA, Wong MT, Chan CM, Guo MY, Ng YH, Djurišić AB, Lee PK, Chan WK, Yu LH, Phillips DL, Ma AP, Leung FC. Mechanisms of antibacterial activity of MgO: Non-ROS mediated toxicity of MgO nanoparticles towards *Escherichia coli*. Small. 2014 Mar 26;10(6):1171–83. doi: 10. 1002/smll.201302434. Epub 2013 Dec 17. PMID: 24344000.
229. Berger TJ, Spadaro JA, Bierman R, Chapin SE, Becker RO. Electrically generated silver ions: Quantitative effects on bacterial and mammalian cells. Antimicrob Agents Chemother. 1996;10:357–58.
230. Nel A, et al. Toxic potential of materials at the nanolevel. Science. 2006;311:622–27. DOI:10.1126/science.1114397.
231. Mahana A, Guliy OI, Mehta SK. Accumulation and cellular toxicity of engineered metallic nanoparticle in freshwater microalgae: Current status and future challenges. Ecotoxicology and Environmental Safety. 2021 Jan 15;208:111662. doi: 10.1016/j.ecoenv. 2020.111662. Epub 2020 Nov 29. PMID: 33396172.

232. Abbaszadegan A, Ghahramani Y, Gholami A, Hemmateenejad B, Dorostkar S, Nabavizadeh M, et al. The effect of charge at the surface of silver nanoparticles on antimicrobial activity against gram-positive and gram-negative bacteria: A preliminary study. Journal of Nanomaterials. 2015;16:53.
233. Wang Y, Zhu X, Lao Y, Lv X, Tao Y, Huang B, Wang J, Zhou J, Cai Z. TiO_2 nanoparticles in the marine environment: Physical effects responsible for the toxicity on algae *Phaeodactylum tricornutum*. Science of the Total Environment. 2016;565:818–26.
234. Duan J, Liang S, Feng L, Yu Y, Sun Z. *Silica nanoparticles* trigger hepatic lipid-metabolism disorder in vivo and in vitro. International Journal of Nanomedicine. 2018;13:7303–18.
235. Sondi I, Salopek-Sondi B. Silver nanoparticles as antimicrobial agent: A case study on *E. coli* as a model for Gram-negative bacteria. Journal of Colloid and Interface Science. 2004;275:177–82.
236. Stern ST, Adiseshaiah PP, Crist RM. Autophagy and lysosomal dysfunction as emerging mechanisms of nanomaterial toxicity. Particle and Fibre Toxicology. 2012 Jun 14; 9:20. doi: 10.1186/1743-8977-9-20. PMID: 22697169; PMCID: PMC3441384.
237. Service RF. Nanotechnology. Nanoparticle Trojan horses gallop from the lab into the clinic. Science. 2010 Oct 15;330(6002):314–15. doi: 10.1126/science.330.6002.314. PMID: 20947742.
238. Kukowska-Latallo JF, Candido KA, Cao Z, Nigavekar SS, Majoros IJ, Thomas TP, Balogh LP, Khan MK, Baker JR. Nanoparticle targeting of anticancer drug improves therapeutic response in animal model of human epithelial cancer. Cancer Research. 2005 Jun 15;65(12):5317–24. doi: 10.1158/0008-5472.CAN-04-3921. PMID: 15958579.
239. Pritt B. Common parasites. Pathology. 2020;52:S49.
240. Rokkas T, Gisbert JP, Malfertheiner P, Niv Y, Gasbarrini A, Leja M, Megraud F, O'Morain C, Graham DY. Comparative effectiveness of multiple different first-line treatment regimens for *Helicobacter pylori* infection: A network meta-analysis. Gastroenterology. 2021, 161, 495507.
241. Imarhiagbe O. Aspects of the Biology and Ethnobotany of Parasitic Angiosperm Species in Nigeria. 2021. Available online: https://www.intechopen.com/chapters/77552.
242. Campbell S, Soman-Faulkner K. Antiparasitic drugs: StatPearls Publishing; 2021.
243. Syed Z, Sonu K, Dongre A, Sharma G, Sogani M. A review on hydroxamic acids: Widespectrum chemotherapeutic agents. International Journal of Biology and Biomedicine. 2020, 14, 7588.
244. Bahuguna A, Rawat DS. An overview of new antitubercular drugs, drug candidates, and their targets. Medicinal Research Reviews. 2020;40:263–92.
245. Raj S, Sasidharan S, Balaji SN, Saudagar P. An overview of biochemically characterized drug targets in metabolic pathways of *Leishmania parasite*. Parasitology Research. 2020;119:2025–37.
246. Adeyemi OS, Sulaiman FA. Evaluation of metal nanoparticles for drug delivery systems. The Journal of Biomedical Research. 2015;29:145–49.
247. MubarakAli D, Thajuddin N, Jeganathan K, et al. Plant extract mediated synthesis of silver and gold nanoparticles and its antibacterial activity against clinically isolated pathogens. Colloids Surf B Biointerfaces. 2011;85:360–65.
248. Lekshmi NCJ, Sumi SB, Viveka S, et al. Antibacterial activity of nanoparticles from *Allium* sp. Journal of Microbiology and Biotechnology Research. 2012;2:115–19.
249. Bhardwaj R, Saudagar P, Dubey VK. Nanobiosciences: A contemporary approach in antiparasitic drugs. Molecular and Cellular Pharmacology. 2012;4:97–103.
250. Das S, Bhattacharya A, Debnath N, et al. Nanoparticle-induced morphological transition of *Bombyx mori* nucleopolyhedrovirus: A novel method to treat silkworm grasserie disease. Applied Microbiology and Biotechnology. 2013;97:6019–30.

251. Rahul S, Chandrashekhar P, Hemant B, et al. In vitro antiparasitic activity of microbial pigments and their combination with phytosynthesized metal nanoparticles. Parasitology International. 2015;64:353–56.
252. Wigginton NS, Titta A, Piccapietra F, et al. Binding of silver nanoparticles to bacterial proteins depends on surface modifications and inhibits enzymatic activity. Environmental Science & Technology. 2010;44:2163–68.
253. Salma AA, Amer HA, Shaemaa HA, et al. The effects of gold and silver nanoparticles on transaminase enzymes activities. International Journal of Cancer. 2011;3:2249–2329.
254. Srivastava M, Singh S, Self WT. Exposure to silver nanoparticles inhibits selenoprotein synthesis and the activity of thioredoxin reductase. Environmental Health Perspectives. 2012;120:56–61.
255. Adeyemi OS, Whiteley CG. Interaction of nanoparticles with arginine kinase from *Trypanosoma brucei*: Kinetic and mechanistic evaluation. International Journal of Biological Macromolecules. 2013;62:450–56.
256. Adeyemi OS, Whiteley CG. Interaction of metal nanoparticles with recombinant arginine kinase from *Trypanosoma brucei*: Thermodynamic and spectrofluorimetric evaluation. Biochimica et Biophysica Acta. 2014;1840:701–6.
257. Chen H, Roco MC, Li X, Lin Y. Trends in nanotechnology patents. Nature Nanotechnology. 2008;3:123–25.
258. Li Y, Duan X, Qian Y, Yang L, Liao H. Nanocrystalline silver particles: Synthesis, agglomeration, and sputtering induced by electron beam. Journal of Colloid and Interface Science. 1999;209:347–49.
259. Raveendran F, Wallen J. Completely "green" synthesis and stabilization of metal nanoparticles. Journal of the American Chemical Society. 2003;125:13940–41.
260. Park EJ, Yi J, Kim Y, Choi K, Park K. Silver nanoparticles induce cytotoxicity by a Trojan-horse type mechanism. Toxicology in Vitro. 2010 Apr;24(3):872–78. doi: 10.1016/j.tiv.2009.12.001. Epub 2009 Dec 5. PMID: 19969064.
261. Leirião P, Rodrigues CD, Albuquerque SS, Mota MM. Survival of protozoan intracellular parasites in host cells. EMBO Reports. 2004 Dec;5(12):1142–47. doi: 10.1038/sj.embor.7400299. PMID: 15577928; PMCID: PMC1299194.
262. Poland CA, Duffin R, Kinloch I, Maynard A, Wallace WA, Seaton A, Stone V, Brown S, Macnee W, Donaldson K. Carbon nanotubes introduced into the abdominal cavity of mice show asbestos-like pathogenicity in a pilot study. Nature Nanotechnology. 2008 Jul;3(7):423–28. doi: 10.1038/nnano.2008.111. Epub 2008 May 20. PMID: 18654567.
263. Jiang W, Kim BY, Rutka JT, Chan WC. Nanoparticle-mediated cellular response is size-dependent. Nature Nanotechnology. 2008 Mar;3(3):145–50. doi: 10.1038/nnano.2008.30. Epub 2008 Mar 2. PMID: 18654486.
264. Meghana S, Kabra P, Chakraborty S, Padmavathy N. Understanding the pathway of antibacterial activity of copper oxide nanoparticles. RSC Advances. 2015;5:12293–99.
265. Nalwade AR, Jadhav AA. Biosynthesis of silver nanoparticles using leaf extract of Datura alba Nees and evaluation of their antibacterial activity. Archives of Applied Science Research. 2013;5:45–49.
266. Ghosh S, Patil S, Ahire M, Kitture R, Kale S, Pardesi K, Cameotra SS, Bellare J, Dhavale DD, Jabgunde A, Chopade BA. Synthesis of silver nanoparticles using *Dioscorea bulbifera* tuber extract and evaluation of its synergistic potential in combination with antimicrobial agents. International Journal of Nanomedicine. 2012;7:483–96. doi: 10.2147/IJN.S24793. Epub 2012 Feb 1. PMID: 22334779; PMCID: PMC3273981.
267. Raffi M, Hussain F, Bhatti TM, Akhter JI, Hameed A, Hasan MM. Antibacterial characterization of silver nanoparticles against *E. coli* ATCC-15224. Journal of Materials Science & Technology. 2008;24:192–96.
268. Habash MB, Park AJ, Vis EC, Harris RJ, Khursigara CM. Synergy of silver nanoparticles and aztreonam against *Pseudomonas aeruginosa* PAO1 biofilms. Antimicrobial Agents and Chemotherapy. 2014;58, 5818–30. 10.1128/AAC.03170-14.

269. Schreurs WJ, Rosenberg H. Effect of silver ions on transport and retention of phosphate by *Escherichia coli*. Journal of Bacteriology. 1982;152:7–13.
270. Conner SD, Schmid SL. Regulated portals of entry into the cell. Nature. 2003 Mar 6;422(6927):37–44. doi: 10.1038/nature01451. PMID: 12621426.
271. Sabella S, Carney RP, Brunetti V, Malvindi MA, Al-Juffali N, Vecchio G, Janes SM, Bakr OM, Cingolani R, Stellacci F, Pompa PP. A general mechanism for intracellular toxicity of metal-containing nanoparticles. Nanoscale. 2014 Jun 21;6(12):7052–61. doi: 10.1039/c4nr01234h. PMID: 24842463; PMCID: PMC4120234.
272. Tait SW, Green DR. Mitochondria and cell death: Outer membrane permeabilization and beyond. Nature Reviews Molecular Cell Biology. 2010 Sep;11(9):621–32. doi: 10.1038/nrm2952. Epub 2010 Aug 4. PMID: 20683470.
273. Rai M, Kon K, Ingle A, Duran N, Galdiero S, Galdiero M. Broad-spectrum bioactivities of silver nanoparticles: The emerging trends and future prospects. Applied Microbiology and Biotechnology. 2014;98, 1951–61. 10.1007/s00253-013-5473-x.
274. Wu Z, Zhang B, Yan B. Regulation of enzyme activity through interactions with nanoparticles. International Journal of Molecular Sciences. 2009 Nov 20;10(10): 4198–209. doi: 10.3390/ijms10104198. PMID: 20057940; PMCID: PMC2790103.
275. Gadek TR, Nicholas JB. Small molecule antagonists of proteins. Biochemical Pharmacology. 2003 Jan 1;65(1):1–8. doi: 10.1016/s0006-2952(02)01479-x. PMID: 12473372.
276. Arkin MR, Tang Y, Wells JA. Small-molecule inhibitors of protein-protein interactions: Progressing toward the reality. Chemistry & Biology. 2014 Sep 18;21(9):1102–14. doi: 10.1016/j.chembiol.2014.09.001. PMID: 25237857; PMCID: PMC4179228.
277. Fanti JR et al. Biogenic silver nanoparticles inducing *Leishmania amazonensis* promastigote and amastigote death in vitro. Acta Tropica 2018 Feb.;178:46–54. doi: 10.1016/j.actatropica.2017.10.027.
278. Nikolova MP, Joshi PB, Chavali MS. Updates on biogenic metallic and metal oxide nanoparticles: Therapy, drug delivery and cytotoxicity. Pharmaceutics. 2023 Jun.;15(6). doi: 10.3390/pharmaceutics15061650.
279. da RA, Sanfelice S, et al. Biogenic silver nanoparticles (AgNp-Bio) reduce *Toxoplasma gondii* infection and proliferation in HeLa cells, and induce autophagy and death of tachyzoites by apoptosis-like mechanism. Acta Tropica. 2021 Oct.;222. doi: 10.1016/j.actatropica.2021.106070.
280. Arruda da Silva Sanfelice R, et al. Biogenic silver nanoparticles reduce *Toxoplasma gondii* infection and proliferation in RAW 264.7 macrophages by inducing tumor necrosis factor-alpha and reactive oxygen species production in the cells. Microbes and Infection. 2022 Jul.;24(5). doi: 10.1016/j.micinf.2022.104971.
281. Machado LF, et al. Biogenic silver nanoparticles reduce adherence, infection, and proliferation of *Toxoplasma gondii* RH strain in HeLa cells without inflammatory mediators induction. Experimental Parasitology. 2020 Apr.;211. doi: 10.1016/j.exppara.2020.107853.
282. Abou Elez RMM, Attia ASA, Tolba HMN, Anter RGA, Elsohaby I. Molecular identification and antiprotozoal activity of silver nanoparticles on viability of *Cryptosporidium parvum* isolated from pigeons, pigeon fanciers and water. Scientific Reports. 2023 Dec.;13(1). doi: 10.1038/s41598-023-30270-2.
283. Oluranti O, Rotimi L, Gideon A, Oluwatayo A, Olusola M, Adedosu T. Biogenic silver nanoparticles from two species of *Malvaceae*: Synthesis, antimalarial, antitrypanosomal, antimicrobial properties and their potential towards HeLa Cell Line. Tropical Journal of Natural Product Research. 2023 Apr.;7(3): 2642–49. doi: 10.26538/tjnpr/v7i3.26.
284. Dkhil MA, et al. Biosynthesized silver nanoparticles have anticoccidial and jejunum-protective effects in mice infected with *Eimeria papillata*. Environmental Science and Pollution Research. 2023. doi: 10.1007/s11356-023-25383-0.

285. Al Quraishy S, et al. Morphological and molecular approaches for identification of murine *Eimeria papillata* infection. Journal of King Saud University – Science. 2022 Aug.;34(6). doi: 10.1016/j.jksus.2022.102164.
286. Ullah I, Cosar G, Abamor ES, Bagirova M, Shinwari ZK, Allahverdiyev AM. Comparative study on the antileishmanial activities of chemically and biologically synthesized silver nanoparticles (AgNPs). 3 Biotech. 2018 Feb.;8(2). doi: 10.1007/s13205-018-1121-6.
287. Kamal M, et al. In vitro assessment of antimicrobial, anti-inflammatory, and schistolarvicidal activity of macroalgae-based gold nanoparticles. Frontiers in Marine Science. 2022 Dec.;9. doi: 10.3389/fmars.2022.1075832.
288. Prasanna P, Kumar P, Kumar S, Mandal D. Synthesis of 7, 8-dihydroxyavone functionalized gold nanoparticles and its mechanism of action against *Leishmania donovani*: 2021, doi: 10.21203/rs.3.rs-37873/v3.
289. Das S, et al. Andrographolide engineered gold nanoparticle to overcome drug resistant visceral leishmaniasis. Artificial Cells, Nanomedicine, and Biotechnology. 2018 Oct.;46(sup1):751–62. doi: https://pubmed.ncbi.nlm.nih.gov/29421940/.
290. Ahmad A, et al. Silver and gold nanoparticles from *Sargentodoxa cuneata*: Synthesis, characterization and antileishmanial activity. RSC Advances. 2015 Aug.;5(90): 73793–806. doi: 10.1039/c5ra13206a.
291. Ghosh S, et al. Dioscorea bulbifera mediated synthesis of novel $Au_{core}Ag_{shell}$ nanoparticles with potent antibiofilm and antileishmanial activity. Journal of Nanomaterials. 2015;2015. 10.1155/2015/562938.
292. Karthik L, Kumar G, Keswani T, Bhattacharyya A, Reddy BP, Rao KVB. Marine actinobacterial mediated gold nanoparticles synthesis and their antimalarial activity. Nanomedicine. 2013 Oct.;9(7):951–60. doi: 10.1016/j.nano.2013.02.002.
293. Foo YY, Periasamy V, Kiew LV, Kumar GG, Malek NA. *Curcuma mangga*-mediated synthesis of gold nanoparticles: Characterization, stability, cytotoxicity, and blood compatibility. Nanomaterials. 2017 Jun.;7(6). doi: 10.3390/nano7060123.
294. Cheraghipour K, et al. Synthesis, characterization, and antiparasitic effects of zinc oxide nanoparticles-eugenol nanosuspension against *Toxoplasma gondii* infection. Heliyon. 2023 Aug.;9(8). doi: 10.1016/j.heliyon.2023.e19295.
295. Gur T, Meydan I, Seckin H, Bekmezci M, Sen F. Green synthesis, characterization and bioactivity of biogenic zinc oxide nanoparticles. Environmental Research. 2022 Mar.;204. doi: 10.1016/j.envres.2021.111897.
296. Lail N, et al. Biosynthesis and characterization of zinc oxide nanoparticles using *Nigella sativa* against coccidiosis in commercial poultry. Scientific Reports. 2023 Dec.;13(1). doi: 10.1038/s41598-023-33416-4.
297. Jan H, et al. Biogenic synthesis and characterization of antimicrobial and antiparasitic zinc oxide (ZnO) nanoparticles using aqueous extracts of the Himalayan columbine (*Aquilegia pubiflora*). Frontiers in Materials. 2020 Aug.;(7). doi: 10.3389/fmats.2020.00249.
298. Costa IN, et al. Biogenic silver nanoparticles can control *Toxoplasma gondii* infection in both human trophoblast cells and villous explants. Frontiers in Microbiology. 2021 Jan.;11. doi: 10.3389/fmicb.2020.623947.
299. Alarfaj NA, Amina M, Al Musayeib NM, El-Tohamy MF, Al-Hamoud GA. Immunomodulatory and antiprotozoal potential of fabricated sesamum radiatum oil/polyvinylpyrrolidone/au polymeric bionanocomposite film. Polymers (Basel). 2021 Dec.;13(24). doi: 10.3390/polym13244321.
300. Rheder DT, et al. Synthesis of biogenic silver nanoparticles using *Althaea officinalis* as reducing agent: Evaluation of toxicity and ecotoxicity. Scientific Reports. 2018 Dec.;8(1). doi: 10.1038/s41598-018-30317-9.
301. Vasanth SB, Kurian GA. Toxicity evaluation of silver nanoparticles synthesized by chemical and green route in different experimental models. Artificial Cells, Nanomedicine, and Biotechnology. 2017 Nov.;45(8):1721–27. doi: https://pubmed.ncbi.nlm.nih.gov/28278585/.

302. Lee YJ, Ahn EY, Park Y. Shape-dependent cytotoxicity and cellular uptake of gold nanoparticles synthesized using green tea extract. Nanoscale Research Letters. 2019;14. doi: 10.1186/s11671-019-2967-1.
303. Nishiyama N. Nanomedicine: Nanocarriers shape up for long life. Nature Nanotechnology. 2007;2(4):203–4.
304. Mody VV, Siwale R, Singh A, Mody HR. Introduction to metallic nanoparticles. Journal of Pharmacy and Bioallied Sciences. 2010 Oct;2(4):282–89. doi: 10.4103/0975-7406.72127. PMID: 21180459; PMCID: PMC2996072.
305. Javed R, Zia M, Naz S, et al. Role of capping agents in the application of nanoparticles in biomedicine and environmental remediation: Recent trends and future prospects. Journal of Nanobiotechnology. 2020;18(172). https://doi.org/10.1186/s12951-020-00704-4.
306. Munyayi TA, Vorster BC, Mulder DW. The effect of capping agents on gold nanostar stability, functionalization, and colorimetric biosensing capability. Nanomaterials (Basel). 2022 Jul 19;12(14):2470. doi: 10.3390/nano12142470. PMID: 35889694; PMCID: PMC9319646.
307. Javed R, Usman M, Tabassum S, Zia M. Effect of capping agents: Structural, optical and biological properties of ZnO nanoparticles. Applied Surface Science. 2016;386:319–26.
308. Arzt E. Size effects in materials due to microstructural and dimensional constraints: A comparative review. Acta Materialia. 1998;46:5611–26.
309. Lifeng, JIA, Tao, HE, Zhipeng, LI and Xue-Mei, LI. Monolayer-protected gold nanoparticle surface-bound catalysts: preparation and application. Chinese Journal of Catalysis. 2010;31(11–12):1307–1315.
310. Devi JM. Simulation studies on structural and thermal properties of alkane thiol capped gold nanoparticles. Journal of Molecular Graphics and Modelling. 2017;74:359–365.
311. Neouze M-A, Schubert U. Surface modification and functionalization of metal and metal oxide nanoparticles by organic ligands. Monatshefte fűr Chemie Chem. Mon. 2008;139:183–95. doi: 10.1007/s00706-007-0775-2.
312. Mousavi-Khattat M, Keyhanfar M, Razmjou A. A comparative study of stability, antioxidant, DNA cleavage and antibacterial activities of green and chemically synthesized silver nanoparticles. Artificial Cells, Nanomedicine, and Biotechnology. 2018;46(3):1022–1031.
313. Hotze EM, Phenrat T, Lowry GV. Nanoparticle aggregation: Challenges to understanding transport and reactivity in the environment. Journal of Environmental Quality. 2010;39:1909–24. doi: 10.2134/jeq2009.0462.
314. El Badawy AM, Luxton TP, Silva RG, Scheckel KG, Suidan MT, Tolaymat TM. Impact of environmental conditions (pH, ionic strength, and electrolyte type) on the surface charge and aggregation of silver nanoparticles suspensions. Environmental Science & Technology. 2010 Feb 15;44(4):1260–66. doi: 10.1021/es902240k. PMID: 20099802.
315. Tejamaya M, Römer I, Merrifield RC, Lead JR. Stability of citrate, PVP, and PEG coated silver nanoparticles in ecotoxicology media. Environmental Science & Technology. 2012 Jul 3;46(13):7011–17. doi: 10.1021/es2038596. Epub 2012 Mar 21. PMID: 22432856.
316. Gorham JM, Rohlfing AB, Lippa KA, MacCuspie RI, Hemmati A, Holbrook RD. Storage Wars: how citrate-capped silver nanoparticle suspensions are affected by not-so-trivial decisions. Journal of Nanoparticle Research. 2014;16:2339.
317. Elzey S, Grassian VH. Agglomeration, isolation and dissolution of commercially manufactured silver nanoparticles in aqueous environments. Journal of Nanoparticle Research. 2010;12:1945–58.
318. Teeguarden JG, Hinderliter PM, Orr G, Thrall BD, Pounds JG. Particokinetics in vitro: Dosimetry considerations for in vitro nanoparticle toxicity assessments. Toxicological Sciences. 2007 Feb;95(2):300–12. doi: 10.1093/toxsci/kfl165. Epub 2006 Nov 10. Erratum in: Toxicol Sci. 2007 Jun;97(2):614. PMID: 17098817.

319. Izak-Nau E, Huk A, Reidy B, Uggerud H, Vadset M, Eiden S, et al. Impact of storage conditions and storage time on silver nanoparticles' physicochemical properties and implications for their biological effects. RSC Advances. 2015;5:84172–85. doi:10.1039/C5RA10187E.
320. Hosamani, GB, Patil RR, Chandrashekhar SS, Benagi VI, Nandihalli BS. Sunlight mediated green synthesis of silver nanoparticles from soybean seed and its bio-efficacy on Callosobruchus chinensis (L.). Journal of Experimental Zoology India. 2020;23(2) :1091.
321. George S, Gardner H, Seng EK, Chang H, Wang C, Yu Fang CH, Richards M, Valiyaveettil S, Chan WK. Differential effect of solar light in increasing the toxicity of silver and titanium dioxide nanoparticles to a fish cell line and zebrafish embryos. Environmental Science & Technology. 2014 Jun 3;48(11):6374–82. doi: 10.1021/es405768n. Epub 2014 May 15. PMID: 24811346.
322. Sangeetha J, Sandhya J, Philip J. Biosynthesis and functionalization of silver nanoparticles using Nigella sativa, Dioscorea alata and Ferula asafoetida. Science of Advanced Materials. 2014;6(8):1681–1690.

Index

T

V

X

Z

For Product Safety Concerns and Information please contact our EU
representative GPSR@taylorandfrancis.com
Taylor & Francis Verlag GmbH, Kaufingerstraße 24, 80331 München, Germany

www.ingramcontent.com/pod-product-compliance
Lightning Source LLC
LaVergne TN
LVHW011116300926
849160LV00008B/410

9781032457581